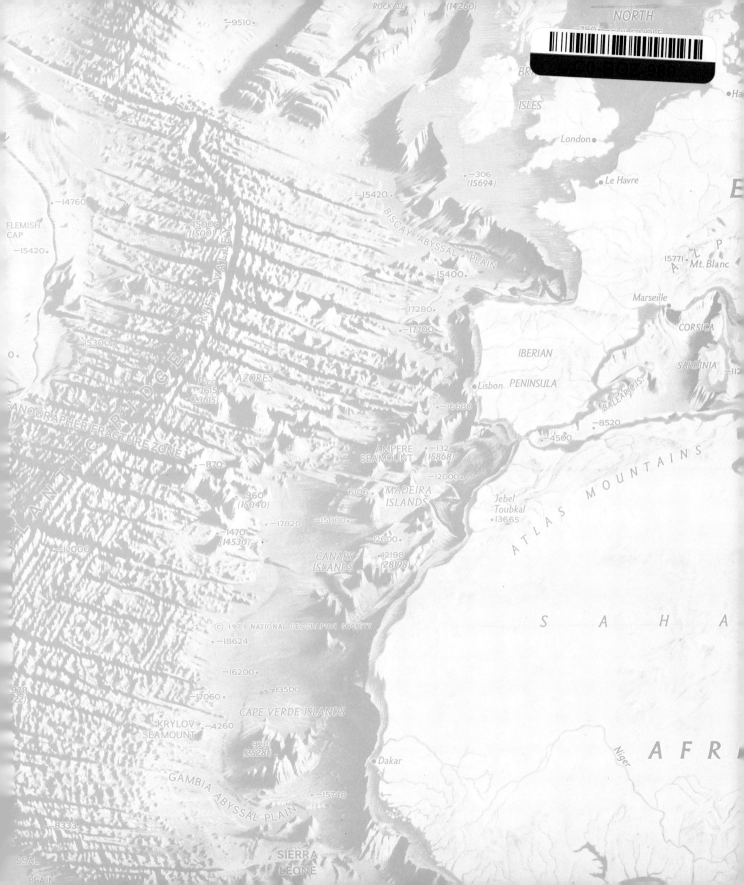

The EARTH and Its ENVIRONMENT

WALTER H. HESSE
California State Polytechnic University, *Pomona*

ROBERT L. McDONALD
Columbia Junior College

DICKENSON PUBLISHING COMPANY, INC. Encino, California Belmont, California

© 1974 by DICKENSON PUBLISHING COMPANY, INC.

All rights reserved. No part of this book may be reproduced, stored in a retrieval system, or transcribed, in any form or by any means—electronic, mechanical, photocopying, recording, or otherwise—without the prior written permission of the publisher, 16561 Ventura Boulevard, Encino, California 91316.

ISBN-0-8221-0115-7

Library of Congress Catalog Card Number: 73-76613

Printed in the United States of America
Printing (last digit): 9 8 7 6 5 4 3 2

Section on astronomy was previously published as *Astronomy: A Brief Introduction*, 1967, Addison-Wesley, Reading, Mass.

Inside cover: Atlantic Ocean floor. © 1973 National Geographic Society.

Production Editor: Linda Malevitz Hashmi; *Copy Editor:* James P. Miller
Designer: James F. Beggs; *Technical Artists:* Charles J. Alessio and Tom Martin
Cover: James F. Beggs

Contents

I INTRODUCTION

1	What Is Science?	3
1.1	Sciences vs. Humanities	4

II ASTRONOMY

2	Introduction to Astronomy	9
2.1	Objectives of Astronomy	9
2.2	Reasons for Studying Astronomy	10
2.3	The Geocentric Theory	11
2.4	The Heliocentric Theory	13
2.5	Structure of the Solar System	17
2.6	Space Science and Society	18
2.7	Summary	21
3	The Sun, the Earth, and the Moon	23
3.1	Formation of the Solar System	23
3.2	Age of the Solar System	25
3.3	The Sun	25
3.4	The Earth	33
3.5	The Moon	38
3.6	Summary	45
4	The Solar System: The Planets, Asteroids, Comets, and Meteoroids	47
4.1	Mercury	49
4.2	Venus	50

4.3	Mars	52
4.4	Jupiter	58
4.5	Saturn	62
4.6	Uranus	62
4.7	Neptune	63
4.8	Pluto	64
4.9	The Tenth Planet	65
4.10	Unanswered Questions	65
4.11	Other Inhabitants of the Solar System	66
4.12	Summary	70
5	Stars	72
5.1	Stellar Distance	72
5.2	Stellar Magnitude	75
5.3	Stellar Temperatures	77
5.4	Stellar Diameters	77
5.5	Stellar Mass and Density	78
5.6	Stellar Motion	79
5.7	Stellar Classification	81
5.8	The Hertzspring-Russell Diagram	82
5.9	Unusual Stars	85
5.10	Stellar Evolution	87
5.11	The Constellations	89
5.12	Summary	90
6	Galaxies	92
6.1	The Galaxy Discovered	92
6.2	Shape, Size, and Population of the Milky Way	93
6.3	Intragalactic Motion	94
6.4	Galactic Structure	95
6.5	Galactic Mass	96
6.6	Types of Galaxies	96
6.7	Galactic Evolution	97
6.8	Galactic Distances	100
6.9	Galactic Motion	103
6.10	Distribution of Galaxies	105
6.11	Radio Galaxies and Quasars	106
6.12	Summary	108
7	Cosmology—Life in the Universe	109
7.1	History of Cosmology	110
7.2	The Evolutionary Universe	111
7.3	The Steady-State Theory	113
7.4	Life in the Universe	114
7.5	Astronomy and Society	119
7.6	Summary	120

III GEOLOGY

8	**Introduction to Geology**	125
8.1	Some Early Principles	126
8.2	The Newly Formed Earth	130
8.3	The Present Earth	132
8.4	The Ocean Basins	135
8.5	Continental Drift	137
8.6	The Earth's Present Atmosphere	142
8.7	Summary	142
9	**The Earth's Materials**	144
9.1	Some Mineral Myths	144
9.2	Minerals Defined	145
9.3	Physical Properties of Minerals	148
9.4	Mineral Classification	153
9.5	The Importance of Minerals	156
9.6	Rocks	158
9.7	Weathering	165
9.8	Economic Importance of Weathering	169
9.9	Summary	171
10	**Mountain Building**	172
10.1	The Mountain-Building Process	174
10.2	Volcanism	177
10.3	Folded and Faulted Mountains	186
10.4	Earthquakes	192
10.5	Mountain Building and Society	198
10.6	Summary	199
11	**Erosion**	201
11.1	The Erosional Work of Gravity	202
11.2	Wind Erosion	210
11.3	Glaciers	215
11.4	Water Erosion	228
11.5	Ground Water	241
11.6	Wave Action	246
11.7	Summary	247
12	**Historical Geology**	249
12.1	The Measurement of Geological Time	250
12.2	The Geologic Rock Record	251
12.3	The Geologic Fossil Record	254
12.4	The Geologic Time Scale	256
12.5	The Precambrian Era	257
12.6	The Paleozoic Era	258

12.7	The Mesozoic Era	262
12.8	The Cenozoic Era	265
12.9	Summary	267

IV THE ATMOSPHERE

13	Introduction to the Atmosphere	271
13.1	Introduction to Meteorology as a Science	271
13.2	The Atmosphere	273
13.3	Weather-Producing Mechanisms	277
13.4	Atmospheric Variables	283
13.5	Clouds	292
13.6	Summary	294
14	Air Movement and Air Masses	298
14.1	Circulation of the Atmosphere	298
14.2	Air Masses	307
14.3	Weather Fronts	310
14.4	Storms	320
14.5	Summary	327
15	Climate and Life on the Earth	329
15.1	The Seasons	331
15.2	Climates of the World	333
15.3	Weather Forecasting	348
15.4	Summary	352
16	Air Pollution and Weather Modification	357
16.1	Air Pollution and Society	357
16.2	Types of Air Pollution	359
16.3	Effects of Air Pollution	362
16.4	Dispersion of Smog	366
16.5	Control of Air Pollution	366
16.6	Weather Modification	368
16.7	History of Weather Modification	368
16.8	Methods of Weather Modification	370
16.9	Problems of Weather Modification	374
16.10	Summary	375

V THE OCEANS

17	Introduction to Oceanography	381
17.1	The Importance of the Oceans	381
17.2	Exploration of Inner Space	389

	17.3	Physical Characteristics of the Sea	394
	17.4	Summary	395
	18	Seawater and Its Properties	397
	18.1	Chemical Properties of Ocean Water	397
	18.2	Physical Properties of Ocean Water	402
	18.3	The Sea and the Climate	410
	18.4	Summary	412
	19	The Restless Sea	414
	19.1	Measurement of Ocean Currents	414
	19.2	Circulation of Ocean Currents	417
	19.3	Forces Causing Sea Circulation Patterns	419
	19.4	Tides	423
	19.5	Wave Activity	429
	19.6	Summary	435
	20	Water Pollution	437
	20.1	Water and Man	437
	20.2	Polluted Waters	438
	20.3	Forms of Pollution	439
	20.4	Solving the Problem	446
	20.5	Summary	450

APPENDIXES

1	Units of Measure: Metric and English Equivalents	453
2	Temperature Scales	455
3	Nomograms of Height, Length, and Temperature	456
4	The Beaufort Scale of Wind (Nautical)	457
5	Mineral Identification Key	459

GLOSSARY	462
INDEX	477

Preface

Scientific progress has accelerated in the last several generations to a degree inconceivable during the time of Galileo and Newton. Our everyday mode of living is influenced by rapid scientific advances with the result that science has become a very important facet of our culture. Whether this is wholly desirable is being vigorously debated, and to debate the subject intelligently requires that the general populace have some understanding of such diverse subjects as water pollution, space research, and landslide activity. This is particularly true of the leaders of the community who are involved in developing policy to which scientific matters are related. It is not possible or even necessary for everyone to be completely knowledgeable in all phases of science. However, it would be most valuable for each citizen to have at least an awareness of the forces that shape and influence his or her environment and life.

This text provides students in the nonscience disciplines with a general background in the earth sciences. The intent is to impart an understanding of the principles of earth science by means of a historical development of these ideas. Wherever possible the influence of natural phenomena upon society is shown by the inclusion of pertinent material. Needless to say it is impossible to cover all aspects of a subject in any one text without becoming encyclopedic, so no attempt to do so has been made. The presentation of the material is such that no mathematical preparation beyond algebra and geometry is required. Such an approach is used to provide the teacher with a basic plan for teaching the earth sciences, and at the same time allow him or her sufficient latitude for making the presentation as rigorous as the students are capable of understanding.

While there is no best way of presenting the various fields of earth science, a nonintegrated text has merit in that it permits the teacher to rearrange the sequence of material to suit his needs without destroying the meaning of the subject matter. It provides sufficient flexibility to allow the inclusion of material of greater interest and use to the student with nonscience interests.

The text is composed of four sections. The first, astronomy, begins with a presentation of the development of man's early thinking on matters pertaining to the universe. This is followed by chapters in which the properties of the sun, planets, and other objects in the solar system are described. Chapters on stellar properties and evolution, galactic structure, and cosmology are included in this section.

The second section deals with geologic phenomena. The first chapter gives a brief summary of the development of geology as a science and concludes with a brief discussion of continental drift. Another chapter deals with rocks and minerals, and two chapters discuss the processes that build up land areas and erode them. The last chapter summarizes the historical changes in landforms and evolution of life as geologists believe it has occurred on the North American continent during the past half billion years.

The third section discusses meteorological phenomena including physical properties of the atmosphere, movement of air masses, weather phenomena, air pollution, and weather modification.

The last section on oceanography gives a brief history of oceanographic research and discusses the benefits we may derive from the sea. It covers properties of seawater and describes the generation of ocean currents, waves, and tidal action. A final chapter on water pollution is included.

Throughout the entire text we have attempted to go beyond a mere description of our environment in physical terms, by showing how our activities are influenced by natural phenomena, and the effect we have on our environment.

In this way we feel we offer a text that differs significantly from others covering the same subject matter, and thereby provide the student with a text much broader in scope and more useful in content.

WALTER H. HESSE
ROBERT L. MCDONALD

Introduction

What Is Science?

Understanding science—what is science?
Relationship of the sciences to the humanities.
What is earth science?

To understand what earth science means we should first look at the meaning of the broader area of science. Science (Latin: "knowledge," from *scire*, "to know") is a creative and dynamic activity. It is an *expression of human experience*—an expression that seeks to provide an understanding of the total environment and the forces that shape it. Science is the process whereby man, individually and collectively, uses this experience to devise a commonly accepted explanation of the workings of the universe around him. Science involves observation and measurement, imagination and hypothesis, communication and criticism, in an endless assault on the unknown or little-known aspects of our surroundings.

The scientist who is involved in such activities observes and measures objects and phenomena of the physical world. To experimental results, he applies imagination in an effort to discern some common action or behavior of matter and energy. He generalizes from the collection of observations and measurements and relationships and laws that have been accumulated, to develop theories which can in some coherent way explain what is taking place. Theories are then used to predict new phenomena and new laws as yet unobserved, and these predictions serve as guides to new experiments and observations.

The scientist maintains continual communication with his colleagues in a variety of ways, subjecting his results to the close scrutiny of his peers. This communication is carried out through the scientific literature, in scientific meetings, and in informal person-to-person seminars and discussions. So important is the role of communication among scientists to the process of mutual

4 criticism that this has led to the assertion by some that modern science is communication.

Introduction

The process or activity of science has developed its rules on the basis of hard and searching experience. Recognizing that science cannot attain the absolute in knowledge, scientists have sought to substitute for the unobtainable absolute, the attainable utmost in objectivity. The scientific tradition, while demanding of each individual the maximum of insight, ingenuity, imagination, discernment, and invention—that is, the utmost in subjectivity—nevertheless wrings as much of the personal out of the equation as possible. This is accomplished by demanding that the individual subject his thoughts and results to the uncompromising scrutiny of his peers. The members of the scientific community accept this tradition, although being human, they may occasionally deviate from this basic philosophy. The acceptance of this tradition gives to science and scientists a unity not only of knowledge, but also of method, one which encircles the world and transcends political divisions. These principles have also had a profound influence upon society, not only in all the new objects of everyday living, but also in our concepts and patterns of thought.

1.1 SCIENCE VS. HUMANITIES

Having briefly discussed the role of the scientists and what science is, we may now compare science to other forms of human endeavor with which most people are more readily familiar.

These endeavors are the humanities, including art, literature, music, and history, to name a few. Each of these, like science, is in its own way an expression of human experience. The artist, like the physicist, considers his environment, interprets it, and presents his view for criticism. The fact that the two see the world in different terms in no way detracts from the basic premise that all these endeavors are primarily based upon the individual's experience and are his interpretations of the universe. In both instances, and in fact in all creative endeavor, the work is presented in one form or another for general review and criticism.

Science need not be a mystery for anyone. It is true that few have the faculty to discover the innermost secrets of science just as there are few who are gifted artists or composers of music. However, we can appreciate and understand the workings of nature when these are revealed to us, just as we can appreciate a symphony or a drama.

Scientists make use of data or facts as do historians, but the collection of facts in either case is not an end in itself. Facts are used to establish or reinforce a theory or to explain the occurrence of an event. Facts serve to illustrate a law or support a hypothesis. The relationship of facts to the development of a theory applies to the interpretation of history as it does in the development of a law in physics. For example, it is a well-established fact that Queen

What Is Science?

Elizabeth I of England never married. This, of course, has led to the development of a number of theories as to the reason. Additional facts need to become available to establish the truth of one or the other of the theories. The same process occurs in science. Everyone recognizes the fact that if one steps out of a second-story window he will fall to the ground, or if an apple comes loose from its stem it will fall to the earth. The "why" of this phenomenon has led to the pronouncement of the universal law of gravitation.

It is not our intent to imply that the humanities and sciences are identical, because this is not the case. There are substantial differences in that some sciences are essentially quantitative, whereas the humanities tend to be more qualitative or descriptive. However, while it is relatively simple to separate, for example, history and physics into either the humanities or sciences category on such a basis, there is no definite dividing line between the sciences and the humanities. To illustrate this we can construct a spectrum of man's intellectual activities, with physics at one end of the array and history at the other. One can readily recognize that astronomy, chemistry, biology, and geology belong with physics, whereas art, music, and drama belong at the other end. As we begin to fill in psychology, anthropology, archeology, sociology, ethics, economics, and other subjects, it becomes increasingly difficult to decide whether these subjects are more closely related to the sciences or to the humanities. The fact that there is no clear-cut distinction then becomes evident.

Returning to earth science, we find this is in reality an assemblage of subjects that deal with various aspects of the earth and its environment. Astronomy, geology, meteorology, and oceanography are most commonly included, but other subject matter may be considered: for example, the newly developing environmental sciences, soil science, climatology, and some aspects of conservation. None of the subjects in earth science are regarded as "pure" sciences, since each includes elements of other sciences. For example, in each of the subjects listed, physics, chemistry, and mathematics play an important role.

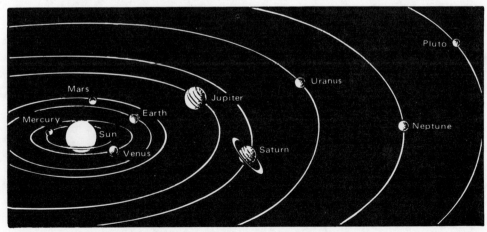

Figure 1.1 Diagram of the solar system.

6   In addition, oceanography includes geology and biology, indicating that considerable integration within the individual earth sciences also occurs.

Introduction   From these relationships, it must soon become evident that earth science deals with the inanimate structures and processes in nature in the same manner that biology deals with living forms in nature. Sciences such as physics and chemistry play a support role for earth science by dealing with the nature of matter, the changes that occur in substances, the energy relationships that cause these changes, and the forces of nature. Mathematics enables the scientist to quantify these phenomena and deal with the forces in a manner that is easily recognizable by other scientists.

II ASTRONOMY

2 Introduction to Astronomy

Goals of astronomy.
Why astronomy was studied by ancient people.
Organization of the geocentric system vs. the heliocentric system.
Kepler's laws of motion.
Astronomic discoveries of Galileo.
Development of the law of gravity by Newton.
Modern view of the structure of the solar system.
Relationship of space science to society.

The astronomer seeks answers to basic questions which nearly all people have asked themselves since time immemorial. What is humanity's place in the scale of things? Is the earth the solitary abode of life in the universe? What is the age of the earth, of the solar system, of the sun? Are there other planetary systems?

2.1 OBJECTIVES OF ASTRONOMY

The objectives of astronomy are to find answers to questions such as those above and to learn about the structure, the behavior, and the evolution of all types of celestial bodies. This includes bodies ranging in size from the smallest (meteorites, comets, planets) to the largest (stars, galaxies). In essence, the astronomer is attempting to learn about matter in space and in the entire universe.

The great names of astronomy are among the greatest discoverers in the whole course of human history. These include Copernicus, Tycho Brahe, Kepler, Galileo, and Newton. The questions that were foremost in their minds have now been answered and today form part of the education of everyone: what are the earth, the sun, the moon, the comets, the planets? Is the earth flat or round? How large is the earth? What is the cause of day and night and the annual variation of seasons? What is the cause of a solar or lunar eclipse? What is a star? What is the Milky Way?

Now the answers to these questions are available, and it is possible to recognize how simple the questions were. Yet gaining these answers ranks among the great triumphs of human inquiry.

Solving the current questions in physics and astronomy requires a different approach than was possible in former ages. Today the laws of physics are derived partly from the study of astronomical objects themselves, partly from the study of laboratory phenomena, and partly from theory. Astronomy has always played a principal part in the development of physical laws and is still one of the chief frontiers of science.

2.2 REASONS FOR STUDYING ASTRONOMY

Many ancient people, particularly the Chinese, Egyptians, and Babylonians of 4,000 to 5,000 years ago, speculated about the nature of the universe. They applied the knowledge they gained toward practical purposes and needs which may be enumerated as follows:

1. *Time:* The alternate daylight and darkness, the regularity of the lunar cycle, and the yearly passage of the stars were recognized and used to mark the passage of time. This was a valuable means of arranging the affairs of man on an orderly basis. It permitted merchants to meet commitments, governments to collect taxes at regular intervals, and farmers to time the planting of their crops. For example, the temples in Egypt were oriented to permit the light from the star Sirius to shine on the altar at dawn in spring. This event signaled the coming of the spring floods of the Nile after which seeds for the new crops were sown.

2. *Religion:* Objects in the sky were looked upon with religious reverence by ancient peoples. In Babylonia, the sign for God was a star. For many primitive civilizations, the sun was recognized as the principal source of light and heat, and for this reason it was worshiped. Many ancient gods and goddesses, prominent in mythology, were represented by various star configurations (later called constellations).

3. *Direction:* The rising and setting of the sun and the location of the polar star permitted travelers to maintain a sense of direction and reach distant destinations by sea as well as by land. Navigation has been dependent upon knowledge of the position of certain stars and still is even to this day.

4. *Prognostication:* The ability to forecast the future stemmed from the ability of astronomer-priests to predict certain events in the heavens such as

an eclipse. Eclipses occurred periodically and were viewed by ancient people with superstitious awe. The Chinese astronomers were adept at such predictions as long as 4,000 years ago, while the Babylonians, 3,000 years ago, left many records of eclipses and movements of the planets. Events in the sky were related to events on earth, and the movement and positions of the planets were considered important in the destiny of individuals and nations. In this manner astrology had its beginning—partly science and partly mysticism—which even in modern times strongly influences the actions of many people.

5. *Knowledge for its own sake:* Not until 2,500 years ago did the Greeks begin to seek answers to astronomical questions for other than practical reasons. Mathematics and physics were applied in the search for truth, although parts of mythology and astrology were retained from their Babylonian source.

2.3 THE GEOCENTRIC THEORY

The Greeks, 2,500 years ago, began to place the objects in the sky into some form of organized system. Pythagorus, Aristotle, and others viewed the system as earth-centered or *geocentric*. That is, the earth, composed of the elements earth, water, air, and fire, occupied the center of the universe. In orbits around the central Earth were the moon, Mercury, Venus, the sun, Mars, Jupiter, and Saturn. These were "the wanderers," or *planets*, which exhibited motions that differed from the stars. According to the geocentric theory, the planets revolved around the earth in perfect circular orbits and moved at uniform velocities. It was thought that the stars were fixed on an outer sphere (the celestial sphere), and that the entire sphere revolved around the earth (fig. 2.1).

Not all the ancient Greeks agreed with this geocentric concept, because certain apparent motions, not consistent with the supposedly harmonious arrangement of the universe, were observed.

These motions or inconsistencies (inconsistent only with the established theory) could not be explained on the basis of the geocentric system. For example, all celestial objects moved across the sky from east to west daily, while over longer periods the planets moved west to east against the background of stars. In addition, when the movements of Mars, Jupiter, and Saturn were plotted against the background of stars, these planets periodically exhibited a *retrograde motion*. The planets moving normally from west to east appeared to reverse their direction at regular intervals for several months and then resume normal travel again (fig. 2.2). Mercury and Venus also changed direction, rising higher in the sky night after night (or morning after morning if in the morning sky) for several weeks or months and then reversing direction again. In addition, these two planets never appeared in opposition to the sun—that is, in a position directly overhead at midnight—which would occur if the planets were orbiting the earth. Periodic variation in the size of the sun and the moon and in brightness of the other planets was another inconsistency

Astronomy

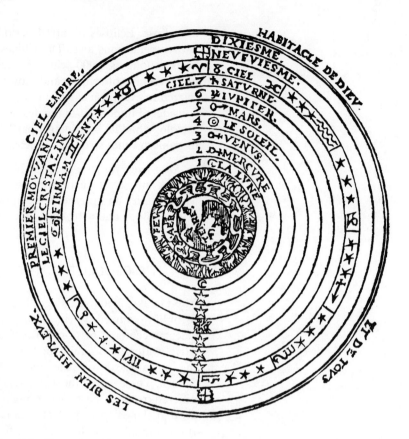

Figure 2.1 The geocentric scheme of the universe as devised by the ancient Greeks. The earth occupied the center and was composed of the four elements: earth, water, air, and fire. Surrounding it were the spheres of the moon, Mercury, Venus, the sun, Mars, Jupiter, Saturn and the outer sphere of the stars. From the DeGolyer copy of Petrus Apianus' Cosmographie, 1551.

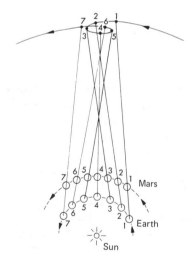

Figure 2.2 Retrograde motion, an apparent motion, is the result of the difference in the orbital speeds of the planets.

that ancient astronomers could not explain. Such action implied variation in distance of the planets from the earth, which was not in keeping with the principles of the geocentric theory.

These inconsistencies caused Aristarchus of Samos (circa 280 B.C.) to suggest that the sun and not the earth was in the center of the universe and that the earth and planets revolved around the sun. However, the consensus of Greek thought was in favor of the geocentric theory, and Aristotle's tremendous influence as a teacher in the fourth century B.C. was principally responsible for fixing the geocentric concept as the acceptable theory of the universe for the next 2,000 years. It was this concept that Copernicus challenged in the early sixteenth century at what may be considered a major turning point in the history of astronomy.

2.4 THE HELIOCENTRIC THEORY

Copernicus (1473–1543), born in Poland, prepared for a career in the church under the guidance of his uncle, but at the same time pursued his intense interest in astronomy. He studied the ancient systems at great length and found that errors, which had supposedly been corrected or accounted for, were constantly recurring. Copernicus objected to the frequent use of artificial devices developed by the ancient astronomers to uphold the uniform circular motion concept for the planets. He expressed the opinion that "... a system such as this seemed neither sufficiently absolute nor sufficiently pleasing to the mind."

Over a period of thirty years, Copernicus made observations and eventually came to the conclusion that movements of celestial objects could be more readily explained if the sun were placed at the center and the earth became the third planet from the sun (fig. 2.3). The planets were arrayed in order of their periods of revolution, with Mercury closest to the sun, followed by Venus, Earth (with the moon as a satellite), Mars, Jupiter, and the most distant, Saturn. The movement of the stars was then recognized as the result of the earth's rotation on its axis. Copernicus did place the sun slightly off-center to the system to explain the planets' variation in brightness from year to year.

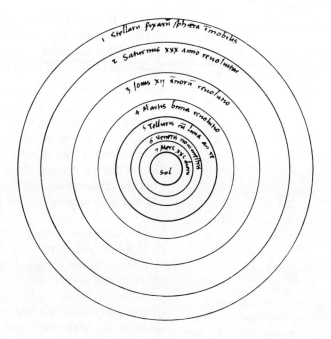

Figure 2.3 The Copernican concept of the heliocentric system. The epicycles used in the system are omitted. From Copernicus' De Revolutionibus, 1543.

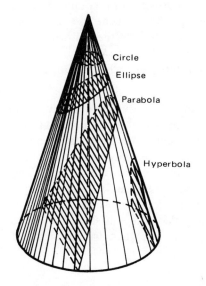

Figure 2.4 Conic sections.

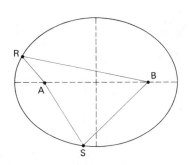

Figure 2.5 An ellipse is a geometric figure for which the sum of the distances of a point on the curve to the focal points will equal the sum of the distances of any other point on the curve to the focal points. Thus, we find that AS + SB will equal AR + RB.

He retained the idea of circular motion and uniform velocity of the planets—a notion that was an important part of the geocentric theory. In this way, the world was presented with the *heliocentric theory* (sun-centered theory), sometimes called the *Copernican system* because of the part played by Copernicus in its development.

The Copernican system was not quite the planetary theory that is recognized today, but it was a major reform that pointed the way. The ideas were not immediately accepted and were much criticized despite their relative simplicity. When Copernicus's book came out, just prior to his death in 1543, it contained an anonymous preface by his friend Osiander proclaiming the system to be an easier and more practical method of calculating the positions of the planets than had theretofore been available.

One of those opposed to the Copernican system was Tycho Brahe (1546–1601) of Denmark. He was an astronomical observer of tremendous skill and gathered vast quantities of accurate data on the movements of planets, particularly Mars. Tycho attempted to test the Copernican system by looking for stellar *parallax.* Parallax is an apparent movement of the nearer stars with respect to the more distant stars when viewed from opposite sides of the earth's orbit. He was unable to detect parallax with his instrument because of the great distance to even the nearer stars and he therefore concluded that the earth did not orbit the sun.

In 1600, the mathematician Johannes Kepler joined Tycho Brahe in Prague. Kepler inherited the vast collection of Tycho's observational material and provided the theoretical complement that resulted in the second great milestone in the history of astronomy: the laws of planetary motion.

Kepler spent many years attempting to relate Tycho's data to the circular orbits of the planets as prescribed by Copernicus. Kepler's faith in Tycho's observations was such that Kepler did not believe the discrepancy between the data on the shape of the Martian orbit and a perfect circle was due to observational error. Kepler also noted, as did Copernicus before him, that the sun did not occupy the center of the solar system but rather a point offset from the center. Eventually he was able to fit the observed orbit to an ellipse and was able to state his first law:

1. *The planets move around the sun in elliptical orbits with the sun at one of the focal points.* An ellipse is a conic section, that is, a geometric figure formed by a plane cutting a cone at an angle to the axis of the cone (fig. 2.4). The curve of the ellipse is such that the sum of the distances from the focal points to a point on the curve is equal to the sum of the distances from the focal points to any other point on the curve. In this manner Kepler described the true shape of the planetary orbit and eliminated the ancient concept of circular orbits (fig. 2.5). The degree of eccentricity of the elliptical orbit is quite small in most cases, and if the orbit were drawn to scale it could hardly be distinguished from a circle.

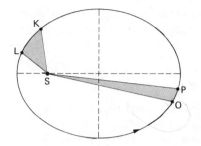

Figure 2.6 An imaginary line from the planet to the sun will sweep over equal areas in an equal period of time. Therefore the area of KLS is equal to the area of OPS.

Kepler also observed that the planets did not orbit the sun at uniform velocity. The planets moved faster when close to the sun (*perihelion*) and slower when at their greatest distance from the sun (*aphelion*). This led to Kepler's second law:

2. *An imaginary line from the center of the sun to the center of the planet sweeps over equal areas in equal periods of time.* This means that a planet travels from A to B in the same period required for it to travel from C to D. However, the imaginary line will cover equal areas because of the differences in the planetary distances from the sun at these two locations (fig. 2.6).

Now it was recognized that planets did not move at uniform velocity, and Kepler had removed another basic provision of the geocentric theory.

Kepler's first two laws were announced in 1609. For the next nine years he labored to develop a relationship between a planet's period of revolution and its distance from the sun. Kepler had nothing other than an intuitive belief that such a relationship existed and in 1618 published his third law, or what was to become known as the *harmonic law*.

3. *The squares of the periods of revolutions of any two planets are in the same ratio as the cubes of their mean distance from the sun.* The periods P of the planets was expressed as time in years and the distance A in astronomical units. The third law may be expressed in the following form:

$$\frac{P_1^2}{P_2^2} \propto \frac{A_1^3}{A_2^3}$$

If the earth is used as the second planet the following relationship exists:

$$P^2 = KA^3$$

where K is a numerical constant whose value varies depending upon the units of time and distance used.

The progress of astronomy to this point had been accomplished without the aid of the telescope, although the instrument was invented while Kepler was making his considerable contribution. Galileo developed his first telescope in 1609, after hearing rumors of its invention in the previous year by a Dutch lens maker, Hans Lippershey. With the aid of the telescope, Galileo made a series of revolutionary discoveries in astronomy, some of which supported the Copernican concept. These discoveries were:

1. The moon had earthlike features such as mountains, plains (which he thought were seas), and craters.

2. The planets were more than structureless points of light—they were finite physical objects with form and detail like the earth. He also observed that Venus exhibited phases like those of the moon, which provided him with evidence that Venus revolved around the sun. Galileo was also able to observe

the rings of Saturn, although he was not able to define them with the instruments he built.

3. Jupiter, and its four large moons (which he discovered), appeared to Galileo as a miniature solar system and added still more proof to him of the validity of the Copernican system. It indicated to Galileo that objects other than the earth could have bodies orbiting around them. The earth was not, therefore, the single center of all objects in the universe.

4. The telescope revealed the existence of sunspots, and Galileo deduced from their motion that the sun rotated on its axis.

5. Turning his instrument to the stars, Galileo found that he could observe many more stars; however, their size was not magnified, a condition he attributed to their great distance. In looking toward the Milky Way he discovered it to be an area of a vast multitude of stars.

Although Kepler and Galileo were contemporaries and confirmed the Copernican system, they essentially ignored each other. They attempted to define the force responsible for motion in the sky but neither was successful. It was left to the genius of Isaac Newton to accomplish this feat in the last half of the seventeenth century. He was able to define the force of gravity in mathematical terms as being directly proportional to the product of the masses of two objects in space and inversely proportional to the square of the distance between them. This may be expressed as

$$F = G\frac{Mm}{R^2}$$

where F is the force, G the gravitational constant, Mm the masses of the two objects and R the distance between them.

Newton also refined the accuracy of Kepler's third law—a modification based on the universal law of gravity. This modification may be stated as follows:

$$P^2(M + m) = A^3$$

where the symbols remain as stated for Kepler's third law, except that M equals the relative mass of the sun and m the relative mass of the planets. If the solar mass is equal to 1.0 then the earth's mass is 3/1,000,000 and $(M + m)$ is essentially equal to the solar mass.

The Newtonian laws remained unchallenged for over two hundred years, but by the twentieth century, observations of certain phenomena did not conform to these laws. It was found that certain motions could not be described by Newtonian mechanics, and it was at this time that Albert Einstein presented his *special theory of relativity* in 1905 and his *general theory of relativity* in 1916. These new concepts were a broader and more accurate description of the motion of objects in space and are now generally accepted. Newton believed, as did others of his time, that the immutable universe served as a fixed standard of reference for measuring position and motion of any object in space. Einstein maintained that no such absolute frame of reference existed and that position and motion of an object could only be described relative to some other object also in motion.

2.5 STRUCTURE OF THE SOLAR SYSTEM

The heliocentric concept first developed by Copernicus gradually led to a more complete understanding of the structure of the solar system. The motion of the planets, according to certain physical laws (those of Kepler and Newton), illustrates the importance of the sun (comprising 99 percent of the mass of the solar system) as the controlling body in the solar system. Not only are the planets thus influenced but also the numerous other objects such as the thousands of asteroids or minor planets which travel around the sun, primarily in orbits between Mars and Jupiter. The first of these asteroids was discovered in 1801 by G. Piazzi, an Italian astronomer. Since then numerous other asteroids have been found, but not all are confined to the main asteroid belt. Some have elongated elliptical orbits that bring them in close to the sun or out near the planet Saturn. In 1781, previous to the discovery of the asteroids, Uranus was found to be located beyond Saturn. In 1846, Neptune was discovered as a result of its influence on Uranus, and in 1930, Pluto was found, completing the list of nine known principal planets.

The orbits of the planets are essentially on the same plane, with most of the orbits inclined less than 3.5° to the ecliptic. The *ecliptic* is the plane of the earth's orbit projected to the celestial sphere. Mercury's orbital plane varies from the ecliptic by 7° and Pluto's by 17°.

The planets exhibit a similarity in direction of motion around the sun. As viewed from the polar star, Polaris, the planets as well as the asteroids revolve in a counterclockwise direction around the sun, or from west to east as seen from the earth. With the exception of Venus and Uranus, the counterclockwise direction is also the favored direction of rotation of the planets on their axes, and the direction of most satellites around their respective planets.

The mean distance of the planets from the sun ranges from 36 million miles for Mercury to 3,700 million miles for Pluto. Each planet is approximately $1\frac{1}{2}$ to 2 times the distance of its nearest internal neighbor from the sun. An easy method for remembering the distances is provided by a relationship known as *Bode's law*, although this is not a physical law in the strict sense. To each number in a series—0, 3, 6, 12 (found by doubling the previous one)—4 is added and the sum is divided by 10. The resulting sequence is the approximate mean distance of the planets to the sun expressed in astronomical units. An *astronomical unit* (A.U.) is the earth's mean distance to the sun and is equal to 92,950,000 miles, generally rounded off to a more convenient 93,000,000 miles. The calculated distances according to Bode's law correspond quite closely to the measured distances of the planets with the exception of Neptune and Pluto (see table 2.1).

Astronomy

Table 2.1

	Bode's Law	Astronomical Units	Distance from Sun 10^6 km
Mercury	0.4	0.3871	57.91
Venus	0.7	0.7233	108.21
Earth	1.0	1.0000	149.60
Mars	1.6	1.5237	227.9
Ceres	2.8	2.7673	414.2
Jupiter	5.2	5.2037	778.3
Saturn	10.0	9.5803	1428.0
Uranus	19.6	19.1410	2872.0
Neptune	38.8	30.1982	4498.0
Pluto	77.2	39.4387	5910.0

It is difficult to envision the magnitude of the solar system from a simple listing of the distances in a table. By constructing a diagram based on a recognizable scale it may be possible to judge the extent of the distance to the earth's neighbors, and we can see the immensity of the space that man plans to explore. If we consider a scale where the earth's distance to the sun (one A.U.) is equal to 1 foot, then Mercury would be almost 5 inches and Venus $8\frac{1}{2}$ inches from the sun. Mars would be 1 foot 6 inches, and the asteroids would be distributed between 1 and 4 feet from the sun. A slight bit more than 5 feet would reach Jupiter, and it would be $9\frac{1}{2}$ feet to Saturn. One would need to go 19 feet from the sun to reach Uranus, 30 feet to Neptune, and Pluto would be 40 feet from the sun. Man has thus far traveled to the moon ($\frac{1}{32}$ of an inch on this scale) and has sent unmanned probes to Venus, Mars, and Jupiter. The task of reaching the outer planets will be monumental, despite the fact that it can be accomplished with existing knowledge and technology.

Continuing in this line of thought leads to consideration of the prospect of reaching stars. The nearest star is so far that billions of miles would be too small a measure. Even the astronomical unit does not suffice, so instead, the light-year is used. The *light-year* represents the distance light travels in one year moving at a velocity of 300,000 kilometers per second. This makes the light-year equal to 9 trillion kilometers, or 62,400 astronomical units. Proxima Centauri, the nearest star to the sun, is part of a triple star system that is $4\frac{1}{3}$ light years from the sun. The possibility of reaching even this nearby group is quite remote with the technology and knowledge available to us at present.

2.6 SPACE SCIENCE AND SOCIETY

As stated in the introduction, astronomy deals with the study of celestial phenomena, a study which has resulted in the development of some of the important laws of physics. Advances in astronomy have been dependent upon such technological developments as the telescope, the spectroscope and the many auxiliary devices presently used. These devices are necessary to obtain the degree of refinement of observation without which progress in astronomy would be impossible.

Ground-based observations for astronomical purposes have some limitations which were recognized by Socrates 2,500 years ago. He said, "We who inhabit the earth dwell like frogs at the bottom of a pool. Only if man could rise above the summit of the air could he behold the true earth." Man now has risen above the air, has landed on the moon and looks toward the planets. These are not scientific feats but rather technological achievements accomplished through the application of knowledge of physical laws that were recognized many decades ago. Without the knowledge imparted to us by Kepler (laws of motion) and Newton (laws of gravity, laws of motion) the movement into space and space travel would not now be possible. This movement has resulted in the creation of a new discipline—space science.

Space science is more all-encompassing than astronomy, but astronomy will be one of the beneficiaries of this tremendous endeavor. Within recent time, man has been able to see the earth from above the atmosphere and found the view fascinating. New insights have been gained about the earth's environment—the atmosphere and beyond (see Chapter 13). A much clearer view of the heavens may be obtained from high altitude or satellite observations. Landings on the moon and probes to Venus, Mars, and Jupiter have resulted in much new information and the raising of many new questions about the solar system. Orbiting solar observatories, from 550 kilometers above the earth, recorded much data about the sun, data which could not have been obtained from the earth's surface because of atmospheric shielding.

Most of the astronomical phenomena observed in visible light from the ground can be fairly well accounted for by existing theories dealing with matter in near-earth conditions. However, deep-space phenomena in the nonvisible portion of the electromagnetic spectrum, discovered during the last decade, are believed to be at least of equal importance in the universe. The existence of solar X-ray and X-ray from other celestial sources was a surprising discovery, and the origin of this X-ray is difficult to explain. Certain nonthermal processes were discovered occurring in the sun and are of interest because they reveal otherwise unobservable reactions taking place in other celestial objects. These include gamma rays arising in thermonuclear reactions involving the transmutation of elements. At the longwave end of the spectrum, radio waves were detected arising from the interaction of very low-density streams of extremely hot matter with very weak but extensive magnetic fields. These magnetic fields included those associated with the structure of the galaxy, thus providing observational evidence and information on a wide range of phenomena in the universe. Astronomy is, therefore, a worthwhile human endeavor, because an essential part of the study of man is his relationship to his environment. The universe may be a rather remote setting for man, but nevertheless it is a part of his total environment.

Astronomy

The astronomer is interested in knowing more about man's celestial neighbors, and this will be accomplished by future investigations of the moon, the other planets, the asteroids, and possibly the comets. The work already begun will continue, but at a slower pace than during the 1960s, the first decade of the space age. This pace will enable scientists and engineers to deal with the multitude of technical problems that must be solved in order to explore the solar system and will allow for a more thorough examination of the data that is returned to earth before the next project is launched.

The establishment of a large space station in orbit around the earth is now in the planning stages. Planned to accommodate up to one hundred persons, it will be used as a launch site for manned probes to other planets. It will permit scientists to view the heavens without distortion by the earth's atmosphere, and will enable physicists to conduct experiments in the vacuum of space that are not possible on earth.

The possibility of life existing elsewhere in the universe is also of interest to both the astronomer and the biologist. Life, it is thought, will normally occur where conditions for it are appropriate. However, most scientists agree that the appropriate conditions need not necessarily resemble those on earth, nor must all life be comparable to earth forms.

A number of life-detection devices are being considered for landing on other planets. These devices, of necessity, are generally small, weighing as little as 1½ pounds, and are primarily designed to detect microbial life.

One such device is *Gulliver*, which, after landing, will fire adhesive cords outward and then reel them in. The dust and other surface material that adheres to the cords will be immersed in a nutrient solution containing radioactive carbon. If earthlike organisms are present, they will ingest and ferment the solution, creating radioactive carbon dioxide. This would register on a geiger counter and the results would be transmitted to earth.

Similar to this is the *Wolf Trap*, which sucks up samples of soil and air and immerses them in a nutrient solution. If organisms are present the solution will undergo changes in acidity and turbidity which would be reported to earth.

Other devices are also being developed, such as the *Multivator*, a miniature laboratory capable of performing a number of life detecting experiments; a TV-wired microscope for sending pictures of microscopic phenomena back to earth; and a TV-wired telescope to survey the landscape for possible moving or swaying objects that may be living things.

Space science, as previously stated, is broader in scope than astronomy and deals with items that have a greater direct impact upon society. Approximately one-third of the total space effort has gone into the space applications program, which includes work in communications, navigation, meteorology, oceanography, earth resources, and geodesy. These areas have achieved a greater degree of visible results and immediate practical benefits than other areas of space research. Some of the results are familiar to everyone. Television programs transmitted by satellite are commonplace, and the gathering of weather data and photographs of cloud formations are used in daily weather forecasting. No one can foresee the many possible benefits that may be derived from these efforts.

Why does man want to explore space? Is it ego that causes man to feel he must conquer the unknown, to overcome the impossible? Or is it the same insatiable curiosity that has motivated man throughout his history to inquire into the mysteries of his world? This motivation, plus the prospect of adventure, caused him to explore the earth in the sixteenth and seventeenth centuries. After a period of consolidation, it appears that man once again is ready to pursue the unknown and satisfy his curiosity about his environment. The exploration of space is not an isolated event in the development of civilization, nor is it a recent adventure. Man has explored the universe for centuries, and the present space program is a continuing effort on man's part to understand his environment.

2.7 SUMMARY

The basic objectives of the study of astronomy are to become acquainted with our total environment and to learn about natural laws that influence our activities. Ancient civilizations practiced rites that were based upon astronomy and studied astronomical phenomena for practical reasons.

In ancient Greece, two basic concepts evolved as descriptions of the organization of the universe. One idea envisioned the stationary earth as the center of the universe, with all celestial objects revolving in a variety of orbits about the earth. The other view saw the sun as the central figure, with the earth as an object moving about the sun. Aristotle's influence caused the earth-centered (geocentric) concept to dominate astronomical thinking for 2,000 years.

In the sixteenth century, Copernicus found that the motions of the planets and stars could be more simply described if the sun were placed at the center of the system. He retained the concept of uniform velocity and circular motion of planets as originated by the ancient Greeks for the geocentric theory. Johannes Kepler, through intensive study, developed laws of motion which eliminated the uniform-velocity-circular-motion concept. Kepler's laws of planetary motion were an extremely important contribution to astronomy and space science.

Galileo, a contemporary of Kepler's, was exploring the heavens with a newly developed instrument—the telescope—in the early seventeenth century. With it, Galileo collected additional evidence that supported the heliocentric (sun-centered) theory. He was able to observe Venus exhibiting phases, a phenomenon that could only occur if the planet orbited the sun between the earth and the sun. Galileo also discovered four satellites orbiting Jupiter and was able to see more clearly the true nature of the lunar surface.

Astronomy

In the latter half of the seventeenth century, Isaac Newton combined the thinking of a number of scientists of that day and mathematically defined the force of gravity in his universal law of gravitation. This greatly contributed to an understanding of the relationships of objects in space. In brief, his law states that every object in space attracts every other object with a force proportional to the products of their respective masses, and inversely proportional to the square of the distance between them.

Newton's law, while still valid, did not explain all aspects of motion of objects in space. This was more fully accomplished by Einstein in the early part of the wentieth century with his special and general theories of relativity.

As now viewed the solar system consists of the sun, which comprises 99 percent of the mass of the solar system, the planets and their satellites, and a variety of other objects such as asteroids, comets and meteoroids. Mercury, Venus, Mars, Jupiter, and Saturn are visible to the unaided eye and were seen by the ancient astronomers. Uranus, Neptune, and Pluto have been discovered since the invention of the telescope and cannot be seen without such an instrument.

Recent events have brought into being a new discipline, space science, much broader in scope than astronomy, but one in which astronomy will play a leading role. A considerable amount of knowledge has been gained in a very short period about the immediate environment of the earth, but most of this has been due to technological, rather than scientific, gains. One of the principal goals of space science is exploration of the planets for the purpose of learning more about their physical cnaracteristics, and to determine if life in any form exists in other parts of the solar system.

QUESTIONS

1. Why was astronomy studied by ancient people?
2. What inconsistencies appeared in the earth-centered concept of the universe as seen by the ancient Greeks?
3. What was Copernicus's model of the universe and what vestiges of the geocentric system did he incorporate into his system?
4. What ideas were stated by Kepler's first and second law of motion and what vestiges of the geocentric theory did these laws eliminate?
5. What relationship between distance from the sun and period of revolution of a planet is expressed by Kepler's third law?
6. List at least five discoveries made by Galileo as a consequence of his development of the telescope.
7. What is the order of the planets starting from the sun? Which of these have been discovered since the time of Copernicus?
8. List at least five benefits derived from the space science program which are not discussed in the text.

3 The Sun, the Earth, and the Moon

Theories on the formation of the solar system.
Age of the solar system.
Physical properties and structure of the sun.
Source of the sun's energy.
Physical dimensions of the Earth.
The atmosphere and magnetosphere.
The Earth's motion and the measurement of time.
Physical dimensions of the moon.
The moon's motions.
Eclipses.
Lunar surface characteristics.

Through the past 2,500 years, knowledge of the solar system has evolved from considering it as an earth-centered structure to the present concept of celestial bodies revolving around the sun. With the invention of the telescope, additional planets were discovered and added to the group already known to ancient astronomers.

3.1 FORMATION OF THE SOLAR SYSTEM

At this point one might ask how the solar system was formed. Although the structure and organization of the solar system were fairly well established in the early eighteenth century, no attempt had been made to theorize on the possible means of its formation. Even men like Newton considered the universe as immutable, and that the small measurable changes taking place were all cyclic and part of a stable cosmos. By the middle of the eighteenth century, however, several ideas were suggested to account for the formation of the solar system.

In 1749, a French naturalist, G. Buffon, theorized that the planets were formed as a result of a collision between a comet and the sun. The theory held that this event scattered material which formed the planets and became known as the *close-encounter hypothesis*. In 1900, T. C. Chamberlain and F. R. Moulton of the University of Chicago modified this idea, suggesting that a visiting star passed so close to the sun that material was drawn from the sun, forming the planets. If this had occurred, the solar system would be unique in the universe, since celestial collisions are extremely rare. Because of problems which could not be satisfactorily explained on the basis of the close-encounter hypothesis, it was discarded.

Another theory, no longer considered seriously, is the *binary-star hypothesis*. According to this theory, the sun was once one of a pair of stars, a circumstance not unusual in the universe. It proposes that a visiting star or some other phenomenon led to the disintegration of the second star, and that part of the debris remained under the sun's gravitational influence to form the planets.

At present the *nebular hypothesis* prevails as the most probable explanation of how the solar system was formed. It was first suggested in 1753 by the philosopher Immanuel Kant and then put in more precise astronomical terms in 1796 by the French astronomer and mathematician, Pierre Simon LaPlace.

The original form of the theory presumed that the solar system was created from a vast cloud of gas. As the gas cooled and contracted, its speed of rotation increased. Bulging at the center as the gases coalesced, a ring was abandoned by the contracting mass. Successively smaller rings were subsequently left behind, each forming a gas globe which had a circular orbit around the main mass. The main mass, which contained most of the material, formed the sun.

This theory leaves much to be desired, because it is known that gases tend to disperse rather than coalesce. All but abandoned, the theory was modified by C. F. von Weizäcker after World War II and later by G. P. Kuiper. In its modified form, the theory removes some of the objections. The cloud was presumably made up of gas, mainly hydrogen and helium, and dust particles of the heavier elements. Gravitational attraction, light pressure from other stars, and the effect of rotation resulted in the formation of a lens-shaped body. The intense heat, generated at the center of the mass by gravitational compression, finally triggered a thermonuclear reaction, and the sun was born. The rotating disk of gas and dust that remained was in a high state of turbulence and broke up into eddies of irregular size. The eddies, which tended to increase in size toward the outer zones, are sometimes referred to as *protoplanets*. The small particles collided, gradually forming larger and larger bodies which swept up more and more material, eventually to form planets. The heat within the planets themselves was produced by gravitational compression and radioactive decay within the mass. Because of this, some believe the planets were originally in a molten state. Others feel that the planets were relatively cool when formed and heated up internally as they increased in size and with passage of time.

This theory, at present, finds the greatest acceptance among astronomers. However, it is probably not the last word on the formation of the solar system.

3.2 AGE OF THE SOLAR SYSTEM

How long ago were the earth and the solar system formed? The system is very complex, yet it has many common characteristics which would indicate that the solar system was formed at one time rather than in piecemeal fashion. Measuring the rate of radioactive decay of certain elements—for example, the decay from uranium to lead or from strontium to rubidium—has provided a method for determining the age of rocks. The oldest rock found on the surface of the earth was dated at 3.3 billion years old. This, then, is the youngest that the earth could be, since the rock containing the elements must be formed before the measured radioactive decay can begin. Measurements on stony meteorites similar in composition to the mantle of the earth, and more recently on lunar rock, reveal an age of about 4.5 billion years. Because of these similarities, it may be presumed that the meteorites were formed at the same time as the stony mantle and iron core of the earth. It probably took millions of years for the relatively cool earth to heat up sufficiently from radioactivity to permit the mantle and core to separate. Therefore, 4.5 billion years is perhaps a lower limit for the age of the earth. If the length of time necessary for the protoplanet to form into the planet is considered, then the age of the earth approaches perhaps 5 billion years.

The sun's formation is thought to have taken place about 1 billion years prior to the formation of the earth, making the age of the solar system about 6 billion years.

3.3 THE SUN

During the height of the ancient Greek civilization, the sun was mistakenly viewed as a huge flaming rock about the size of Greece. Actually the sun is a unique body in the solar system, in contrast with the cold unlighted planets orbiting around it. The sun is a heaving, turbulent body of hot gas emitting huge quantities of energy. It is about 1,400,000 kilometers in diameter (about 109 times the diameter of the earth) and includes about 99 percent of the total mass in the solar system. The sun is 1,000 times more massive than Jupiter and 332,000 times more massive than the earth. However, the sun's density is only about $\frac{1}{4}$ that of the earth, being 1.4 grams per cubic centimeter compared to the earth's 5.52 grams per cubic centimeter. The sun also rotates on an axis, a fact first deduced by Galileo from the movement of sunspots across the face of the sun. By careful observation, astronomers found that all parts of the sun do not rotate at the same speed. The sun's equator rotates once in about 25 days, while the period of rotation is 28 days at approximately 35° north and south of the equator, and 34 days, 75° from the equator. This variation in the speed of rotation is additional evidence that the sun is composed mainly of gaseous material. From sunspot movement it has also been determined that the solar equator is inclined 7° from the ecliptic.

The sun is of interest because its mass governs the movement of all other members of the solar system, and its radiation is the primary source of energy in the solar system. The sun is actually a star, appearing large because it is relatively close to the earth; for this reason it yields much information on the nature of stars in general.

The sun is not by any means one of the larger stars. Others are vastly greater in size, as, for instance, Betelgeuse, which has a volume 27 million times that of the sun. Actually, Betelgeuse is so large that the earth's orbit could fit entirely within the body of the star. On the other hand, many stars are smaller than the sun. Some are 100,000 times as bright as the sun, some are 1/10,000 as bright. The sun could, on this basis, be considered a typical, average star.

The bulk of the gas of which the sun is composed is hydrogen, estimated to be about $\frac{3}{4}$ of the sun's mass, helium about $\frac{1}{4}$ of the mass; only 1 or 2 percent of the mass is composed of the heavier elements, mostly in an ionized form. Of the 92 naturally occurring elements, about 70 have been identified on the sun. None have been found on the sun which do not also occur on earth, although helium, discovered on the sun in 1868, was not found on the earth until 27 years later.

3.3A The Sun's Structure

Although the sun is a gaseous body, its structure can, for convenience of discussion, be divided into several parts (fig. 3.1).

The *interior* of the sun contains nearly all its mass and is inaccessible for observation. Studies that apply the laws of physics to known data of the sun reveal some characteristics of the solar interior. These studies indicate that the pressure at the center of the sun is approximately 15 billion pounds per square inch and temperatures are about 14,000,000° K*. The tremendous energy being radiated by the sun is generated by the conversion of hydrogen to helium in the sun's interior. This will be discussed further in the next section.

The *photosphere* is the bright visible disk of the sun through which energy from the interior escapes. The temperature of the photosphere is about 5,800° K at the center of the disk and about 300° less at the sun's edge or *limb*. The light from the sun is noticeably brighter from the center than from the limb. This

---

*Kelvin (K) = Centigrade + 273°.

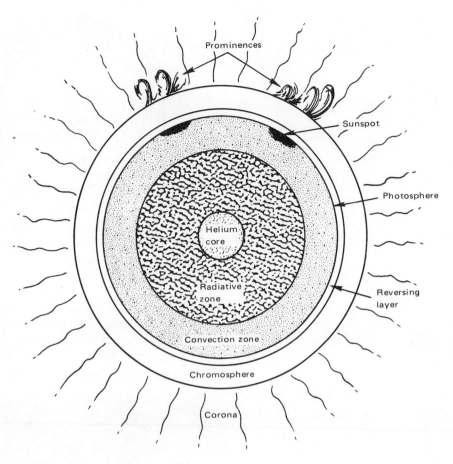

Figure 3.1 Cross-sectional view of the sun.

is because energy from the center of the disk travels directly toward the earth through the solar atmosphere. However, at the outer edges, the light crosses obliquely through the solar atmosphere and more of the energy is absorbed. Because the energy is absorbed, one is unable to see as deeply into the photosphere at the limb. The visible light at the limb, therefore, comes from a higher and cooler region of the photosphere, thus causing the *limb darkening* and lower temperatures (fig. 3.2).

Normally the sun appears as a smooth, luminous disk without noticeable surface features. However, high-altitude photographs have revealed much detail and show the solar surfaces to be quite mottled (fig. 3.3). This mottling has a fine-grained, cellular appearance called *solar granulation*. It is believed the granules are the turbulent motion of hot gases rising to the surface from the interior and are from 150 to 900 kilometers in diameter. The darker, cooler areas between the granules are the cooler gases returning to the solar interior.

Astronomy

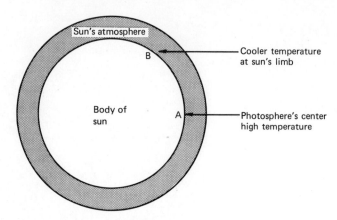

Figure 3.2 Light from the center of the photosphere (a) as viewed from the earth comes from a greater depth of the solar surface, where the gases are considerably brighter and hotter than light from the limb (b).

Sunspots are another feature of the photosphere that are of interest to the astronomer. Sunspots vary in size, most being relatively small—1,000 kilometers in diameter—and lasting only a few days, while a few may reach a diameter of many thousands of kilometers and last several weeks. Sunspots have a dark center, the *umbra*, and a lighter edge, the *penumbra*. The center of

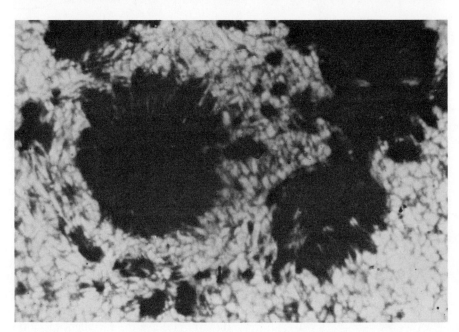

Figure 3.3 Dark areas are sunspots, which may last for days or weeks. Magnetic lines of force can be seen in what appears to be the rim of the sunspots. NASA.

a sunspot is not truly black but is dark by comparison to the surrounding area of the dazzlingly bright sun. The temperature of a sunspot is about 1,000° K lower than the normal surface of the sun.

In the early part of the nineteenth century, Heinrich Schwabe, a German druggist who had a great interest in astronomy, began keeping a systematic record of the sun's surface. He was looking for the planet Vulcan, purported by the French astronomer Urbain Leverrier, to be in an orbit between Mercury and the sun. No such planet existed, but Schwabe's detailed record eventually led him to the discovery of the cyclic nature of sunspot occurrence. He found that sunspot activity varied from minimum to maximum and back to minimum in a cycle that took a period of about 10 to 11 years. He announced his findings in 1843, but was to work another 10 to 15 years before his efforts were recognized. Later it was found that the sunspots varied in position on the sun as the cycle progressed. As the new cycle started at minimum, a few spots would appear about 30° north and south of the equator. As the cycle progressed, new spots, larger and more numerous than earlier in the cycle, appeared closer to the equator until, at maximum activity, most sunspots appeared in a zone around 15° north and south of the equator. At the end of the cycle, a few small spots appeared from 5° to 10° above and below the equator, while at the same time, new spots appeared 30° from the equator. The reappearance of spots in this location signals the start of a new cycle. The cyclic nature of sunspot activity appears to be related in some way to magnetic activity on the sun. If the cycle of activity continues as in the past, the minimum activity in 1964 means that the minimum level will reoccur in 1974 or 1975, followed by maximum activity in 1980.

In 1908, further investigation on the properties of sunspots led George Hale of the Mount Wilson observatory to the conclusion that sunspots had a magnetic field. Hale made use of a discovery by Pieter Zeeman in 1896. Zeeman had found that the spectrum lines of glowing sodium vapor broadened when a tube of glowing vapor was placed between the poles of an electromagnet. The normally single spectrum lines became double or triple under the influence of the magnetic field, and this became known as the Zeeman effect. Hale observed this effect in light from sunspots. By means of a polarizing attachment on the spectroscope, Hale was then able to show that the leading spots in a group had one polarity, and the trailing spots had the opposite polarity. He also found that sunspot groups in the northern and southern hemispheres had reversed polarities. With the start of a new cycle, the polarities in both hemispheres were reversed, thus indicating that the point of minimum sunspot activity was the real start of each new cycle. If the polarity of sunspots were considered along with sunspot activity, then the length of the cycle would be 22 years.

The *chromosphere* consists of a low-density layer thought to be about 3,000 kilometers thick that lies above the photosphere. Temperatures of the chromosphere increase with elevation and are about 15,000° K at the upper levels.

The exact mechanism of this increasing temperature is not wholly understood, but it is thought to be due to the occurrence of shock waves caused by convection currents in the photosphere. The energy of the shock waves is dissipated as heat, resulting in the high temperatures measured in the chromosphere and corona (see below).

The chromosphere is characteristically red in color due to the bright, strong emission line in the spectrum of hydrogen (Hα line). In the past, the chromosphere could only be seen during a solar eclipse when the moon covers the photosphere. Today an instrument called a *coronagraph* permits study of the chromosphere at any time. The coronagraph is a device whereby the image of the photosphere is covered at the focal plane of a telescope.

Occasionally, bright outbursts of energy appear in the chromosphere. These *solar flares*, as they are called, appear to be associated with sunspot activity and are generally of short duration (fifteen minutes to several hours). They are thought to be caused by masses of hot, glowing gas brought up with considerable force from beneath the photosphere. Ultraviolet radiation from such outbursts causes the disruption of the earth's ionosphere, resulting in the interruption of radio communication. Ionized particles from flares also react with gases of the earth's upper atmosphere and cause the auroral displays in the extreme northern and southern skies. Solar flares are also of concern to those involved in manned space flight. Radiation dosages from solar-flare activity is extremely high, even at one astronomical unit, and could be harmful to persons traveling in space beyond the protection of the earth's atmosphere.

The *corona* is the outer and most spectacular visible layer of the sun, seen only during a total eclipse or with a corongraph (fig. 3.4). The corona is an extensive and extremely rarified gaseous envelope surrounding the sun, reaching hundreds of thousands of kilometers into space. Its light is only about one-half that of the full moon and is completely masked by the sun's brilliance. The temperature is about $1,000,000°$ K, due possibly to the shock waves that increase the temperature of the chromosphere.

Solar *prominences* are great flamelike projections of the sun that are best visible along the limb during a total eclipse. Before 1860, prominences were thought to be appendages of the moon, since they could be seen only during an eclipse. Since then, the coronagraph has enabled observers to study and photograph prominences at any time.

Prominences occur tens of thousands of kilometers above the sun's surface and appear to originate in the corona. They appear as great sheets of flame which move at speeds up to 700 kilometers per second. Prominences evidently are cooling gases in the corona, the ions of which react with electrons and emit light. These gases apparently move along magnetic lines of force associated with sunspot activity. At times the luminous material appears to pour down to the photosphere from an "empty" region. In fact the gases are flowing both up and down but, for some yet obscure reason, only those flowing downward are at times radiating in visible wavelengths.

The Sun, the Earth, and the Moon

Figure 3.4 Eclipse of the sun on March 7, 1970. The planet Venus was visible during the eclipse. Courtesy of B. O. Lane.

The energy generated in the sun and the variety of activities taking place thereon result in a constant flow of particles from the solar surface. This particle flow or *plasma* is frequently called the fourth state of matter. That is, matter exists as solid, liquid, gas, or plasma. Plasma is defined as a collection of charged (ionized) particles in sufficient bulk so that its behavior is much the same as a fluid. It is this material emanating from the sun that is called the *solar wind*. Satellite exploration in the vicinity of the earth has made it possible to determine characteristics of the solar wind at one astronomical unit from the sun. At this distance the solar wind has a velocity of approximately 480 kilometers per hour, a flux* of about 5×10^8 protons per square centimeter per second and a density of 10 protons per cubic centimeter. The particles are mostly protons (hydrogen ions) and some helium.

The exact source of the solar wind is not known, but the best guess is that it originates in the corona, perhaps one to two solar radii from the sun's surface. How far out into the solar system does the solar wind extend? Calculations made on the basis of the strength of the solar wind at one astronomical unit indicate that it may extend at least to a distance of 50 astronomical units. This would imply that the planets are orbiting within the physical body of the sun, if we include the solar wind as an integral part of the sun. These findings are restricted to the region close to the ecliptic. Little has been done to determine the properties of space above and below the ecliptic.

*Flux: The number of particles flowing through a given area in a given time.

3.3B The Sun's Energy

The enormous amount of energy released by the sun is the primary source of energy on earth. Early man, viewing the sun, probably concluded that the warmth was the result of a burning process. Burning or combustion, as it is commonly understood on earth, is a process involving the breakdown of organic compounds in the presence of oxygen, resulting in a release of energy and the formation of carbon dioxide. While oxygen and carbon are to be found on the sun, the temperature of the sun is far too hot to allow these elements to combine into compounds, except in sunspots. Carbon is thought to be involved in the release of energy in some massive stars and to a very small degree in the sun, but not in the manner described above. Since compounds or combinations of elements cannot exist on the sun because of the high temperature, the sun is made up of a mixture of atoms of the various elements and subatomic particles.

The atom, or, more precisely, the binding energy of the atomic nucleus, is the actual source of energy on the sun. There are two possible means whereby energy bound up in the atom could be released: nuclear fission and nuclear fusion. In the nuclear fission process, a heavy element is broken down into two lighter elements whose combined mass is slightly less than the mass of the original element. The difference in mass is energy, which is released as in an atomic bomb. The heavy elements in the quantities required for this type of reaction are not present on the sun. However, there is an abundance of hydrogen, which makes the fusion process the likely mechanism for the release of energy. In this process, hydrogen nuclei are combined to form helium through a series of reactions called the proton-proton chain. The first step in the reaction is the fusing of two protons into deuteron or heavy hydrogen isotopes and the emission of a positron:

$$_1H^1 + {}_1H^1 \longrightarrow {}_1H^2 + {}_1e^0$$

The positron subsequently combines with an electron to form a gamma ray. The deuteron encounters and combines with another proton to form a light helium isotope:

$$_1H^2 + {}_1H^1 \longrightarrow {}_2He^3$$

The light helium isotopes fuse to form a stable helium atom and two protons:

$$_2He^3 + {}_2He^3 \longrightarrow {}_2He^4 + 2{}_1H^1$$

Each of these reactions results in the release of a small amount of energy which is represented by the loss in mass. Four hydrogen atoms (or protons), each with an atomic weight of 1.008 and a combined atomic weight of 4.032, will yield one helium atom with an atomic weight of 4.003. The difference represents the mass which is converted into energy radiated by the sun.

The total amount of energy radiated by the sun, or its *luminosity*, can be determined by measuring the strength of the solar radiation received on the earth's surface. In making such a measurement, energy absorbed or reflected by the earth's atmosphere must be taken into account. Careful studies have shown that the amount of energy striking the earth is 1.97 cal*/cm²/min or 1.37×10^6 ergs†/cm²/sec, which is known as the *solar constant*. Since it can be assumed that the sun radiates uniformly in all directions, the total amount of energy given off by the sun should be the solar constant multiplied by the surface area of a sphere with a radius of 1 A.U. This area is equal to 2.82×10^{27} cm², so the total output of energy by the sun is 3.8×10^{33} ergs/sec. This is energy which has been formed at the expense of some of the sun's mass.

How much mass is being expended can be determined by making use of Einstein's formula, $E = mc^2$. The energy (E) output per second is known, and c is the speed of light. From this, the mass (m) can be calculated. It is found that more than 4.6 million tons of matter are converted into energy every second. As large as this rate may be, it has very little effect on the sun's activity. In a year, the sun loses by radiation about one-15,000-billionth of its total mass, but this does not mean that the sun will last for 15,000 billion years, because it will not continue to function in its present manner to the point of complete annihilation. Since only about 0.7% of the mass in the hydrogen-to-helium conversion becomes energy, the sun has a theoretical life expectancy of about 100 billion years. This figure is too large, however, because the temperature at the center of the sun necessary for continuance of this reaction cannot be maintained after about 10% of the available fuel has been exhausted. A realistic life expectancy for the sun is, therefore, about 10 to 11 billion years. The age of the sun is estimated to be about 6 billion years; hence the sun's eventual demise is predicted for some 4 to 5 billion years in the future.

3.4 THE EARTH

The earth is the third planet from the sun. It is neither the largest nor the smallest planet (see table 4.1), but it is the only one in the solar system thus far known to support life. Studying the earth as a planet has the advantage of permitting data obtained from other planets to be interpreted in earthlike terms, since it is assumed that the physical laws of science which apply on earth will apply to all parts of the universe.

There are some problems in studying the earth in its entirety because only a small portion is visible at any one time to a single observer on the surface. Certain aspects of the earth could be viewed better by an observer on the moon. From such a vantage point it would be relatively easy to observe the physical shape and astronomical motions of the earth.

3.4A The Earth's Shape

Measurements have shown that the earth is not a perfect sphere as the Greeks thought. It is an oblate spheroid, slightly compressed at the poles and bulging at the equator as a result of the earth's rotation. The polar radius of the earth is

---

*Calorie: Heat necessary to raise temperature of one gram of water one degree centigrade.
†Erg: One calorie equals 4.18×10^7 ergs.

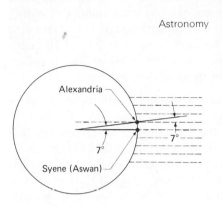

Figure 3.5 Eratosthenes' method of measuring the circumference of the earth.

6,356.6 kilometers, and the equatorial radius is 6,378 kilometers. The circumference at the equator is about 40,100 kilometers. Measuring the radius and the circumference of the earth is not a recent accomplishment. In the third century B.C., a Greek astronomer, Eratosthenes, was able to measure the circumference of the earth. He noted that on the longest day of the year in Alexandria, a stake in the ground cast a shadow of slightly more than 7° of arc at noon (fig. 3.5). On the same day in Syene (Aswan), the sun was directly overhead and cast no shadow. The angular difference between the location of the two cities was 1/50 of a complete circle and the distance between the two cities was 5,000 stadia (920 kilometers). From these facts, Eratosthenes calculated the circumference of the earth and found it to be equivalent to about 250,000 stadia or 46,000 kilometers. This estimate is somewhat higher than the present-day figure of 40,100 kilometers but Eratosthenes did not have the advantage of modern precision equipment.

3.4B The Earth's Atmosphere and Magnetosphere

An observer on the moon would find the surface of the earth partially obscured by a layer of clouds (fig. 3.6). This cloud layer is part of the atmosphere of gases which surrounds the earth, a more detailed description of which may be found in Chapter 13.

The existence of magnetism has been known for more than 2,000 years, but the fact that the earth itself was a huge magnet was not demonstrated until 1600, when Sir William Gilbert explained how the magnetic compass may be used for navigation. The earth's magnetic field was viewed much like that shown (fig. 3.7) in which the small circle represented the earth and the lines represented the magnetic field.

Since the advent of satellite exploration in 1957, the view of the earth's magnetic field has changed, and the term *magnetosphere* has been coined to identify the region of earth's magnetic influence in space (fig. 3.8). The *magnetopause* is the boundary which separates the region close to the earth, where there are strong and oriented magnetic fields derived from the planet itself, from the region outside, where there are weak and fluctuating fields derived from the sun. The solar wind impinging on the earth's field causes field lines from the polar regions to be swept back in a direction opposite the sun to form the earth's *magnetic tail*. Field lines in the tail below the plane of the magnetic equator point predominantly away from the sun, those above the plane point toward it. Between the two at the equatorial plane is a neutral sheet in which field reversal is accomplished.

The flow of the solar wind through interplanetary spaces is supersonic. Therefore, a shock wave, the "bow shock," is created upstream of the earth's field in the same manner that a shock wave is created upstream of an obstacle in a river or a boat plowing through water. Between the bow shock and the magnetopause is a region of turbulence insofar as field direction and particle motion are concerned. This region of turbulence is called the *transition zone*.

Figure 3.6 Photo taken from Apollo 11 about 150,000 kilometers from earth shows Africa and portions of Asia. NASA.

Within the confines of the magnetosphere are the *Van Allen radiation belts* discovered by James A. Van Allen in 1958. The radiation belts were described by him as two concentric belts of charged particles, one approximately $1\frac{1}{2}$ radii from the earth and the other about 3 to 4 radii. Some scientists now prefer to speak in terms of a single radiation zone with a slot between them where radiation levels are lower, rather than two well-defined belts.

Both belts contain protons and electrons trapped by the earth's magnetic field. Proton energies in the outer belts are too low to contribute to the particle count, and this led early observers to conclude that the outer belt was composed primarily of high-energy electrons.

Recent satellites have continued the exploration of the earth's radiation belt and mapped it in greater detail. The more the belts are studied the more complex

Astronomy

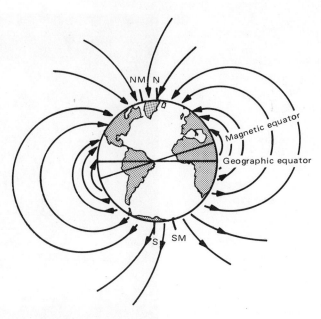

Figure 3.7 Old view of earth's magnetic field.

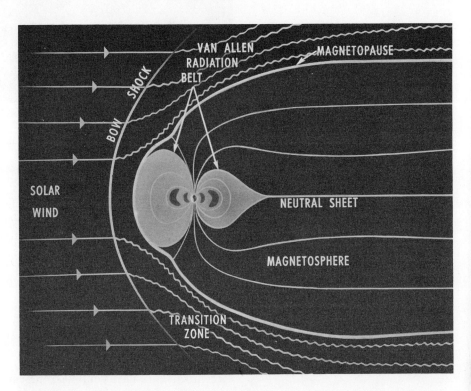

Figure 3.8 A diagram of the magnetosphere established by satellite experiments. NASA.

they seem. First, they are strongly affected by the magnetic field of the earth, and second, both the belt and the magnetic field are influenced by the activities of the sun. These interrelationships continue to be studied to obtain a greater knowledge about our near-earth environment.

3.4C Movement of the Earth

The earth moves in a variety of ways within the confines of the solar system. Some of the motions readily recognized include *rotation* (turning of the earth on its axis), *revolution* (movement of the earth in orbit around the sun), and *precession* (conical motion of the earth's axis due to the attraction of sun and moon on the earth's equatorial bulge).

The rotation of the earth is evidenced by the daily movement of celestial objects (sun, stars) from east to west across the sky. This motion was incorrectly attributed by the Greeks to the movement of these objects in space, but ample proof now exists that the earth moves. One of the first of these proofs was provided in the nineteenth century by the French physicist, J. B. L. Foucault, who demonstrated the rotation of the earth in the Pantheon in Paris by means of a huge, swinging pendulum. A 25-kilogram weight attached to a metal thread about 65 meters in length was suspended over a calibrated disk to indicate the direction of swing. When the pendulum was started swinging the weight slowly veered westward, indicating the earth's eastward rotation.

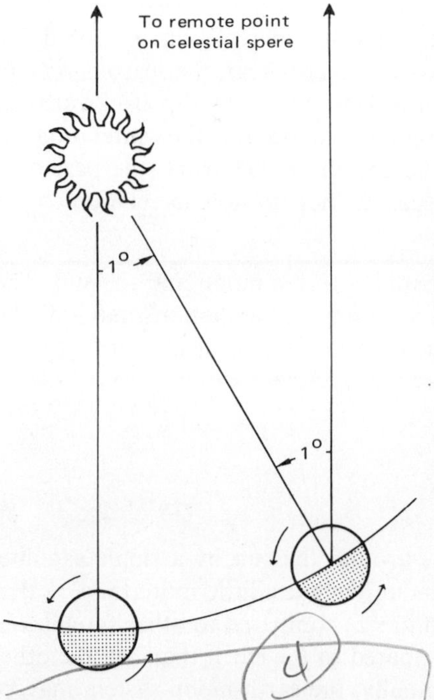

Figure 3.9 A sidereal day represents a 360° rotation of the earth on its axis. A solar day represents the time necessary for a point on the earth's surface to rotate once and return to the same position relative to the sun at the end of one rotation. This represents a rotation of about 361°.

The earth's rotation is used as a convenient means of telling time, each rotation being equivalent to one day of 24 hours. There are, however, two ways in which this period may be measured. One is by selecting a distant star and noting when it crosses the celestial meridian.* The interval of time from one transit of the star across the meridian to the next transit is one rotation. This interval is called the *sidereal* (star) *day* and represents the true rotation of the earth through 360°. The second means of measuring earth's rotation makes use of the interval between one transit of the sun across the meridian until the next transit. This interval is the *solar day*. Since the earth moves in orbit (fig. 3.9) each day, the solar day is approximately 4 minutes longer than the sidereal day.

Thus

1 sidereal day = 24 h sidereal time
or 23 h 56 m 4.091 s/of mean solar time

1 solar day = 24 h of mean solar time
or 24 h 3 m 56.55 s/sidereal time

The earth's revolution around the sun is used as a measure for one year. It requires nearly 365¼ days to make one complete orbit. For this reason, three calendar years out of four have 365 days, and the fourth year (leap year) has 366 to compensate for the accumulated quarter days.

---

*Celestial meridian: An imaginary line (great circle) on the celestial sphere passing directly overhead (zenith) and crossing the viewer's horizon at 90°.

The precession of the earth, a movement not so readily discernible, resembles the gyrations of a spinning top or gyroscope. This movement is so slow that it is barely visible even to close observation, and only over long periods of time. The axis of the earth is inclined 23½° from an imaginary axis perpendicular to the ecliptic plane. Precession is the movement of the earth's axis at this angle around the imaginary axis perpendicular to the ecliptic plane. This movement causes a shifting of the earth's axis relative to the stars, so that while Polaris is the present polar star, Alpha Draconis was the polar star in 3,000 B.C., and Vega will be the polar star in 14,000 A.D. To return to the present position after a complete circle will require 26,000 years.

Because of precession, the earth's period of revolution is also measured in two ways. The interval of time required for the earth to make one complete (360°) orbit around the sun is the *sidereal year*. The length of the sidereal year is 365 d 6 h 9 min 10 s, and this is considered the true period of revolution. The *tropical year*, used for calendar purposes, is measured between vernal equinoxes. The *vernal equinox* is that moment on approximately March 21 when the sun is directly over the equator, apparently moving from south to north. Because of precession, the vernal equinox shifts slightly westward each year causing the tropical year to be 20 m 23.42 s shorter than the sidereal year, or 365 d 5 h 48 min 46 s. By adapting the calendar to the tropical year, the seasons remain in phase with the calendar. Otherwise the seasons would slip backwards in relation to the calendar, eventually causing spring to occur in the winter months, with other seasons following suit. At the end of 26,000 years (the period of precession), the seasons would again occur as they do now in relation to the calendar.

There are other motions of the earth shared with the moon and the sun. The total effect is quite complex and the speed of any of the motions may only be given relative to that object against which the movement of the earth is measured. This is more fully described in section 3.5A.

3.5 THE MOON

The earth is accompanied in its travels around the sun by a single satellite, the moon. The moon is 3,476 kilometers in diameter, little more than ¼ that of earth. Although the moon ranks fifth in size compared to other satellites in the solar system, it is more massive compared to the earth than are the other satellites compared to their planets. Actually, the earth-moon system may be more characteristic of a double-planet system than a planet-satellite system.

The mean distance from the center of the earth to the center of the moon is 384,404 kilometers or, more conveniently, 380,000 kilometers. It requires 1.28 seconds for light to travel from the moon to the earth. The moon does not revolve around earth in a perfect circle: Its orbit is elliptical and slightly eccentric. At its closest approach (perigee), the moon is 356,555 kilometers from the earth, and at its greatest distance (apogee), 406,863 kilometers from the earth.

3.5A   Lunar Motion

The motion of the earth has been discussed previously without reference to the effect of the moon on this movement. What is called the earth's orbit is, in reality, the orbit of the center of mass, or *barycenter*, of the earth-moon system. Slight shifts of the earth, occurring on an approximately monthly cycle, show that the system's center of mass is 4,660 kilometers from the center of the earth on a line between the center of the earth and the center of the moon. This places the barycenter within the earth. As a result, the earth's orbit is not a smooth path, but rather the earth and the moon follow a slightly serpentine course around the sun (fig. 3.10).

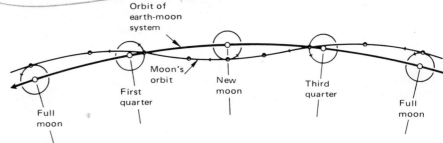

Figure 3.10   The earth and the moon follow a serpentine path relative to the sun as they orbit.

Knowledge of this type is important in that the data permits one to calculate the lunar mass if the mass of the earth is known (it is), and the relative distances of the centers of the earth and moon to the barycenter are known. The mass and volume data of the moon permits determination of lunar density, which is 3.3 grams per cubic centimeter. This type of information serves as a basis for speculating on the lunar structure as compared to the structure of the earth.

The moon's revolution around the earth takes about one month. However, there are two ways of measuring the moon's period of orbit. The *sidereal month* is the true period of revolution, being the time interval between two successive conjunctions of the moon's center with the same star, as seen from the earth. The time it takes the moon to travel this 360° orbit around the earth is a sidereal month of 27 d, 7 h, 43 m, 11.5 s, or about $27\frac{1}{3}$ days. The *synodic month* is the interval between successive conjunctions of the moon and the sun, or from new moon to new moon. This month is longer than the sidereal month by more than 2 days, being 29 d, 12 h, 44 m, 2.8 s, or about $29\frac{1}{2}$ days. The difference is due to the fact that during the moon's revolution about the earth, the earth in its turn has moved along its orbit around the sun as illustrated in figure 3.11.

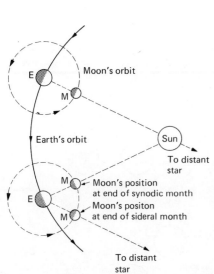

Figure 3.11   The sidereal month is measured as the time it takes the moon to go 360° around the earth. The synodic month is the time it takes the moon to make a successive conjunction with the sun, and it is slightly more than two days longer than the sidereal month.

The moon's eastward motion, resulting from its movement in orbit around the earth, causes the moon to fall behind the rotation of the earth so that the moon returns over the same meridian an average of 50 minutes later each day. This is referred to as the *daily retardation* of the moon.

The moon rotates on its axis in the same length of time in which it revolves around the earth, namely, the sidereal period of $27\frac{1}{3}$ days. Because of this, the moon presents the same hemisphere toward the earth at all times, permitting viewers on earth to see only a portion of the lunar surface.

One of the first celestial phenomena to be viewed and understood by man was the phases of the moon (fig. 3.12). The moon itself is dark and only reflects light from the sun as the moon moves around in its orbit. The alternately increasing and then decreasing lunar areas reflecting sunlight toward the earth are the phases of the moon. When the moon passes between the earth and the sun, the hemisphere of the moon facing the earth is dark. This is the new moon phase. On the following night a slight crescent of light will appear signaling the beginning of a new lunar month. The crescent of light increases until about

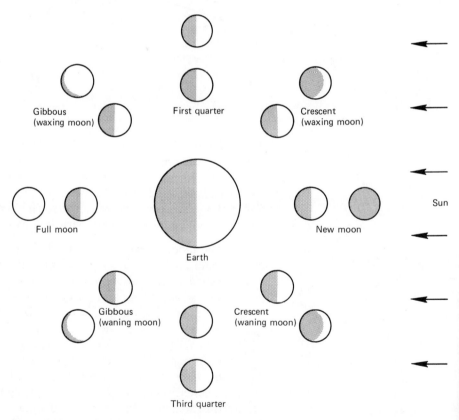

Figure 3.12 The phases of the moon.

half the moon's face is lighted, which represents the first-quarter phase. Then follows the gibbous phase: The lighted portion of the moon increases nightly until the position of the moon is opposite that of the earth from the sun and the full moon is visible. The phases are then repeated in reverse order through gibbous, last quarter, and crescent, to new moon again.

The boundary between the bright and dark portions of the moon, the *terminator*, appears as a smooth, regular line. However, on closer examination with magnification, the line is seen to be irregular due to the mountainous nature of the lunar surface.

3.5B The Eclipse

Until recently it was not possible to view the outer edges of the sun except during a total eclipse. This event made possible the examination of prominences and corona. In the pattern of movements of the sun, earth, and moon, there are always two to five solar eclipses during a year. Some are partial, some annular (that is, a thin ring of the sun's disk is left visible), and some total. On the average, only two total eclipses occur on earth every three years.

A *total eclipse* of the sun can occur only when the moon is new, that is, when it is between the earth and the sun. Furthermore, it occurs only when the moon is at or near its closest approach to the earth. Since the moon's shadow is 373,520 kilometers long, it is obvious that an eclipse could not occur under average conditions. However, because of its elliptical orbit, the moon's shadow is sometimes longer than the distance from the moon to the earth, allowing the shadow to fall on the earth, causing a total eclipse. One other factor that must be considered is the proximity of the moon to the ecliptic plane. The moon's orbit is inclined 5° to the ecliptic, which means during some new moon episodes the lunar shadow will fall above or below the earth (fig. 3.13). Because of the orbital inclination, it is necessary for the moon to be within one earth radius of the ecliptic in order for the eclipse to occur on earth.

It is possible to see a total eclipse when the true shadow of the moon, the *umbra*, passes overhead. The umbra produces a round shadow never more than 274 kilometers in diameter at or near the point on the earth directly beneath (i.e., nearest) the sun. As the umbra approaches the grazing position, the shadow becomes more and more elongated. The *penumbra*, which surrounds the umbra like an inverted cone, does not completely hide the sun, and observers located on any point in the lightly shaded area in figure 3.14 will see only a partial eclipse. The penumbra forms a circle about 6,500 kilometers in diameter.

At the moment of totality, when the moon completely covers the disk of the sun, it is possible to see the solar prominences. Also, the darkened sky is lightened by the corona, which may extend for more than $1\frac{1}{2}$ million kilometers from the sun's surface. A total eclipse will last only a few minutes, then a sliver of light reappears and the eclipse is over.

Astronomy

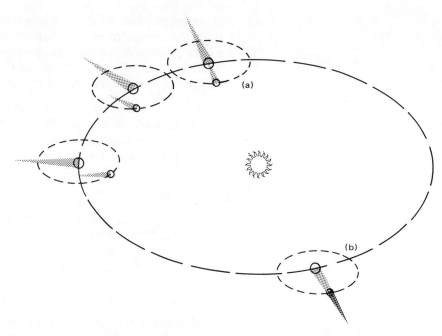

Figure 3.13 A solar eclipse can occur only when the moon is near the ecliptic during the new moon phase (a). A lunar eclipse occurs at full moon with the moon at or near the ecliptic (b).

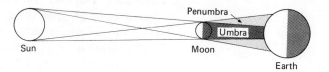

Figure 3.14 An eclipse of the sun.

As previously mentioned, the distance from the earth to the moon varies so that the umbra of the moon does not always reach the earth. In this case, an observer directly under the tip of the dark cone will see a thin ring of light surrounding the moon, the thin ring of light of the *annular eclipse*. This has a diameter of about 50 kilometers wider than a total eclipse. Surrounding this area is a partial eclipse about 6,500 to 9,500 kilometers wide.

Eclipses can be predicted quite accurately for some time into the future. This is accomplished through our detailed knowledge of the motions of the moon and the sun relative to the earth. Predictions for solar eclipses in the near future are:

    June 20, 1974: Southern Indian Ocean, Antarctic Ocean, southern Australia
    Oct. 23, 1976: East Africa, across Indian Ocean and Australia
    Oct. 12, 1977: Mid-North Pacific, southeastward into northern South America

Feb. 26, 1979: North Pacific Ocean, northwest tip of United States, across Canada into central Greenland

Feb. 16, 1980: Atlantic Ocean, across central Africa, Indian Ocean, India, southern China

The moon may be eclipsed whenever the earth passes between the moon and the sun. The distance between earth and moon is not a factor, because the earth's shadow extends well beyond the moon's orbit. However, the moon must be in a full-moon position and must be within one earth radius of the ecliptic in order for a lunar eclipse to occur. When totally eclipsed, the moon will still be visible because the red wavelengths of sunlight are bent sufficiently when passing through the earth's atmosphere to reach the moon. This light is reflected back to earth, giving the moon a dull reddish appearance.

3.5C The Lunar Surface

When seen through a telescope, the most obvious feature of the moon is its cratered surface. The craters measure up to 240 kilometers in diameter and are formed by meteorite impact or volcanic activity.

At one time the moon was thought to be a cold, dead world but more than 400 "transient lunar events" have been observed in the past few centuries. In 1787, Sir William Herschel saw a red area near the crater Aristarchus glowing like "slowly burning charcoal thinly covered with ashes." Over the centuries, many moon watchers have seen such flashes of light and glowing red spots. More recently, infrared maps of the moon have revealed many hot spots which may be lava flows or some other manifestation of lunar volcanic activity. One of the samples returned from the moon by *Apollo 12* had many characteristics of volcanic ash, although no direct volcanic activity was observed at the site of the landing.

Numerous impact craters have been observed on the moon even on what appears to be the relatively smooth *mares* or "sea" areas. These areas were called seas by early astronomers (after Galileo) because they looked like bodies of water through the telescopes used in the early seventeenth century.

Typical of the seas are the landing sites of *Apollo 11* (Sea of Tranquility) and *Apollo 12* (Ocean of Storms). The *Apollo 12* landing site was among a group of small craters ranging from 45 to 360 meters in diameter, similar to those shown in figure 3.15. The ground surface was covered by particles ranging in size from microscopic to boulders several yards across, and beads of glassy material were found on and in the ground. In addition to the glass, plagioclase, olivine, pyroxene, troilite, ilmenite, sanidine, and iron were positively identified as minerals in the rock. The rocks were mainly crystalline igneous and some breccia. Chemically, the *Apollo 12* samples contained more iron, magnesium, and nickel than *Apollo 11* samples, but less titanium, potassium, and rubidium. The age of the *Apollo 12* samples (1.7 to 2.7 billion years) was about one billion years less than those of *Apollo 11*. However, one rock from the Ocean of

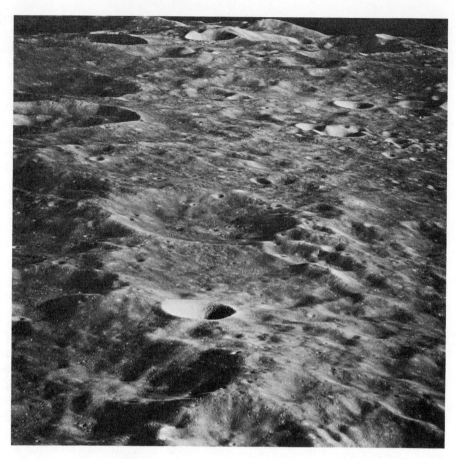

Figure 3.15 Photograph of craters on the far side of the moon taken from Apollo 10. NASA.

Storms contained significantly higher quantities of uranium, thorium, and potassium than other samples collected, and radioactive dating techniques indicated this sample to be 4.6 billion years old.

Stratification of the loose surface material was detected in the core sample taken at the *Apollo 12* landing site. This was attributed to separate layers of ejected material being deposited each time a meteorite impacted and formed a crater. The craters differ widely in age, a factor determined from the variation in crater shape and the change in the ejected material.

Although there is no detectable atmosphere on the moon, some of the rock samples did contain small amounts of the noble gases. The fines (particles > 1 cm in diameter) and the breccia were found to contain a relative abundance of helium originating from solar wind activity. The crystalline rocks generally yielded smaller quantities of gas because of radioactive decay.

Apollo 14 and *Apollo 15* were missions to more mountainous regions. Hills at the *Apollo 14* site resulted from material ejected by the impact that formed Mare Imbrium. Material returned by *Apollo 14* led to the theory that the impact occurred about 700 million years after the formation of the moon. At that time, the theory holds, a planetesimal, possibly another earth satellite about 150 kilometers in diameter, was swept up by the moon at an impact velocity of no more than 6 kilometers per second.

The Apennine Mountain range, landing site of *Apollo 15*, was also formed by the Imbrium impact and by subsequent faulting. Seismic data from *Apollo 15* seems to indicate that the moon has a crust and a mantle (see section 8.3).

The *Apollo 16* mission landed in the Descartes region in the Central Highlands. This area was thought to have been formed by volcanism but now appears to be a mixture of volcanic and impact activity.

The *Apollo 17* landing site was Taurus Littrow, where volcanic activity was examined, and lunar samples were collected for further determinations of the moon's age. This site also represented a location removed from previous landing sites which permitted further geophysical measurements of the moon in conjunction with scientific equipment left at other sites.

The absence of a lunar atmosphere is responsible for the wide range in temperatures on the moon's surface. Temperatures vary from $100°C$ at lunar noon to $-150°C$ during the lunar night. High temperatures and the lack of atmosphere may have precluded the development of life on the moon. No organisms were found in the lunar samples, nor was there any evidence of fossil material. Thus, it would appear that the first extraterrestrial body in the solar system to be explored by man is devoid of life. This may be an advantage, for if the moon is colonized at some time in the future, the lack of alien organisms will be one less hazard man will have to contend with.

3.6 SUMMARY

The solar system consists of the sun, which contains 99% of the mass of the solar system, the planets and their satellites, and various other objects. Mercury, Venus, Mars, Jupiter, and Saturn are visible without the aid of a telescope. Uranus, Neptune, and Pluto have been discovered since the invention of the telescope. It is believed that the solar system was formed 5 to 6 billion years ago from a cloud of dust and gas.

The sun, the most important object in the solar system, is the primary source of energy—energy produced by the conversion of hydrogen to helium. The sun is a star, like the multitude of stars seen at night. Other bodies in the solar system are visible only because they reflect light received from the sun.

The earth and its near environment are profoundly influenced by the sun's activity. The solar wind affects the earth's magnetosphere as well as the radiation belts formed by protons and electrons trapped within the magnetosphere.

Several motions of the earth are readily recognized: namely, rotation, revolution, precession, and movement around the barycenter. Some of these movements are useful in providing a convenient method for measuring the passage of time.

The earth is accompanied by a single natural satellite, the moon. A great deal is known about lunar motion and the relation of this motion to solar and lunar eclipses. Recent landings on the moon have provided much information about the moon, and indirectly, much about the nature of extraterrestrial bodies.

QUESTIONS

1. Describe the formation of the solar system according to the nebular hypothesis as suggested by von Weizsäcker and Kuiper.
2. Describe sunspot activity on the sun during the course of a complete cycle.
3. Describe the process whereby energy is generated in the sun.
4. What is the difference (aside from 4 minutes) between the solar day and the sidereal day?
5. What circumstances must be satisfied in order for a total eclipse of the sun to occur?
6. Describe the motion of the earth and the moon with respect to the barycenter.
7. How does the tropical year differ from the sidereal year?
8. Define the following terms: solar limb, corona, vernal equinox, precession, daily retardation, terminator.

4 The Solar System: The Planets, Asteroids, Comets, and Meteoroids

Difference between a star and a planet.
Properties and characteristics of the individual planets.
Asteroids.
Comets.
Meteoroids, meteors, meteorites.

The sun, the primary source of energy and the controlling force in the solar system, is circled by nine planets. Of these planets, only the earth, third planet from the sun, is presently known to be inhabited, but the other planets are of interest because they are objects in space which can be compared closely with the earth. Much of what is known about the planets is of a statistical nature and may readily be shown in tabular form for easy study (table 4.1). Despite extensive studies of the planets, relatively little is known about them—in fact, more may be known about distant stars many light years away.

A star is a self-luminous body which radiates great quantities of light. The analysis of the light tells much about the composition of the star's surface layer. In some stars there is enough turbulence, or churning and mixing, inside the star to permit the assumption that the spectrum of light coming from the surface of the star represents the composition of a fraction of the star's interior. The planets, on the other hand, are solid bodies surrounded by gaseous atmospheres of varying densities. The planets only reflect light they receive from the sun and do not radiate light as the sun and stars do. They are, therefore, not subject to the same type of analysis. Of equal importance to the self-luminous nature of stars is the fact that stars of all ages are found in the sky.

Table 4.1 Some Planetary Data

	Mercury	Venus	Earth	Mars	Jupiter	Saturn	Uranus	Neptune	Pluto
Discovery	Ancient	Ancient	—	Ancient	Ancient	Ancient	Herschel 1781	Adams, Leverrier, 1846	Tombaugh 1930
Diameter (km)	4,680	12,200	12,742	6,648	139,500	116,340	47,500	44,800	6,000 (?)
Period of rotation	59 d	243 d (?)	23h 56m	24h 37m	9h 50m	10h 15m	10h 45m	12h 48m	6.4 d
Period of revolution	88 d	225 d	365¼ d	687 d	11.8 y	29.6 y	84 y	166 y	248 y
Mass (Earth=1)	0.054	0.814	1.000	0.107	317.4	95.0	14.5	17.6	?
Density (gm/cm)	6.03	5.11	5.52	4.16	1.34	0.68	1.55	2.23	?
Atmosphere	Very tenuous	Carbon dioxide water vapor (extensive)	Nitrogen, oxygen, carbon dioxide, argon, water vapor (extensive)	Carbon dioxide water vapor (thin)	Methane ammonia (extensive)	Methane ammonia (extensive)	Methane ammonia (probably extensive)	Methane ammonia (probably extensive)	?
Satellites	None	None	1	2	12	10	5	2	?
Orbital velocity (km/sec)	47.9	35.0	29.8	24.1	13.1	9.6	6.8	5.4	4.7
Escape velocity (km/sec)	4.2	10.4	11.1	5.0	60	35.5	22.6	24.2	?

The Solar System: The Planets, Asteroids, Comets, and Meteoroids

Stars that are young, middle-aged, old, and nearly extinct have been discovered. Study of these stars in the last few decades has added much to the knowledge of the complete life cycle of a stellar body. On the other hand, very little is known about the history of the planets, other than that they are all approximately the same age, about 4.5 billion years old. Exploration of the moon and probes sent to Venus and Mars have revealed some new details about the nature of the planets but have also raised new questions. These questions will be studied by carefully designed experiments conducted by those who have a broad understanding of the earth as a planet.

The planets can be divided into two groups: the terrestrial (earthlike) planets and the major or Jovian planets. Mercury, Venus, Earth, and Mars have similar densities, whereas the major planets have a much lower average density. Why these differences exist is one of the intriguing questions currently being pursued. The ninth planet, Pluto, appears to be similar to the terrestrial planets, but because of its great distance, not much is known about its physical characteristics.

4.1 MERCURY

Mercury, named after the messenger of the gods in Roman mythology, is the closest to the sun and has the shortest period of revolution. It is little more than a third of the diameter of the earth and has an average density of about 5 grams per cubic centimeter, slightly less than the earth's average density.

Mercury's period of revolution around the sun is 88 days. Its period of rotation was also thought to be 88 days, but recent radar observations and careful calculations in 1965 have shown that Mercury rotates on its axis in 59 ± 3 days. The exact rotation period is not known, but it is suspected to be $\frac{2}{3}$ of the orbital period (58.64 days). Such a rotation amounts to 3 rotations for every 2 orbits around the sun and is the result of the sun's force of gravity upon Mercury's rotation. The earth is not influenced in the same manner because of its greater distance from the sun. But the moon, being so near the earth, is affected to a much greater degree by the earth's gravitational field. This has resulted in a one-to-one relationship between the moon's period of rotation and revolution.

Mercury was always thought to be a body without an atmosphere because of the great amount of energy in the form of heat absorbed by the side that faced the sun. Some indirect evidence suggests that the planet may, in fact, have an atmosphere, possibly composed of small amounts of argon. However, Mercury's albedo (amount of sunlight reflected from a planet's surface) and spectroscopic studies of the planet indicate no atmospheric gases.

The mean distance of Mercury from the sun is 36 million miles. However, the planet's orbit is elliptical, making the distance to the sun 43.5 million miles at *aphelion* (point in orbit of greatest distance from the sun) and 28.5 million miles at *perihelion* (closest approach to the sun). Little is known of the surface, but radar studies in 1970 have shown the presence of several large rough areas and one smooth area.

4.2 VENUS

Venus, named for the Roman goddess of beauty, is sometimes called the earth's twin because of a similarity in size and mass (table 4.1). Venus is about 0.7 astronomical units from the sun, has the most nearly circular orbit of the planets, and requires 225 days to complete one revolution around the sun. The planet is thought to rotate slowly on its axis with a period of 247 days in a retrograde direction (opposite to that of the earth). This has not been confirmed due to the dense atmosphere surrounding the planet, which hides its surface from view. The atmosphere is far denser than the earth's and contains particles of high reflectivity, causing Venus to appear as a brilliant object in the sky.

Reliable measurements of the atmospheric cloud cover have been made when light from stars, passing directly behind Venus, has been studied. Values from 85 to 120 kilometers have been obtained for the altitude of the cloud tops, a figure confirmed by the 1962 *Mariner II* probe to Venus. Data from this probe indicated that the cloud layer is about 25 kilometers thick, starting 70 kilometers from the Venusian surface and extending to an altitude of about 95 kilometers. The highest measured elevation for clouds on earth are noctilucent (night glowing) clouds at approximately 85 kilometers; the cloud layer on earth is not continuous as appears to be the case on Venus.

The composition of the Venusian atmosphere has not yet been completely determined. However, carbon dioxide is positively identified as a major constituent, comprising 90 to 95% of the total. Very small amounts of water vapor (about 0.1%) have also been detected. The determinations were first made from earth. In 1967, *Venera 4*, a Soviet probe to Venus, parachuted a capsule with a gas analyzer to the surface of Venus and verified these findings. *Venera 4* also had equipment for measuring temperatures and atmospheric pressure. At an altitude of 25 kilometers above the surface (where equipment ceased to function), the temperature was 280° C (540° F) and the pressure about 20 atmospheres. The American *Mariner V* flyby, which arrived at Venus at the same time as *Venera 4*, sent signals through the Venusian atmosphere back to earth. These signals verified the temperatures and pressure data of the Soviet spacecraft. When the data from the two spacecraft was extrapolated downward, a surface temperature of 425° C and a surface pressure of about 75 atmospheres was obtained. The temperature readings confirmed findings made by radio astronomers in 1956 and received from the *Mariner II* flyby in 1962.

On July 22, 1972, the Soviet *Venera 8* made a soft landing and survived the environmental rigors of the Venusian surface for about 50 minutes. Data returned to earth indicated a surface temperature of 470° C (900° F) and a pressure of 90 atmospheres. The atmosphere is composed primarily (97%) of carbon dioxide and a little less than 2% nitrogen, with less than 1% water vapor detected in the clouds. Traces of ammonia were also detected in the cloud layer. Despite the presence of the thick cloud layer, some light did filter down to the surface of Venus.

Wind movements were recorded with velocities of 50 meters per second (110 mph) noted at an altitude of 45 kilometers, but tapering down to 2 meters per second at the surface. The surface appeared to be composed of loose material similar to material on the earth's surface and had a density of 2.5 grams per cubic centimeter. It was compared to the composition of terrestrial granite.

Information received on the Venusian temperature prompts the asking of the question: Why is the surface of Venus so hot? Carbon dioxide and water vapor in the atmosphere absorb heat radiated by the planet. The heat may initially come from solar energy or escape to the surface from the interior of the planet. In the case of the earth, most of the heat escapes into space because of the small percentage of carbon dioxide and water vapor in the atmosphere. In the case of Venus, with the large amounts of atmospheric carbon dioxide (70,000 times more than on earth) to absorb planetary infrared radiation, the heat is trapped, raising the surface temperature considerably above the level attainable if Venus were an airless body. This heating process is called the "greenhouse effect."

Next, one might ask about the source of atmospheric carbon dioxide on Venus. Earth also has a considerable amount of carbon dioxide, but it is tied up as carbonate rock. The atmospheric carbon dioxide reacted with calcium and magnesium released from silicate rock by action of water. Since there appears to be little water on Venus, this reaction could not take place, and the carbon dioxide remains as a gas in the atmosphere.

Another question that may be asked is, how much water is on Venus? Water on the earth would average about 3.5 kilometers deep if spread uniformly over the entire surface of the globe. This water is believed to have surfaced from the earth's interior as a result of volcanic activity during its 4.5 billion-year history. Venus, as the earth's twin, is believed to have a similar supply of water trapped in its interior and gradually released to the surface. The best evidence to date, however, indicates that Venus has the equivalent of no more than a few inches of water in its atmosphere, and this exists in vapor form at the high temperatures that prevail on Venus.

The scarcity of water on Venus may be the result of the photodissociation process, whereby water is broken down to hydrogen and oxygen by radiant energy from the sun. The same reaction occurs on earth but at a reduced rate because of earth's greater distance from the sun. The hydrogen from this reaction escapes from the Venusian atmosphere into space, while the oxygen is thought to react with methane and carbon monoxide to form carbon dioxide.

52

Astronomy

The surface of Venus, never seen despite the fact that Venus is the closest planet to earth, is thought to be a hot, dry desert, possibly with hurricane-like winds blowing dust high into the atmosphere. At the high temperatures that exist, Venus must be devoid of life and an unlikely prospect for future manned exploration.

4.3. MARS

Skipping the earth, we come to Mars, which is approximately 80 million kilometers farther from the sun than the earth. Mars rotates on its axis once every 24 hours and 37 minutes and circles the sun in 687 days (table 4.1).

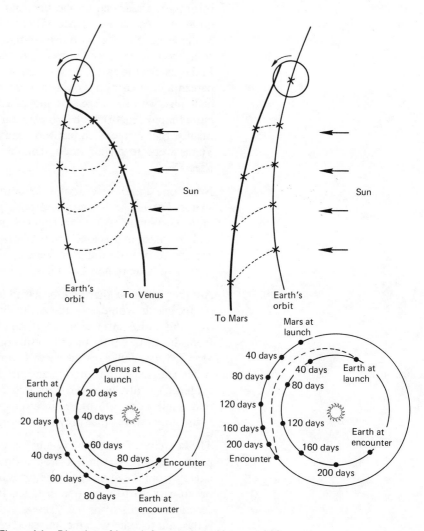

Figure 4.1   Direction of launch for a probe to Venus and Mars.

The Solar System: The Planets, Asteroids, Comets, and Meteoroids

Two questions with respect to Mars are foremost in the minds of astronomers. Are there canals on Mars? Is there life on Mars? Markings that looked like canals were originally observed by Giovanni Schiaparelli in 1877 and have since been seen by some astronomers and not by others. The possible existence of canals lead to speculation on the presence of intelligent life on Mars, and these controversies are part of today's thinking about the red planet.

Another question asked about Mars relates to the changing surface colors. Seasonal darkening of the Martian surface as seen from the earth occurs each Martian year as the polar cap shrinks in size. The darkening spreads from the polar cap toward the equator. This action has been variously interpreted as growth of vegetation as water is released from the polar cap during the Martian spring and summer, or as a change in the color of certain surface minerals as moisture from the polar cap is released into the atmosphere. As yet no generally accepted explanation for this darkening phenomenon is available.

Several probes to Mars have added to the store of new knowledge on Mars but have not yet settled any of these questions. The first probe, *Mariner IV*, encountered Mars on July 15, 1965, after a flight of 307 days. The probe came within 10,000 kilometers and transmitted its findings 217,000,000 kilometers back to earth.

The 21 *Mariner IV* pictures clearly showed about 100 craters along the narrow strip of the planet it surveyed. The craters seemed remarkably moonlike in their mode of formation. Some even have the peculiar mound in the center of the crater that is characteristic of many lunar impact craters. The Martian surface appears very old. The rate of weathering can be guessed by studying the craters, but the estimates vary widely. Weathering is certainly faster than on the airless moon but undoubtedly much slower than on earth, where most craters are all but obliterated by erosion within a few million years of their formation.

Canals certainly were not obvious, although some straight-line features in these *Mariner IV* photographs could be associated with the canallike structures seen from earth. The linear features seen by *Mariner IV* appear to be well-weathered, pockmarked geologic features, possibly natural cracks in the planet's crust caused by meteoroid impact. Nothing was observed in the photographs suggesting any form of life on Mars.

The other experiments on *Mariner IV* disclosed additional important features of the planet. Mars has no significant magnetic field of its own and no radiation belts. The atmosphere is thinner than deduced from terrestrial observations: 4 to 7 millibars, mostly carbon dioxide and argon. (Atmospheric pressure on earth is approximately 1,000 millibars).

In 1969, two simultaneous probes, *Mariner VI and VII*, completed a successful voyage to Mars and returned more detailed data to earth than did the previous probe. *Mariner VI* took photographs across the equatorial zones of the planet to include many known light and dark features of its surface. *Mariner VII* photographed the surface in an approximately north-south course, intersecting that of *Mariner VI* and continuing over the south polar cap. The two tracks were designed to cross at different times of the day to obtain a range of lighting conditions in the same area.

Experiments concerning the Martian atmosphere confirm the *Mariner IV* observations. Carbon dioxide is a major constituent and may comprise up to 95 percent of the atmosphere. Argon was present, and a very small amount of oxygen was detected in the upper atmosphere, with hydrogen occurring in the exosphere, or Martian corona. No nitrogen appears to be present, but small amounts of water were detected.

The atmosphere of Mars lacked any indication of clouds except for a trace over the polar cap which could be clouds or fog of solid carbon dioxide. A thin layer of haze was detected by the final far-encounter sequence of *Mariner VII* photographs. This haze layer, possibly composed of tiny particles, is approximately 8 to 16 kilometers thick and varies in altitude above the surface from 16 to 50 kilometers.

Mariner IV revealed that the surface of Mars was heavily cratered and resembled the moon more than the earth. *Mariner VI and VII* confirmed this and showed that there were several different types of Martian terrain. One type of terrain included many craters from 50 to 80 kilometers in diameter, and a few as large as 500 kilometers. Another form of surface was a chaotic region described as irregular and jumbled with a topography similar to that of a large landslide on earth. A featureless area akin to a dry lake bed on earth was also photographed. No mountain systems like those that occur on earth were detected.

A prominent surface feature on Mars, and one of great interest to astronomers, is the polar cap. The polar caps have been a subject of much discussion on whether the "snow" was frozen water or frozen carbon dioxide. The evidence now available favors carbon dioxide snow; however, water ice may be present in small amounts. Evidence favoring the carbon dioxide hypothesis is based on spectrum analysis which positively identifies solid carbon dioxide over most of the south polar cap. Temperatures measured over this region are sufficiently low to permit carbon dioxide to freeze at the low atmospheric pressures that exist on Mars. Atmospheric pressure measurements by the later probes ranged from 4 to 7 millibars, the same figures that were obtained by *Mariner IV*.

The spacecraft found temperatures on the Martian surface to be quite variable, ranging from $-125°$ C over the polar cap to $16°$ C at noon at the equator. This is close to the temperatures obtained for the equatorial region from earth-based observations. Nighttime temperatures at the equator, unobservable from the earth, are as low as $-75°$ C.

Some of the surface markings seen through earth-based telescopes were not observed in the Mariner photographs. For example, the seasonal darkening, often visible from earth, was not evident, nor were there any topographic

The Solar System: The Planets, Asteroids, Comets, and Meteoroids

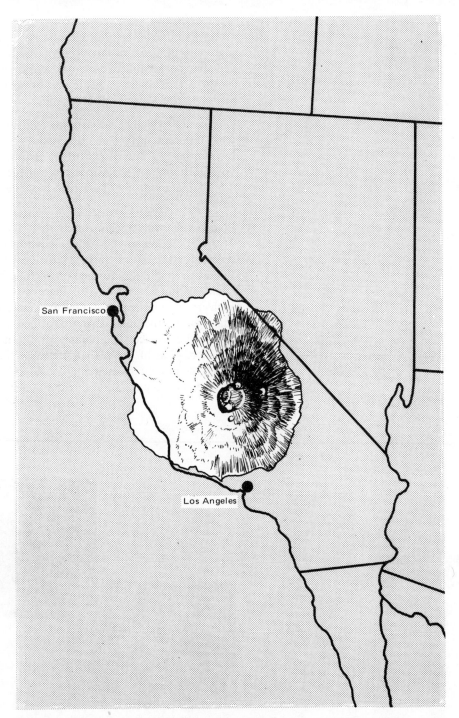

Figure 4.2 Gigantic volcano on Mars—called Nix Olympics—is 500 kilometers across the base. The main crater at the summit is 65 kilometers in diameter. The volcanic structure is shown superimposed on a drawing of California. JPL/NASA.

features visible in the photograph that could be related to the seasonal darkening. Only in a few instances were any linear features visible that might conceivably be identified as canal markings from the long distance to earth. Nothing that might indicate the presence of a vast network of canals was visible in the Mariner photographs.

Mariner IX, which began orbiting Mars in November 1971, has revealed somewhat different surface characteristics for the planet than were photographed by *Mariner VI* and *VII*. Surface features included huge geologically young volcanoes and an equatorial plateau region with faults and rifts as much as 4.5 kilometers deep. These rifts indicate the possibility of intensive tectonic activity.

Planet-wide windstorms raised dust that initially restricted the mission of *Mariner IX*. When the wind died down and the dust settled, photographs revealed evidence of surface erosion by wind, and sand-dune formations similar to sand dunes formed on earth. Channels were also photographed that could readily be explained if there was a plentiful supply of water to cause such erosion on the planet. One explanation that would allow for water erosion suggests that the polar caps are formed of water ice covered by a thin layer of carbon dioxide ice. The polar caps alternate in size every 50,000 years due to the precession of the planet. Between these events there are periods during which the polar caps melt and considerable rain results, causing erosion of the surface and accumulation of the water as ice at the alternate pole.

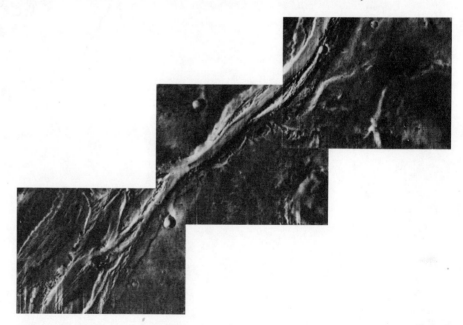

Figure 4.3 A channel thought to have been formed by running water in Mars' geologic past is seen in mosaic of three pictures taken by Mariner IX. Flow of the channel is from lower left to upper right. This small segment is about 75 kilometers in length. JPL/NASA.

The Solar System: The Planets,
Asteroids, Comets, and Meteoroids

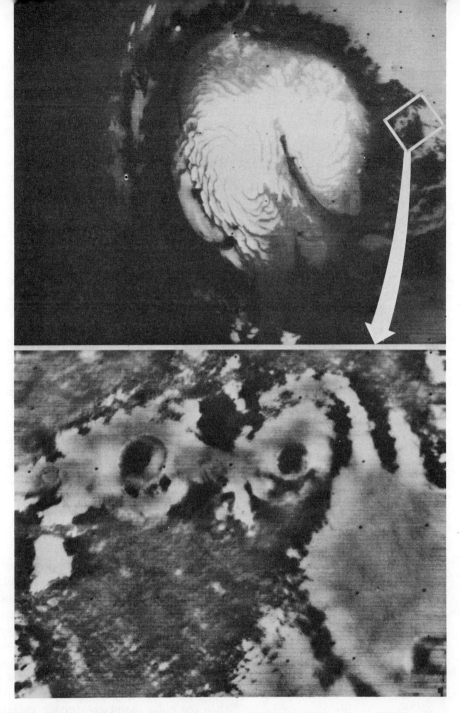

Figure 4.4 The picture shows the polar cap at its minimum extent—about 1,000 kilometers across. The curved patterns in the interior of the frost cover are formed on outward facing slopes receiving more direct sunlight than the flat areas and defrosting earlier. The various shades of gray correspond to surfaces of very bright carbon dioxide and, perhaps, water ice, and bright and dark rock debris. JPL/NASA.

Thus far no data received from any of the Mariner probes to Mars indicates that life exists on this planet. Some believe if life exists at all it is probably microbial. The lack of atmospheric nitrogen and the scarcity of water on the Martian surface are serious limiting factors. Water seems only to occur as vapor or ice on Mars and not as a liquid. No earth organisms could survive in such a dry environment, although it is possible that life forms have evolved that could utilize water in the form of ice or vapor. The intensity of ultraviolet radiation from the sun that strikes the surface of Mars is another factor that reduces the prospect of life. Ultraviolet radiation is more intense on the Martian surface than on earth because Mars does not have the benefit of an insulating atmosphere. Despite this discouraging outlook, the search for life will be continued until its presence or absence is definitely established.

In the mid-eighteenth century, both Voltaire and Jonathan Swift referred to two Martian moons in their writings, although the moons were not discovered until 1877. In that year Mars was at a favorable opposition when Asaph Hall, an American astronomer, discovered the two tiny satellites. Phobos (Fear), closest to the planet, is 6,000 kilometers above the surface and has a period of revolution of 7 hours 39 minutes. Deimos (Panic), 20,200 kilometers from the Martian surface, requires 30 hours 18 minutes to orbit the planet. Deimos is the smaller of the two satellites and may only be about 8 kilometers in diameter. Phobos appears to be about 16 kilometers in diameter. The size of the satellites is only approximate, since they are too small to be measured directly. It is assumed that the albedo or reflectivity of the satellites is the same as that of Mars. By determining the brightness of the two satellites, some estimate of their size can be made.

Neither of the satellites has been examined in any detail by the recent Mariner probes, but no doubt this will be accomplished on future probes.

4.4 JUPITER

Next, beyond Mars and the asteroid belt, is the orbit of Jupiter, largest planet in the solar system. Jupiter is a major planet, eleven times the diameter of the earth and 318 times its mass. Jupiter comprises about 70 percent of the solar system's mass not incorporated in the sun. Jupiter requires 11.8 years to complete one revolution of its orbit. As seen from earth, the planet's period of rotation varies with latitude, being 9 hours and 50 minutes at the equator and slower toward the poles, indicating the extensive nature of the gaseous atmosphere. The disk of Jupiter is somewhat flattened or oblate, causing the equatorial diameter (see table 4.1) to be approximately 9,200 kilometers greater than the polar diameter—further evidence of the partially gaseous structure of the planet.

What one observes of Jupiter through a telescope is the swirling, turbulent surface of the atmosphere. The atmosphere is composed primarily of hydrogen, methane, and ammonia, and possibly some water and helium. Water and

The Solar System: The Planets,
Asteroids, Comets, and Meteoroids

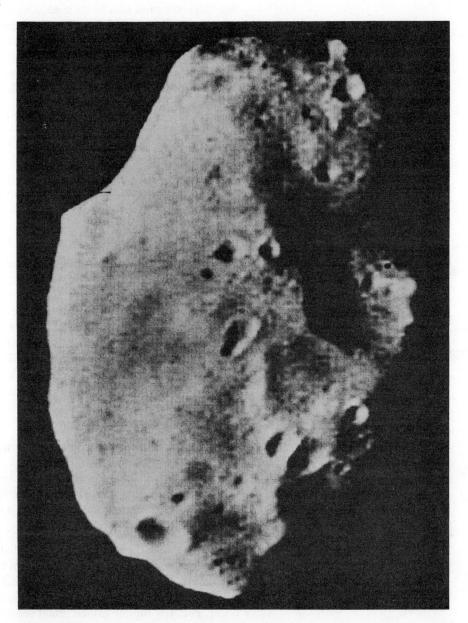

Figure 4.5 A computer-enhanced photograph of Phobos, the innermost moon of Mars, taken at a distance of 5,540 kilometers from Mariner IX. The profusion of craters suggests that Phobos is very old and possesses considerable structural strength. JPL/NASA.

ammonia may be in a frozen state, since the temperature measured at the atmospheric surface is below $-135°$ C. When viewed from earth, the atmospheric clouds—possibly composed of frozen ammonia—show bands and zones of varied colors, and a spot which, over a period of years, ranges from a barely visible pink to a fairly conspicuous red. The "Great Red Spot," as it is known, has been seen for several hundred years and is elliptical in shape. It is about 50,000 kilometers long and 25,000 kilometers wide. Despite its large size, the spot is not fixed but shifts in longitude as though it were floating in the atmosphere. Many interesting suggestions have been made to account for this feature, but its exact nature is as yet undetermined.

The solid surface of Jupiter is completely hidden from view by the dense atmosphere. The reflected sunlight from Jupiter comes from only a shallow layer of the planet's atmosphere and provides very little information concerning the lower layers of the atmosphere or of the solid surfaces beneath. From the observed motions of Jupiter's satellites it has been determined that the solid surface of the planet does not lie close beneath the cloud layer. It is also thought that the density of the gases increase with depth to a point where moving through the atmospheric gases would be similar to walking or swimming through water.

An overall density of 1.33 grams per cubic centimeter indicates that Jupiter does have a core. One astronomer has suggested that the pressure becomes so great that the planet's interior may be composed, at least in part, of metallic hydrogen. A number of questions have yet to be answered about the solid portion of Jupiter. How large is the core and what is its composition? Is the planetary surface solid in the sense that it can sustain shear stresses,* or is it in a fluid state like the earth's core? What is the nature of the transition zone between atmosphere and surface? It is also necessary to determine if heat is generated internally. The answers to these questions are fundamental to an understanding of the interior structure of Jupiter, and to an understanding of its origin and evolution.

Jupiter is within a factor of 30 of being massive enough to be a star. If Jupiter were 30 times more massive, the pressure and temperatures at its center, produced by its own weight pressing on its interior, would raise to the threshold for activating nuclear fuel. This would make Jupiter a star. If this had occurred at the time the solar system was formed it would be a double-star system.

In recent planetary research on infrared radiation from the planets, it has been found that the infrared energy radiating from Jupiter is 4 times greater than the energy the planet receives from the sun. This raises the question as to the source of so much energy. Is it nuclear? This seems unlikely in view of the fact that Jupiter has only $\frac{1}{30}$ the mass required for nuclear reactions to occur. Perhaps Jupiter gained much energy when it first condensed out of the

---

*Shear stress: Distortion of a solid resulting from two forces acting in an opposite direction and tangentially to each other on the solid.

solar nebula as parts fell on one another under the force of their mutual gravitational attraction. Perhaps the planet was heated to so high a level that it is still releasing to space the remains of that primordial energy. As yet the source of so much energy is unknown, but its existence is one of the most interesting planetary discoveries made in the recent past.

Jupiter definitely has belts of the Van Allen type which extend out to several planetary radii. The presence of these belts suggests the existence of a strong planetary magnetic field. There is some indication that the Jovian magnetic pole does not coincide with the pole of rotation but rather is some 9° away. Jupiter is the only planet other than the earth known to possess a magnetic field. The existence of the field and some of its characteristics has been inferred from radio telescope studies. More information is expected from *Pioneer 10*, a probe sent to Jupiter early in 1972.

Jupiter has twelve satellites, four of which were discovered by Galileo and are quite large (2,900 to 4,800 kilometers in diameter), and eight ranging in size from 25 to 250 kilometers in diameter. The four Galilean satellites, plus one other, that is the closest to the planet, have nearly circular equatorial orbits. The other seven have eccentric, highly inclined (to the Jovian equator), and distant orbits. The four outermost of these are about 14 million miles from Jupiter and revolve around the planet in retrograde direction in slightly over two years.

The Galilean satellites, $\frac{1}{4}$ million to one million miles from Jupiter, are interesting bodies to study in their own right. Io is extremely orange and there is indirect evidence that it may have an atmosphere. Ganymede and Callisto have somewhat lower densities than Io and Europa or the earth's moon. These characteristics will make interesting subjects for future studies.

The question of whether or not life could occur on Jupiter has been discussed. At first it was thought that Jupiter was a sorry place for the evolution of life due to the low temperatures and poisonous gases. However, according to basic laws of physics, the temperature of the atmosphere should rise below the cloud tops. The temperature of the earth's atmosphere rises from $-45°C$ at cloud top altitude of 9,000 meters to $15°C$ at the ground surface. The temperature on Jupiter probably also rises in the same manner, and calculations indicate that at some point in the atmosphere or on the surface of the planet, if such exists, there is a region where temperature suitable for the support of life.

Jupiter has an abundance of the atmospheric gases believed necessary for the origin of life. The gases are hydrogen, methane, ammonia, and water vapor, and because of their presence there is speculation as to the possibility of life forming at some level below the cloud surface. These are speculations based on what can be observed from earth of a planet almost 400 million miles away. Future probes to Jupiter will no doubt include experiments for the detection of life.

4.5 SATURN

Saturn is in many ways similar to Jupiter. The atmosphere shows banding and cloud variations similar to those observed on Jupiter, but on the whole does not appear to be as active. Saturn, one of the major planets, has the lowest density of any planet in the solar system. It is the second largest planet, surpassed only by Jupiter in size. It is approximately twice the distance of Jupiter from the sun and requires $29\frac{1}{2}$ years to make one revolution around the sun (see table 4.1).

The major gases observed in Saturn's atmosphere are hydrogen and methane. Because of Saturn's lower temperature as compared with Jupiter, most of the ammonia may be in solid (ice) form and not appear in the atmospheric spectrum. Nothing is known about Saturn's interior, but it may be assumed that the same characteristics found on Jupiter may also exist on Saturn.

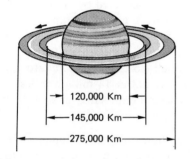

Figure 4.6 Saturn and its ring structure.

A unique feature of Saturn is its spectacular system of rings (fig. 4.6), first seen by Galileo and more accurately described by Christian Huygens in 1655. The system is composed of three concentric rings of which the center is the largest and the brightest, and therefore simply called the *bright ring*. The inner ring, or *crepe ring*, is faint and comes to within 11,300 kilometers of the surface of the planet. The *outer ring* is separated from the bright ring by a 4,800-kilometer space known as *Cassini's gap*. The outside diameter of this ring is 275,000 kilometers. The rings are not solid but rather are formed by billions of small particles of frozen gas which orbit the planet like so many small satellites. The particles close to the planet revolve more rapidly than those further out, which is in accordance with Kepler's third law. When oriented edge-on toward the earth, the rings are barely visible; thus, the system cannot be very thick. The rings may possibly be a few miles thick at most, although some astronomers think the thickness may be only a few feet. The rings, no doubt, are quite thin, for stars can occasionally be seen through them.

Saturn has ten satellites associated with it. All but Titan are less than 1,600 kilometers in diameter. Titan is 4,600 kilometers in diameter and has the distinction of being the only satellite in the solar system definitely known to have an atmosphere—probably of methane. Most of the other satellites revolving around Saturn have low densities and possibly are composed of ice.

4.6 URANUS

Uranus, just barely visible to the naked eye, and the first planet to be discovered after the invention of the telescope, is almost 20 astronomical units from the sun. William Herschel, an English astronomer, discovered the planet in 1781 while examining a region in the constellation Gemini. At first he thought the disklike object was a comet, but after observing it for several months and calculating an orbit, he concluded that it was a planet located beyond Saturn in the solar system.

Not a great deal is known about Uranus because of its great distance. However, it does have a hydrogen-methane atmosphere similar to Jupiter and Saturn, and it is considered one of the major planets despite its being only about $\frac{1}{3}$ Jupiter's diameter (see table 4.1). The planet has a greenish tinge, thought to be due to the relatively high amount of methane in the atmosphere, but few markings of any importance can be seen. For this reason, the rotation of the planet, which is the reverse of the other planets, was determined by spectroscopic examination. In doing so, one other peculiarity of the planet was noted, namely that the axis of rotation is almost parallel to the orbital plane (fig. 4.7). The planet's equatorial plane is inclined 82° to the plane of the orbit, giving the planet the appearance of rolling along as it revolves around the sun.

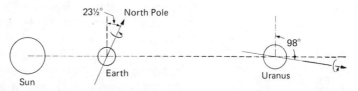

Figure 4.7   The axial inclination of Uranus from a perpendicular to the ecliptic plane is 82°.

Five known satellites orbit the planet, all in the equatorial plane. Herschel discovered two, and the most recent was found in 1948. All are fairly small—1,600 kilometers or less in diameter—and within 595,000 kilometers of the planet.

4.7   NEPTUNE

Neptune, a pale green dot discovered in 1846, appears as a "twin" of Uranus, since it is essentially the same size and has the same atmospheric characteristics. Before its initial discovery, Neptune was detected by an Englishman, John Adams, and a Frenchman, Urbain Leverrier, as a result of the influence the unseen planet had on Uranus. Uranus arrived at its 1821 position ahead of schedule, leading the two astronomers to suspect the existence of an unseen planet. In 1846, Leverrier communicated his findings to Johann Galle at the Berlin Observatory, who then observed the new planet almost in the exact location suggested by Leverrier. This was the same location calculated by Adams, but the astronomer to whom Adams sent the results did not have a detailed sky chart of the proper region and so failed to make use of the information. Nevertheless, Adams was given credit for his part in finding Neptune.

Not a great deal is known about Neptune (see table 4.1). So distant is this planet that light requires almost four hours to travel from Neptune to the earth.

It is slightly smaller than Uranus but has a greater mass and the highest density of the major planets. The green color is thought to be due to the prominence of methane in the atmosphere. This seems to be the only gas present in abundance due to the low temperatures ($-200\,°C$) on Neptune.

Neptune does have two satellites, the larger of which, Triton, is closest to the planet and orbits the planet in retrograde direction. It has been suggested that Triton will eventually crash into Neptune or break up to form a ring around the planet like those around Saturn. Triton is now about 350,000 kilometers from Neptune. The other satellite, Nereid, travels in a normal direction but in an eccentric orbit from one to six million miles from the planet.

4.8 PLUTO

Certain small irregularities in the movement of Uranus could not be explained entirely by the presence of Neptune and this led some astronomers to consider the possibility of a "trans-Neptunian" planet. The existence of such a planet was first mentioned in print in 1879 by a Frenchman, Camille Flammarion, although it was Percival Lowell in the early twentieth century who, from his Arizona observatory, made the first intensive search for the object. By studying the behavior of Uranus and Neptune, Lowell hoped to locate the object he called "planet X." His search lasted almost ten years (from 1905 until his death), during which time he suggested two possible locations for the perturbing planet.

Meanwhile, a similar investigation was being conducted by W. H. Pickering, who mathematically arrived at approximately the same position in orbit for the unseen planet. Photographs taken at the Mount Wilson Observatory in 1919 based on Pickering's determinations failed to reveal the new planet, but only because the image of the planet happened to fall directly on a small flaw on the plate.

Finally in 1930, twenty-five years after Lowell initiated the search for planet X, the planet was discovered by Clyde Tombaugh, a young American astronomer at Lowell Observatory. Tombaugh made use of a "blink" microscope whereby two photographs, taken at different times of the same region of the sky, are rapidly shifted back and forth so the observer sees a constant but flickering picture. If one object in the sky has moved during the interval between the time the photographs were taken, that object would seem to appear and disappear or "blink" as the photographs are shifted. In this way, Tombaugh recognized the new planet quite close to one of Lowell's suggested positions. Ironically, Lowell was in the process of comparing two photographs with the blink microscope at the time of his death. Had he been able to finish, he might have discovered the planet he searched for so long.

The discovery of Pluto may not, in the final analysis, have been the result of meticulous measurement and patient pursuit. Rather it may have been an entirely fortuitous search covering the right area of the sky for the wrong reasons. Pluto appears to be much smaller and of less mass than Lowell's calculations would indicate. As a result, many astronomers consider the finding of Pluto at that point in space a lucky coincidence. Only more precise information on the physical characteristics of the planet will resolve this question.

Pluto is the most distant planet, being 39.5 astronomical units (on the average) from the sun. However, Pluto's orbit has the highest eccentricity of any of the planets, causing it to travel 55 million kilometers inside Neptune's orbit at perihelion. At this time, Pluto is 29.5 astronomical units from the sun, while at aphelion its distance is 50 astronomical units. Pluto will be at perihelion in the year 2010. There is no likelihood of Neptune and Pluto colliding because of their orbital characteristics. The inclination of Pluto's orbit is 17° to the ecliptic, which means that the two planets never approach each other by less than 390 million kilometers. Some astronomers have suggested that Pluto was once a Neptunian satellite and that Pluto's escape has resulted in this peculiar orbital configuration.

Pluto appears to have no atmosphere and is extremely cold, possibly −210° C or lower. While it does have a low temperature as the major planets do, Pluto seems small by comparison (6,000 kilometers in diameter) and is more like the terrestrial planets. The period of rotation is 6.39 days, a finding based on a slight regular variation in the brightness of the planet. Pluto's period of revolution is 248.4 years.

4.9 THE TENTH PLANET

In 1972, evidence of the existence of a tenth planet was obtained from calculations of the orbit of Halley's comet. Halley's comet is one of the most studied of the comets that periodically revolve around the sun. Deviations from its normal orbit indicate that the comet is affected by a mass equivalent to three times that of Saturn, located approximately 65 astronomical units from the sun. The hypothetical planet would have a retrograde orbit requiring 512 years to complete one revolution, and the orbit would be inclined 60° to the ecliptic. The proposed direction of revolution and the inclination of the orbital plane to the ecliptic is at variance with that of the other planets, making this planet an unusual member of the solar system, if it does exist. A subsequent search for the planet in the area where it was predicted to be has thus far failed to reveal its existence.

4.10 UNANSWERED QUESTIONS

The first question that comes to mind deals with the prospect of additional planets as yet undiscovered in the solar system. Some attempt has been made by Clyde Tombaugh to answer this question, but his efforts have thus far been fruitless. He has explored a region of the sky far beyond the orbit of Pluto without results. Because of the vastness of space, it is easy to overlook a tiny, dark object in the remote areas of the solar system.

None of the planets have as yet been directly explored by man, and all the information thus far gathered has been obtained through intensive study of these various bodies from earth. Models of planetary atmospheres and surface phenomena are developed from observations. These observations may later be rejected as new information is received and new models are constructed. For example, Venus has been variously described in the past fifty years as a swampy jungle, a hot desert, a world covered by oceans, and a planet covered by oil or other hydrocarbons. Although Venus is now thought to be extremely hot at the surface, a conclusion reached on the basis of fairly reliable data, there is a degree of uncertainty about the Venusian surface that can be resolved only by direct exploration by man or machine.

Other questions are constantly being raised, such as how the massive atmosphere of Venus developed. There must have been unique processes of an unknown nature to provide an earthlike planet with such a considerable atmosphere as compared to Earth and Mars. Could a form of life have developed and been lost on Venus, a life with processes attuned to an environment so different that it is not possible to speculate upon its form?

Does the failure to find evidence of volcanic gases in noticeable concentration indicate much less volcanism on Mars than has occurred on Earth? Has Mars already lost its secondary atmosphere? Mars, Venus, and Earth, when examined as sister planets, present important problems that cannot presently be explained. The outer planets are cold, and what is known of their atmospheres fits well into theories of planetary evolution.

4.11 OTHER INHABITANTS OF THE SOLAR SYSTEM

The planets and their attendant satellites are not the only objects orbiting the sun. Thousands of particles, ranging from asteroids (the largest is Ceres, 770 kilometers in diameter), to the tiny meteoroids, revolve around the sun in continuous orbits. These orbits are not regular but vary as the objects are influenced by the gravitational forces of the larger planets. On occasion these objects may be captured by the planets to be drawn in to crash on the surface or to take up residence as a satellite, changing orbits from one around the sun to an orbit around the planet. The outer satellites of Jupiter may have been asteroids that were captured by the large planet.

4.11A Asteroids

Ceres, the largest of the asteroids,* was discovered by Guiseppe Piazzo, in 1801, in an orbit beyond Mars at 2.8 astronomical units from the sun. Within a

---
*Also called minor planets or planetoids.

few years, several more were found in generally the same orbit—Pallas (1802), Juno (1804), and Vesta (1807). These have diameters in excess of 150 kilometers and seem to be the only large asteroids. The majority of the asteroids have diameters between 8 and 50 kilometers, with about 50 having diameters greater than 50 kilometers. It is estimated that some 55,000 of these small bodies, large enough to be seen with the 100-inch telescope, revolve around the sun, although the orbits of only about 1,600 have been accurately determined.

The majority of the asteroids occur in an orbit between Mars and Jupiter (2.8 A.U.) and revolve around the sun in 3 to 6 years. Another small group called the Trojan group move so as to form a vertex of an equilateral triangle with Jupiter and the sun. Part of the group precedes Jupiter, forming one such triangle, and part follows Jupiter to form a second triangle. A small number of asteroids also have orbits which take them in close to the earth. These asteroids, known as the Apollo group, have been useful in determining the length of the astronomical unit. This is due to the fact that parallactic measurements (see section 5.1) of good accuracy can be made with these close-approaching objects. Most of these asteroids are quite small, being only a mile or so in diameter. Hermes, discovered in 1937, comes to within 1 million kilometers of the earth and moves across the sky so rapidly it is difficult to photograph.

Most of the larger asteroids are spherical, but many of the small ones are irregular in shape, a fact deduced from their varying brightness. This has led to the thought that at one time several large asteroids moved in approximately the present orbit of the asteroids. The strong gravitational pull of Jupiter may have influenced the movement of the asteroids, causing them to deviate from their normal orbits and collide. The resulting fragments would revolve around the sun in orbits of different eccentricity and inclination than the original asteroids. Such an event has been found to be statistically possible, but there is no certainty that this was the manner in which the present asteroids formed.

4.11B Comets

Aside from an eclipse, no astronomical phenomenon has created so much consternation during man's history as has the comet. Comets have for centuries been viewed as omens of disaster, causing their movements to be noted with great care. The large comets are bright objects with long tails that appear to hang motionless in the sky as if to point an accusing finger at man's foibles.

Numerous theories have been suggested to explain the origin of the comets, but none have been proved or disproved. The comet-cloud hypothesis proposes that comets come from a comet-cloud (composed of billions of comets) that revolves around the sun in an orbit extending out to 150,000 astronomical units. These comets have orbits that are nearly parabolic, of high eccentricity and inclined at every possible angle to the ecliptic plane. Because of the extremely elliptical orbits, comets may have periods of thousands or perhaps millions of years, and therefore only a few are seen at any one time. Only a very tiny fraction of the total number of comets comprise the short-period comets, which are those that appear at regular, short intervals. Halley's comet,

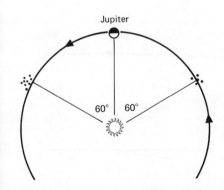

Figure 4.8 The Trojan asteroids precede and follow Jupiter in its orbit forming an equilateral triangle with the sun.

one of the most famous of these, revolves around the sun every 76 years, and its appearance has been traced back to 240 B.C. Its next visit is expected in 1986.

The short-period comets were once all thought to be members of the comet-cloud. As some of these comets made a rare appearance near the sun, they were captured by the gravitational force of the large planets, thus having their orbital period reduced and causing them to be visible at regular intervals. Others were perturbed into hyperbolic orbits and presumably left the solar system.

The comet is composed of three parts: the nucleus and the coma, which make up the head of the comet, and the tail. Approximately 99 percent of the comet's mass is concentrated in the nucleus, which appears through the telescope like a star seen through a fog. The nucleus is relatively small—1 to 10 miles in diameter—and composed of solid particles (meteoroids), frozen gases (methane, ammonia, carbon dioxide), and water ice. Some have described the comet as a "dirty snowball."

Surrounding the nucleus is the coma, a large luminous envelope that gives the nucleus its fuzzy appearance. The coma may occassionally be so bright that it conceals the nucleus entirely. The coma is composed of gases and dust particles that are being sloughed off from the nucleus. This action causes the coma to grow to a considerable size, in some cases up to 500,000 kilometers or more in diameter. The sun's energy evaporates the frozen gases, thereby increasing the size of the coma. This causes the nucleus to be gradually reduced in size as the comet orbits the sun.

The tail is an extension of the coma, and it is difficult to determine where one ends and the other begins. Generally, tails are associated only with large comets, as these are the comets that are losing sufficient matter to give substance to a tail. The tail may be extremely long, up to one to two astronomical units in length, and it is extremely tenuous, permitting stars to be seen through it. Because of its tenuous nature, the tail is influenced by the solar winds, which cause the comet's tail to be pointed away from the sun (fig. 4.9).

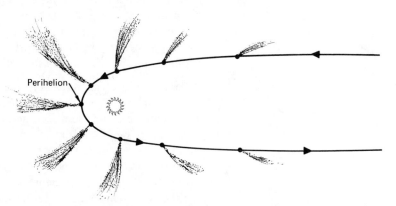

Figure 4.9 Cometary tails are repulsed by the solar winds.

Because a comet is losing matter as it orbits the sun, it cannot last indefinitely. It has been suggested that approximately 1 percent of the nucleus is sloughed off during each passage around the sun, and that the average life of a comet is about 70 passages. This lends support to the theory that comets are captured by the large planets to become short-period comets, replacing those that expire.

4.11C Meteoroids, Meteors, and Meteorites

Countless small particles of matter called *meteoroids* move in elliptical orbits around the sun. Their origin is not definitely known, but several good theories regarding their source have been proposed. Meteoroids may be fragments from the asteroid belt or the debris of comets scattered over the cometary orbit, or meteoroids may be primary solar system material remaining from planetary formation.

Occasionally, the meteoroid particles encounter the earth and are captured by its gravitational force. Meteoroids entering the atmosphere are traveling at high velocities ranging from 11 to 70 kilometers per second. The particle heats up and disintegrates, disrupting the atmospheric gases in the zone through which the particle travels. This results in a streak of light in the sky called a *meteor* or *shooting star*. The light that is seen is not the incandescent particle, which may be 80 to 120 kilometers from the observer and too small to be visible. What is seen is light emitted by the reaction of the gases in the zone through which the particle travels. The light is only briefly visible, usually lasting less than a second, although the larger particles leave a faintly luminous wake behind them that is seen for a longer period. On rare occasions, one may see a bright *fireball* indicating the passage of a large particle through the atmosphere.

At times, a greater-than-normal number of meteors are sighted, seemingly coming from a definite point in the sky called a *radiant*. These *meteor showers*, as they are called, occur at regular intervals, and all meteoroids from a particular shower move in common orbits. These orbits have been identified with those of comets, and it is believed that the meteoroids causing the meteor shower are debris from existing or expired comets. For example, the Orionids, one of the showers, occur on approximately October 21 of each year and are associated with Halley's comet. The name of a shower is taken from the constellation (in this case, Orion) that is in the area of the sky from which the shower appears to be coming.

Most of the meteoroids entering the atmosphere from space are completely vaporized or else are too small to heat up and thus they filter down to the earth's surface as dust. The larger meteoroids survive the plunge and reach the ground in one piece or in fragments. These are called *meteorites*.

Meteorites are recovered whenever possible, since they were, until recently, the only extraterrestrial matter available for study. Meteorites are classed as iron (siderite), stony-iron (siderolite), or stony (aerolite). The iron meteorites are composed of about 90% iron, 5% nickel, and smaller quantities of other metallic elements. They appear dark, metallic, irregular in shape, have smooth thumbprint-like depressions, and are of high density. Iron meteorites are easy to recognize but are the least common of the three types. They are believed to be similar in composition to the earth's core.

Stony-iron and stony meteorites are more common than iron meteorites, but they are more difficult to identify because of their similarity to rock material on earth. Their surfaces are usually brown or black, with a crust formed by its surface melting as they passed through the atmosphere. The stony meteorites may contain 10 to 20% metal, whereas the stony-irons have a metal content intermediate between iron and stony meteorites. Stony meteorites are similar to material on the earth's surface.

Large meteorites are not common; there are only about three dozen catalogued as weighing more than a ton. The largest, an iron meteorite, is the Hoba, which lies partially buried in southwest Africa. This meteorite measures $2.7 \times 3 \times 1$ meter and weights about 55 metric tons.

The presence of large impact craters is evidence that much larger meteorites have crashed into the earth's surface at very infrequent intervals during the earth's 4.5 billion-year history. One such crater, the Barringer meteorite crater in Arizona is 1,280 meters in diameter and 175 meters deep. The meteorite that formed the crater, possibly 50,000 years ago, is estimated to have weighed about one million metric tons and to have been about 60 meters in diameter. The crater was formed by the explosive expansion of gases heated by the impact. Aerial and space photography have recently revealed what appear to be "fossil craters" of meteoritic origin in Canada. Other craters, where no meteoritic material has been found, have impact characteristics such as fractured rock patterns, known as *shatter cones*, and a silica material, *coesite*, formed only under extremely high pressures. One such structure, the Vredefort Ring in Transvaal, South Africa, is approximately 210 kilometers in diameter, too large to be a volcanic crater. It rivals the larger lunar craters in size.

4.12 SUMMARY

The sun, located at the approximate center of the solar system, controls the movement of nine planets in their orbits. The planets range from Mercury, at 58 million kilometers from the sun, to Pluto, which is a mean distance of 5.9 billion kilometers from the sun. A great deal is known about the planets with respect to motion, periods of rotation and revolution, number of satellites, size and mass. Except for earth, little is known about the nature of planetary atmospheres, surface conditions, and whether or not life has existed or does exist on these planets. These are the problems for which answers are currently being sought, for they will help man to better understand his environment.

QUESTIONS

The Solar System: The Planets, Asteroids, Comets, and Meteoroids

1. Rank the planets in order of size, mass, and density. Are there any relationships revealed by these rankings?
2. What is a reason given for the high carbon dioxide content of the Venusian atmosphere?
3. What has been the history of "canal" observations on Mars?
4. What would the motions of the Martian moons be like as seen from the surface of Mars?
5. What are some of the questions being asked about the planet Jupiter?
6. Describe the "rings" of Saturn.
7. What characteristics of Pluto suggest that it once was a satellite of Neptune?
8. What is one theory on the source of the comets?
9. What is a meteor shower?

5 Stars

The measurement of stellar distance. Triangulation and parallax.
Apparent and absolute magnitude.
Measurement of stellar temperatures.
Diameters of stars.
Measurement of mass and density of stars.
Radial and proper motion.
OBAFGKM
Meaning of the H-R diagram.
Types of Clusters.
Giants and supergiants, variable stars, white dwarfs, nova and supernova, pulsars and neutron stars.
Life cycle of a star.
The constellations.

From the earth at night, it is possible to see many hundreds of stars which generate energy in much the same manner as the sun.

The stars range widely in brightness, a factor that may in part be attributed to their differences in size or distance. They differ in color, indicating differences in surface temperatures. By analyzing the light from the stars using telescopes and spectrographs, a great deal of information has been gathered about individual stars. This information has permitted a characterization of the entire stellar population.

5.1 STELLAR DISTANCE

Knowledge of stellar distance is one of the fundamental needs of the astronomer, for it is on the basis of distance that the determination of many other physical properties of stars depends. The ancient Greeks, while they understood the principle of parallax (explained below), were unable to detect the parallactic displacement of stars with the instruments available to them. This is not surprising when it is realized that the angle of parallax of the nearest star is less than 1 second of arc, using the earth's orbital radius (150,000,000 kilometers) as a base. The first serious effort to detect and measure parallax was made by William Herschel in the eighteenth century. He unsuccessfully attempted to accomplish it by studying the relative movement of pairs of stars seen in his telescope.

It was not until 1838 that a German astronomer, F. W. Bessel, made the first reliable measurement of stellar distance. He chose 61 Cygni, a star just visible to the unaided eye that had previously been seen to move through space at an angular distance of 5 seconds of arc per year. Because of this angular motion, he deduced that the star was relatively close to the earth.

The method of measuring parallax is similar to the method of triangulation used by surveyors to find the distance to some point not easily accessible (fig. 5.1). To find the distance from H to C, a line AB is laid out and carefully measured. Angle A and angle B can be determined, and from this information, the distance HC can be calculated.

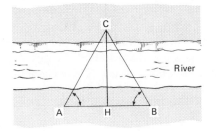

Figure 5.1 Triangulation is a means of finding the distance to an inaccessible site.

A similar technique is used for calculating stellar distance. The diameter of the earth's orbit is used as a baseline (fig. 5.2), and the position of the star is noted at six-month intervals against the background of more distant stars that change their relative position only slightly. The displacement is a measure of the parallax. By common consent of astronomers, the earth's orbital radius is used as the baseline, and, therefore, angle SCB is known as the *angle of parallax*. With this information it is possible to calculate SC. While this represents the distance from the star to the sun, it can also be used as the distance from the star to the earth, because in measuring stellar distances, the earth's orbital radius is insignificant.

The measurement of the extremely small angles involved (0.76 second of arc for the nearest star) is tedious and requires great patience. Recordings must be made at six-month intervals over several years in order to obtain a correct set of measurements. This extended period of observation is necessary because the star being studied is in motion, and a correction for this motion must be made. Furthermore, the solar system is in motion and thus the observer's position changes, necessitating a corresponding correction. Even refraction

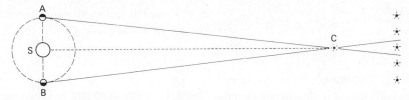

Figure 5.2 Triangulation is used to find the distance to a nearby star. To obtain the parallactic angle at C, measurements are made when the earth is at A and six months later when it is at B.

of light by the atmosphere must be considered, as well as the peculiarities of the instrument being used. Bessel included these factors in his calculation and measured the angular parallax of 61 Cygni as slightly under 0.3 second of arc. From this he determined the distance to 61 Cygni as approximately 696,000 A.U.

It can be seen that distances to the stars are great, and the units of measurement used within the confines of the solar system become too cumbersome for stellar distances. For this reason, a simple unit of measure, easily definable, is necessary. The light-year fits these requirements. It is defined as the distance traveled in one year by a light beam which we know has a velocity of 300,000 kilometers per second. This makes the distance from 61 Cygni to the earth 11 light-years, a much easier quantity to handle and remember. Another unit of measure used for stellar distances is the *parsec*. The parsec is the distance to a point in space where the angle of parallax is equal to 1 second of arc. This distance is equal to 3.26 light-years or 206,265 A.U. This would make the distance to 61 Cygni about 3.4 parsecs. A simple formula gives the relationship between the angle of parallax in seconds of arc and distance:

$$D(\text{parsec}) = \frac{1}{p(\text{seconds of arc})}$$

After Bessel reported his first stellar distance, scarcely fifty more were determined in the next fifty years. When photography was introduced as an astronomical technique, this endeavor was greatly speeded up, so that several thousand stellar distances have now been determined by parallactic displacement. Because of the very small angular measurements, the use of parallax is limited to measuring the distance to stars which are relatively close to the earth. For all practical purposes, the point where error limits the use of parallax for measuring stellar distances is at about 100 parsecs. Beyond this point, distances are measured by indirect means.

Several methods are available for determining stellar distances indirectly, and all are dependent on the apparent brightness of the light coming from a star. Apparent brightness is a function of the amount of light coming from the source and the distance the source is from the observer. The relationship is such that at double the distance, the light is only $\frac{1}{4}$ as bright, while at $\frac{1}{2}$ the distance, the light is 4 times as bright. In figure 5.3, a beam of light striking a screen 1 square foot in area at 1 foot from the source would be 4 times as bright per square foot as the same beam of light striking a screen 2 feet from the source. The same amount of light energy is covering 4 times the area. This fact provides a simple law which states that the apparent brightness of a light source is inversely proportional to the square of the distance from the source or, mathematically,

$$L \propto \frac{1}{d^2}$$

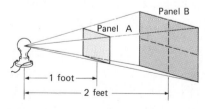

Figure 5.3 Panel A is one-half the distance from the light source that panel B is and receives four times the light energy.

where L is the apparent brightness, and d is the distance. This method involves having some knowledge of the properties of stars. As will be shown later, stars may be categorized into groups on the basis of similar properties. Thus, a star whose distance from the earth has been established by measuring its parallax may be similar in class to a very distant star. The comparative brightnesses can be measured photoelectrically, and from this information it is possible to calculate the distance to the more distant star with a satisfactory degree of accuracy. A star measured to be $\frac{1}{4}$ as bright as a similar star can be regarded as being twice the distance from the observer.

A group of stars very useful for indirectly determining distances are the pulsating stars known as the Cepheid variables. These vary in brightness at regular intervals, a discovery made by Henrietta Leavitt in 1908. It was found that the brightness of these stars is related to the length of the cycle (the period) through which the brightness varies (fig. 5.4). Cepheid stars having the same period have the same brightness, and the longer the period, the brighter the star. Therefore, from Cepheids of known distance, it is possible to determine the distance to Cepheids of similar period which are too far away to permit measurement by parallax.

5.2 STELLAR MAGNITUDE

Hipparchus was the first astronomer to catalog the stars on the basis of brightness. He assigned numbers, ranging from 1 to 6, to about a thousand stars to indicate *apparent magnitude*, with 1 for the approximately twenty very bright stars, and 6 for those stars just barely visible. This system worked admirably until the invention of the telescope, when many more stars were revealed and numbers had to be added to include these new stars in the system. It can be seen that in this system the larger magnitude numbers apply to the fainter stars. Measurements made of the radiation of various stars revealed that stars of magnitude 1 were about 100 times brighter than stars of magnitude 6. This led in the nineteenth century to the adoption of a magnitude scale in which two stars having a ratio in brightness of 100 to 1 differ by 5 magnitudes. This means that a star of the first magnitude is 100 times brighter than a star of the sixth magnitude. By the same token, a star of magnitude 8 is 100 times brighter than one of magnitude 13. The brightness ratio between two stars differing by 1 magnitude is the fifth root of 100, or 2.512. In other words, a star of the third magnitude is 2.512 times brighter than a star of the fourth magnitude. It can be seen from the foregoing that the magnitude of a star is not proportional to the brightness. The difference in brightness between two stars may be found by raising 2.512 to a power equal to the difference in magnitude of the two stars. For example, star A of magnitude 4 is $(2.512)^3$ times brighter than star B of magnitude 7.

This system supplies the apparent magnitude of a star as seen by an observer on earth. A star may be extremely large and extremely bright but because of its great distance appear faint to an observer. On this type of scale, based on apparent magnitude, the sun is the brightest object in the sky; it is much brighter

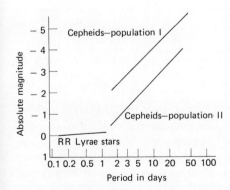

Figure 5.4 Period-luminosity curve for Cepheid variable stars.

than stars of the first magnitude. This has brought about the necessity for using negative magnitudes. The sun, therefore, has an apparent magnitude of -26.7, and the full moon, which is the second brightest object in the sky, has an apparent magnitude of -12.7. Next in order of brilliance is the planet Venus, with a magnitude of -4.2 when brightest. The brightest star in the sky is Sirius (Alpha Canis Majoris), with an apparent magnitude of -1.4. Magnitudes of other stars may be found in table 5.1.

Table 5.1 Properties of Certain Prominent Stars

Name	Distance, Parsecs	Apparent Magnitude (m_v)‡	Absolute Magnitude (M_v)‡	Spectral Class
Sun	—	−26.7	4.9	G2
*Alpha Centauri	1.3	0.0	4.4	G2
†Barnard's Star	1.8	9.5	13.2	M5
*Sirius	2.7	−1.4	1.5	A1
†61 Cygni	3.4	5.2	7.5	K6
*Procyon	3.5	0.3	2.6	F5
Altair	5.1	0.8	2.6	A7
Formalhaut	6.9	1.2	2.0	A3
Vega	8.0	0.0	0.5	A0
Arcturus	11.0	−0.1	−0.3	K2
Pollux	11.0	1.2	1.0	K0
*Castor	14.0	1.6	0.9	A1
*Capella	14.0	0.1	−0.6	G2
Aldebaran	16.0	0.9	−0.7	K5
Achernar	20.0	0.5	−3.0	B5
*Regulus	26.0	1.4	−0.6	B7
Canopus	30.0	−0.7	−4.0	F0
Bellatrix	40.0	1.6	−2.0	B2
*Spica	80.0	1.0	−2.0	B1
Betelgeuse	150.0	0.7	−5.5	M2
*Rigel	250.0	0.1	−6.8	B8
Deneb	430.0	1.3	−6.9	A2

*Multiple star systems.
†Accompanied by dark companions.
‡The abbreviations mv and Mv stand for "magnitude visual." It is customary to use a lower-case "m" for apparent magnitude and an upper-case "M" for absolute magnitude.

Many times it is desirable to compare the brightness of several stars without the complicating factor of distance. For this purpose the concept of *absolute magnitude* has been introduced. Absolute magnitude is the magnitude a star would have if it were at a standard distance of 10 parsecs from the observer on earth. This means that stars at great distances (greater than 10 parsecs) would appear much brighter at the standard distance, while stars that are relatively close would be dimmer. The sun, with an apparent magnitude of

−26.7, would have an absolute magnitude of 4.9 at the standard distance, thus becoming one of the fainter stars visible to the unaided eye.

Magnitude determined by eye is somewhat different from magnitude obtained from a photographic plate, because the reaction to light of the photographic emulsion is different from that of the retina of the eye. The difference between these two values for a given star results in a quality called the *color index* of a star. There exists a relationship between the color index and temperature, which makes the visual apparent magnitude and photographic apparent magnitude valuable in determining the surface temperatures of distant stars. The same technique may be applied when one uses absolute magnitudes.

5.3 STELLAR TEMPERATURES

Several methods of measuring the temperature of distant bodies are available. The thermocouple has been found to be most useful in measuring energy in the infrared part of the spectrum. The spectroscopic determination of the wavelength of most intense radiation can also be used to calculate temperature, while the use of visual and photographic magnitudes (see preceding paragraph) has provided an indirect means of temperature measurement. By combining and comparing data obtained by the several methods, values for temperatures of many stars have been obtained. Temperatures of some representative stars are given in table 5.2.

5.4 STELLAR DIAMETERS

The diameter of the sun can be readily calculated because its angular diameter and distance can easily be measured. Stars, on the other hand, appear as pinpoints of light in even the most powerful telescopes because of the great distances involved, making diameter measurements very difficult. However, several ingenious methods of measuring stellar diameters have been devised, so that the sizes of many stars are now known. A. A. Michelson made the first such determination in 1920, using the 100-inch telescope on Mount Wilson and a device called an interferometer to measure the angular diameter of some of the nearer stars. Michelson found the angular diameter of Betelgeuse to be 0.047 second of arc, and the parallax 0.018 second of arc. Having determined the distance, he then computed the diameter to be 400 million kilometers. This method has been used for fewer than twenty of the nearer stars because of the cumbersome equipment required for more refined measurement. The data even for these few have been of great importance, nevertheless, since this has made it possible to verify the correctness of the diameters of stars calculated by indirect methods.

Table 5.2 Characteristics of the Spectral Classes

Spectral Class	Color	Temperature, °C	Principal Characteristics	Representative Stars
O	Blue	>25,000	Lines of ionized helium and other elements are present. Hydrogen lines are weak	10 Lacertae
B	Bluish-white	11,000–25,000	Stronger hydrogen lines than in Type O. Neutral helium and some ionized elements present	Achernar, Spica
A	White	7500–11,000	Strong hydrogen lines. Ionized calcium, iron, magnesium, and others. No helium lines	Sirius
F	Yellow-white	6000–7500	Hydrogen lines weaker than in Type-A stars. Calcium lines are prominent along with some metals	Procyon
G	Yellow	5000–6000	Strong calcium lines and metals are prominent. Hydrogen lines are weak	Sun
K	Yellow-orange	3500–5000	Lines of metals are dominant. Molecular TiO is present. Hydrogen very weak	Arcturus
M	Red	<3500	Strong lines of metals and strong molecular bands of TiO	Betelgeuse
R	Yellow-orange	3500–5000	Same characteristics as Type-K star except molecular bands of carbon molecule are present	
N	Red	<3500	Same characteristics as Type-M star except molecular bands of carbon molecule are present	
S	Red	<3500	ZrO present instead of TiO as in Type-M star	

5.5 STELLAR MASS AND DENSITY

The mass of a celestial body is difficult to measure unless it is possible to determine its gravitational effect on a nearby body. Within the solar system, the mass of planets can be found when there are satellites, but the problem becomes more complex with those planets that do not have satellites. Knowledge of stellar masses is, for the most part, based on information obtained from *binary stars*, where the effect of the two components upon each other can be studied.

Binary stars may be compared to the earth and the moon, where the system is made up of two bodies revolving around a common center of gravity. *Visual binaries* are those in which the two stars of the pair are distinguishable to an observer on earth.

While stars vary greatly in such properties as brightness and diameter, the range in mass compared with that of the sun is, for the most part, rather narrow. There are stars with a mass 100 times that of the sun, but these are extremely rare. The heaviest known star is HD 698, which has a mass about 113 times that of the sun, while Luyten 726-8, one of the lightest, has a mass of about 0.04 that of the sun. For the great majority of stars, the mass is between 0.1 and about 5 times that of the sun.

Mass may also be determined for spectroscopic binaries—those binary stars that are so close that their components cannot be resolved with a telescope. These stars have high orbital velocities which can be detected by spectroscopic methods—hence their name. The masses of the components of spectroscopic binaries are obtained from indirect data but, as a statistic, serve a useful purpose.

When the mass of a star has been computed, it is possible to determine the density of the material of which the star is composed provided the size of the star is known. *Density* is the weight of a given volume of substance. Since the mass and volume of the sun is known, the mean density can be calculated and is found to be 1.4 grams per cubic centimeter. The density of the earth by comparison is 5.52 grams per cubic centimeter. While the variations in mass of the stars are generally limited to a narrow range, the density varies widely because of the wide range in stellar sizes. Many stars have densities much lower than that of water. In fact, densities for some stars have been determined to be lower than for the best vacuum obtainable on earth. Such a star is sometimes called a "hot vacuum." Epsilon Aurigae, the largest known star, has a density about one hundred-millionth that of water. Extremes of high densities have also been discovered. One of these is for Sirius B, which has a mass approximately equal to that of the sun (0.95), but is only about 1/30,000 the size. This means that the same amount of matter which makes up the sun is compressed into a much smaller volume. The density of Sirius B, therefore, is almost 1 ton per cubic inch. This is not the densest star known. This record is held by a white dwarf star about 2,700 kilometers in diameter, which has a density of almost 1,600 tons per cubic inch.

5.6 STELLAR MOTION

Since the discovery of stellar motion by Edmund Halley in the eighteenth century, the motions of many stars have been studied. The measurements require great precision and are complicated by the rotation of the earth and the

Astronomy

revolution of the earth around the sun, the earth's precession, and the movement through space of the entire solar system.

The speed and direction of a star is found by measuring its *radial velocity* and *proper motion*. Radial velocity is the speed with which a star is moving toward or away from the earth. Radial velocities of more than 10,000 stars have been measured and the velocity is expressed in miles or kilometers per second. It is designated positive if the star is moving away from the sun, and negative if it is moving toward the sun. It must be remembered that a complicating factor in making such a measurement is the fact that the sun is also moving. Thus, what is being measured is the speed with which the star and the sun are moving toward or away from each other. Stars, of course, vary in radial velocity but 32 kilometers per second is quite common. The highest velocity is over 650 kilometers per second.

The radial velocity is not the true velocity of a star, nor does it indicate the true direction in which a star is moving. The proper motion, defined as the angular change in direction of a star in seconds of arc per year, must also be considered. Great precision is required, because the annual movement is usually quite small. At the present time measurements of this type are made photographically by comparing star photos taken a number of years apart. Periods of from twenty to fifty years are not unusual. The displacement of a star over a period of years is a measure of the motion that has taken place.

The star with the highest known proper motion is Barnard's star, so named after its discoverer. At a distance of almost 6 light-years, it is the second closest star to the sun (after Alpha Centauri), and it is invisible to the naked eye. The star's high velocity causes it to change its position by 10.25 seconds of arc per year. In figure 5.5, A represents a star as originally photographed from the earth. The proper motion is the displacement that has caused the star to appear to move from A to B. The movement from A to C represents radial motion. The actual direction taken by the star is represented by the movement from A to D. The velocity of motion from A to D can be calculated, but first it is necessary to determine the tangential velocity, or the velocity of the star in moving from A to B. For this purpose it is necessary not only to know its proper motion, but also the distance of the star from the observer on earth.

So far, this discussion has been on the basis of a star's motion relative to the sun as though the sun were motionless, which is not the case. That the sun moves* with respect to other stars was first realized by Herschel in the early nineteenth century. Astronomers at Greenwich had already measured the movement of thirty-six stars, but it was Herschel who caught the significance of the sun's motion, and by careful study was able to show that it was in the direction of the constellation Hercules. Now, with better instruments, radial velocity and proper motion have been used to verify the sun's motion.

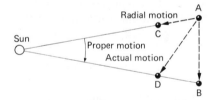

Figure 5.5 Relationship of proper motion (AB), radial motion (AC), and actual motion (AD) of a star with respect to the solar system.

---

*The sun's motion in this context is that of the entire solar system.

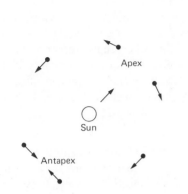

Figure 5.6 The average proper motion of stars surrounding the sun provides a clue to the direction of motion of the sun through space.

The proper motion of stars lying in the path of the sun, either at the apex of the sun's motion (direction toward which the sun is traveling), or the antapex of the sun's motion (the point away from which the sun is moving), will be little influenced by solar motion. The stars at the apex would appear to be moving to the left and right as the sun approaches, much as the trees in a forest do as they are approached (fig. 5.6). At the same time, the stars at the antapex would seem to be closing in behind as the sun moves away from them. The maximum effect of proper motion can be observed at points at right angles to the sun's direction of motion. Only those stars with velocities greater than that of the sun will show a proper motion in the same direction as the sun moves. Since these are few, the great majority will show a proper motion in a direction opposite to that of the sun; in other words, toward the antapex.

The radial motion of stars moving at right angles to the direction of the sun's motion is not influenced by solar movement. However, stars at the apex and antapex do reflect the sun's motion. Stars at the apex have an average radial velocity of -20 kilometers per second and those at the antapex average $+20$ kilometers per second. This is interpreted to mean that the sun is approaching the star Vega (at the apex) in the constellation Lyra at a velocity of 20 kilometers per second.

The proper and radial motion of the neighboring stars does not indicate their motion around the center of the galaxy. Rather it indicates small differences in motion with respect to the sun. The movement is analogous to skaters on a rink. All are generally moving in one direction around the center, but the movement of each individual is slightly different from the movement of any other individual.

5.7 STELLAR CLASSIFICATION

By the turn of the twentieth century, great progress had been made in astronomy, and attention was turned toward the classification of stars. The systematic organization of data on animal and plant life and on rocks and minerals, based on similarities, led inevitably to the question of similarities between stars. Were the stars all different, or were there some common characteristics under which they could be assembled into logical groups? In 1863, Angelo Secchi classified the stars into four groups according to the arrangement of the dark lines on their spectra. Henry Draper later modified this system to include description of variations in star spectra not indicated by Secchi. E. C. Pickering of the Harvard Observatory developed a new method for obtaining improved images with the spectrograph, which made it necessary to refine the classification in order to account for greater detail. This requirement led to the Harvard system of spectral classification, which was based on the examination of more than 200,000 spectra. So many types were found that the system required the use of almost every letter in the alphabet. However, upon completion of the task it was found that 99 percent of the types fitted into about seven main spectral classes, and these constituted a temperature scale when properly rearranged. The colors of the stars ranged from blue, indicating high temperature, to red, for low temperature. According to the letter designation originally ascribed to the different types, the spectral classes are denoted by the letters *O, B, A, F, G, K*, and *M* when placed in order of highest to lowest temperature. Spectral classification has proved to be a valuable device. Simply identifying the color of a star makes it possible to infer other characteristics such as temperature and prominent lines of certain elements in the spectrum of that star. Each of the spectral classes is divided into 10 subclasses and identified by a number from 0 to 9 to indicate placement within each class. Thus, *F5* indicates a position halfway between *F* and *G*. The sun is designated *G2*, indicating its placement between *G* and *K*.

While the majority of the stars fit into the spectral classes just discussed, there are several additional groups prominent enough to be mentioned. For the most part these are similar to the stars in the main spectral classes, but certain characteristics place them in separate groupings which are branches of the main spectral classes. Thus, *R* stars are a branch of class *K*, while *N* and *S* stars are a branch of class *M*. The spectral classes may be shown by symbols as follows:

$$\begin{matrix} & & & & & & R & N \\ & & & & & & / & / \\ O & - & B & - & A & - & F & - & G & - & K & - & M \\ & & & & & & & & & & & & \backslash \\ & & & & & & & & & & & & S \end{matrix}$$

An examination of table 5.2 could lead to the erroneous conclusion that stars are composed of different materials. Stars are made up primarily of hydrogen, with small amounts of helium and very small amounts of other elements. The diverse temperature conditions on stars result in different intensities of spectral lines, causing certain elements to be prominent at various places in the spectral scale. The fact that certain elements are prominent does not preclude the presence of the others.

5.8 THE HERTZSPRUNG-RUSSELL DIAGRAM

In 1905, the Danish astronomer Einar Hertzsprung considered the possibility of a correlation between the luminosity of a star and its spectral classification. By comparing a star's color and its luminosity, he found that, in general, white and blue stars had high luminosity and red stars low luminosity. He found some exceptions to this rule: A few red stars were also highly luminous, indicating great size. In 1913, an American astronomer, Henry Russell, made a similar discovery, and a combination of the two efforts resulted in the Hertzsprung-Russell or H-R diagram (fig. 5.7). The abscissa of the diagram shows the spectral types which correspond to the temperature and color of a

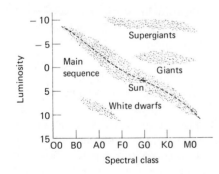

Figure 5.7 The Hertzsprung-Russell diagram as it appears when approximately 1,200 local stars are plotted according to spectral class and absolute magnitude.

Stars

star, and the ordinate shows luminosity as related to the sun and the absolute magnitude.

Since each star was represented by a dot on the diagram, it soon became evident that the great majority of the stars of known distance fitted within a narrow band ranging from the hot, highly luminous stars to the relatively cool, dimmer stars. This band is called the *main sequence*. In the upper right-hand side of the diagram is a group of stars red in color but highly luminous, indicating great size. These have been designated as *giants* or *supergiants*, depending on the degree of luminosity. Around 1925, an additional group of stars, the *white dwarfs*, were discovered and placed in the lower left-hand quadrant of the diagram. It was felt that these stars must be small because, in spite of their white color, they yielded very little light. Approximately 89% of the stars within a radius of 150 light-years are on the main sequence, 9% are white dwarfs, 1% are giants or supergiants, and the remainder are variable stars, which will be discussed later.

The H-R diagram is more than an interesting array of facts based on luminosity, distance, and temperature. It is a device that enables astronomers to make valid comparisons of the stars and learn something about the life history of a star. For example, all yellow stars on the main sequence have been found to have similar mass; stars that are white to blue have greater mass than yellow stars, and red stars have less mass. This evidence led to a new relationship between mass and luminosity; namely, that stars with greater mass are more highly luminous. The degree of luminosity is not directly related to mass, however, for if the mass is twice as great, the luminosity may be increased tenfold. A star that is 5 times the mass of the sun may be 50 or more times as bright as the sun and use up its available fuel 50 times as fast. From this it is possible to conclude that bright, massive stars are short-lived compared to the sun. Information such as this makes it possible to estimate stellar age.

In recent decades, the observation of certain star clusters has aided in studying the life histories of stars. Star clusters are homogeneous groupings of stars which are acting as a unit in that they are moving in the same direction and at the same velocity. Two main types of clusters are recognized. The *open* or *galactic cluster* is usually composed of a few to a thousand or so stars rather widely spaced, so that individual members are easily definable. Typical among the galactic clusters are such well-known groups as the Pleiades and the Hyades. There are about 500 such clusters known in the Milky Way galaxy. The other type is the *globular cluster*, much more compact, made up of about 100,000 stars, and much more distant from the earth than the galactic clusters. It is not possible to see them with the unaided eye, with one exception, Messier 13 in the constellation Hercules. About 120 globular clusters have been located, some within the galaxy, some beyond it. Messier 13, which is about 25,000 light-years away, contains upward of 100,000 stars in a compact mass. Although the stars appear to be an almost solid mass, the average distance between them is about one light-year and, even in the dense center, the stars are thousands of astronomical units apart. The probability of collision is thought to be practically zero, although their close proximity could disturb the orbit of outer planets if any exist.

Studying the stars in a cluster has several advantages. First, it may be assumed that the stars in a cluster were formed at approximately the same time and are, therefore, all about the same age. The second advantage is the fact that stars in a cluster are all about the same distance from an observer on earth. Therefore, a star which appears 100 times brighter than another in the same cluster must have a greater mass, since distance is not a factor. An H-R diagram of a cluster may be somewhat different from the H-R diagram for local stars in figure 5.8. For example, an open cluster in the constellation Serpens, Messier 16,

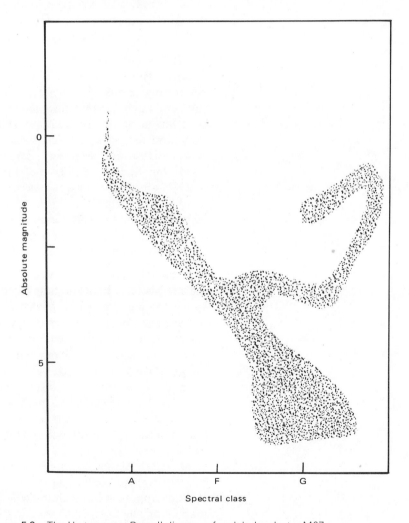

Figure 5.8   The Hertzsprung-Russell diagram of a globular cluster M67.

shows a preponderance of massive blue stars but no white, yellow, or red stars. Since stars of high mass are short-lived, existing for as little as 25 to 30 million years, it may be concluded that the stars (and the cluster) are as yet very young. On the other hand, an open cluster in the constellation of Cancer, Messier 44, lacks the massive blue stars in the H-R diagram and contains only white, yellow, and red stars. The only stars left are those which last a long time, which indicates that these are middle-aged stars. Messier 13 in Hercules, previously mentioned, is older yet, showing no blue or white stars. Only the yellow stars (like the sun) and the red stars are still present. Those stars which remain are very old, and all the large bright stars have disappeared.

5.9 UNUSUAL STARS

The sun may be considered a normal star, one about which a great deal is known and one which has no unusual characteristics. Most of the stars in the main sequence may be considered normal, differing only in luminosity, mass, and temperature, but conforming fairly well to a recognizable pattern. Now some atypical stars should be commented on before we turn to a discussion of stellar evolution.

Several unusual types appear on the H-R diagram, among them the red giants and supergiants located in the upper right-hand quadrant of the diagram. These are mostly type K or M stars, lying above the main sequence because of high luminosity. Betelgeuse and Antares are examples of supergiants, and Arcturus and Capella are typical of giants.

White dwarfs are extremely small stars, having a color similar to spectral type A stars but only about 1/100,000 as bright. These stars are placed below and to the left of the main sequence on the H-R diagram. Their luminosity is so low, due to their presumably small size, that they are invisible to the naked eye. White dwarfs are noted for their fantastically high density* which is due to their small size (planetary) and large mass (nearly solar). Only a few hundred have been discovered, and most of these were found because of their influence upon larger stars associated with them in a binary system. They are composed of *degenerate* gas, i.e., base atomic nuclei and free electrons.

Certain stars fluctuate in brightness and have become known as the *variable* stars. The variable characteristic may be due to one of several things. In some instances, binary stars (two stars revolving around a common center of gravity) will have the plane of their orbit edge-on toward the earth. Each of the pair will eclipse the other periodically, thus causing a variable effect. These eclipsing variables are not true variables in that the individual stars do not actually fluctuate in brightness.

---

*A cubic inch of such a star may have a mass of one ton or more and yet remain a gas, though it no longer obeys the usual gas laws.

Pulsating variables are those that fluctuate in brightness because they alternately expand and contract in size and consequently become cooler and hotter. The first such star was discovered in 1784 by an English astronomer, John Goodricke. He found that the star Delta Cephei varied in magnitude between 4.1 and 5.2 in a period of 5.4 days. More were discovered and studies of these stars revealed that there were several categories of pulsating variables based on their period-luminosity relationship. *Cepheid I* stars have an absolute magnitude of approximately -2 to -6, and periods ranging from 1 to 100 days. These stars brighten rapidly and dim gradually. The *Cepheid II* stars are about 1.5 magnitudes dimmer than Cepheid I types, and have periods ranging from 5 to 50 days. The brightening and dimming of these stars occur at a more gradual rate and the light curve is more irregular than that of the type I Cepheids. *RR-Lyrae variables* are short-period variables having periods of less than a day, typically from 7 to 17 hours.

Some of the giant red stars are long-period pulsating variables that vary in brightness from 2.5 to 7 magnitudes and have periods of 80 days or longer. Stars in this category with the larger periods are also less regular in their cycle and vary in brightness within wide limits. Betelgeuse is classed as one of the semiregular variables. Still other variables are completely irregular.

A more spectacular form of variable is the *eruptive variable* of which the *nova* is an example. A nova is an existing star which erupts suddenly and violently, increasing in brightness many thousands of times. The brightness declines to prenova stage at a much slower rate, in some cases taking years to return to normal. The exact cause of the outburst is unknown, but it appears that an outer layer of gas is ejected at velocities reaching 1,000 miles per second. This expanding shell of gas is responsible for the increase in brightness. A few novas have been observed to have more than one outburst without apparent damage to the star. There is no way of predicting when one will occur. On the average, two or three novas have been discovered telescopically each year, but many may escape detection since a constant vigil is not kept on all parts of the sky at all times. Most of the novas have occurred in the spectral type *O* or *B* stars rather than in the relatively cool type *G* stars, such as the sun. This does not completely eliminate the prospect that a type *G* star may become a nova, but the chances seem much reduced.

Another type of eruptive variable is the *supernova*, much rarer than the nova but much more cataclysmic. Whereas the nova flares up to thousands of times its normal brightness, the supernova will become tens or even hundreds of thousands of times brighter than normal. The gas shell expands away from the star at velocities up to 3,000 miles per second, and the material ejected in this shell may constitute a major portion of the star. Three supernovas have been observed during the past 1,000 years: One was recorded by the Chinese in 1054, one was observed by Tycho Brahe in 1572, and one was seen by Kepler and Galileo in 1604. Now supernovas are observed in other galaxies and are thought to occur in a galaxy about once every hundred years.

Another unusual type of object, first discovered in 1967, is the *pulsar*, of which about fifty are known at present. A pulsar is a small object which emits radio signals at short intervals in bursts of less than 2 seconds duration. The exact nature of the object is not known, although several theories have been proposed. One theory is that the bursts of energy come from a rapidly rotating *neutron star*. A neutron star, although never positively identified as such, is thought to exist as an extremely dense mass of neutrons (more dense even than a white dwarf) possibly compressed to this state by a supernova. A neutron star with the same mass as the sun would theoretically have a diameter of approximately 10 kilometers. Very strong magnetic fields associated with a rapidly rotating neutron star could be the energy sources of the pulses. White dwarfs and sources of extremely high temperatures have also been suggested as possible pulsars, but certain objections have caused these to be rejected.

Interestingly, further contraction has been hypothesized permitting the density theorized for a neutron star to be exceeded as a result of uninhibited gravitational collapse. The collapse could continue until the object reaches a limiting size depending upon its mass. For the sun, this limit is a diameter of approximately six kilometers. Now, in theory, the object may decrease to less than this limit and it would then become a *black hole*. In this instance the gravitational force would be so strong that no matter would escape, nor could any radiation including light. Thus the hypothetical object would be removed from the observable universe.

5.10 STELLAR EVOLUTION

All the bits and pieces of stellar information that culminated in the H-R diagram have contributed to the development of theories as to the life history of stars. The first interpretation of the diagram given by Russell himself was essentially as follows: The star initially formed from a dark cloud of dust and gas into a red giant. The new star contracted and became steadily brighter as it moved up the main sequence to yellow, white, and blue positions on the diagram. The star gradually used up its nuclear fuel, reversed its position on the diagram as the temperature slowly declined, and the star ended up as a red dwarf at the lower right-hand end of the main sequence. This explanation implied that the sun was either a relatively young star ascending toward the bright end of the sequence or a relatively old star descending toward the lower end of the sequence. Neither is correct, and this interpretation is no longer accepted.

A current theory on stellar evolution has stars forming in gas-rich regions of space and evolving through a life cycle, going through young, mature, and old-age stages. At the outset, a nebula, a cloud of dust and gas, begins to condense and collect as a result of mutual gravitational attraction. As the gas contracts, more material is attracted to it, and the entire mass begins to take

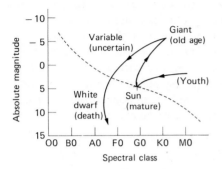

Figure 5.9 Pathway of possible stellar evolution shown on H-R diagram.

on a spherical shape. With continued contraction resulting from the embryonic star's own gravitational force, temperatures and densities rise. Part of the energy produced is converted to heat in the star's interior and part is radiated as light, heat, and radio waves. Eventually the star's central temperature is high enough to initiate a thermonuclear reaction. Temperature and pressure continue to rise until the heat generated is equal to the energy radiated from the star, and the internal pressure is sufficient to counter gravitational compression. The star is now reasonably stable: The forces are in equilibrium and the star is on the main sequence (fig. 5.9).

The place occupied by the star on the main sequence is dependent on its mass. Stars of high mass (relative to the sun) will take a position in the upper left-hand portion of the H-R diagram and become hot, short-lived stars, while stars of small mass will be relatively cool, long-lived stars at the lower end of the main sequence. A star spends the major portion of its life on the main sequence; its hydrogen is being converted to helium with a release of energy.

When the hydrogen fuel is diminished by perhaps 10 to 15 percent, the star becomes unstable. Temperatures at the core cool and the star contracts. This contraction again increases the temperature and initiates nuclear reactions in a shell around the core, causing the star to expand greatly to become a giant or supergiant, depending on its initial mass. This shifts the placement of the star to the right of the main sequence on the H-R diagram. Interior temperatures increase to perhaps 100 million degrees. The helium core, formed as a result of the conversion of hydrogen but inert up to this point, ignites at this high temperature to form a new source of fuel. In this new reaction, three helium nuclei are thought to combine to form carbon. This activity results in additional changes in the star and is looked upon as the end of the red-giant stage in the life cycle.

There is considerable uncertainty as to what occurs next, but it is thought that the star's position on the H-R diagram is reversed and that it returns to a main sequence position, although somewhat higher than its previous position; at this point, it becomes a variable star. During this period, other heavier elements are being formed as a result of the extremely high temperatures in the star's interior. The star continues to shrink, its position dropping below the main sequence, and it becomes a white dwarf. The star may possibly go through a nova or supernova stage before becoming a white dwarf. This is thought to be true for stars having a mass greater than 1.4 times that of the sun. The excess mass is shed in a series of nova eruptions, or else the star is almost completely destroyed as a supernova. White dwarfs are formed of "degenerate matter" made up of atoms in which the electrons have been uncoupled from the atomic nuclei, allowing very close packing of the particles. The nuclear fuel is exhausted, and the feeble light of the white dwarf is dependent on the residual heat from the interior, which may last for as long as a billion years. The light from a white dwarf becomes progressively dimmer until nothing remains but a dense, dark, burned-out stellar clinker.

The debris resulting from a nova or supernova again becomes part of the dust and gas found in interstellar space. Having once been part of a star, the debris consists of the heavier elements as well as the residual hydrogen and helium. This material combines with interstellar hydrogen so that subsequent stars formed from this enriched dust and gas will have a greater proportion of heavy elements than stars formed in the past. Old stars in the galaxy have a lower proportion of heavy elements than younger stars, a fact leading some astronomers to the conclusion that the original stars in the galaxy were composed of pure hydrogen. The build-up of heavy elements from these original stars and the formation of new stars from this material has resulted in second- and possibly even third-generation stars in the galaxy. Heavy elements are detected in the sun, and since these elements could not yet have been formed by the sun at this stage, it is considered to be at least a second-generation star, formed at least partly from the debris of previously existing stars.

5.11 THE CONSTELLATIONS

Many stars are easily visible to an observer on the earth. They appear to be scattered haphazardly across the sky. Upon closer examination, certain patterns become evident, and these patterns led the ancient astronomers to conceive of the constellations. The constellations, representing figures that are figments of man's imagination, have been handy points of reference for locating particular celestial bodies. Today's professional astronomer, aiming his telescope to locate a particular object or star which he may not even be able to see, must have more precise data. But for the amateur, the constellations still serve this purpose very nicely.

The geometric patterns in which the stars are arranged were named after gods and goddesses and mythological animals, but they have little or no resemblance to these figures.

Some people think that no more resemblance was intended than exists in modern times when a geographic location is named after some person. No record exists as to who named the constellations, but it is believed that the forty-eight constellations listed in Ptolemy's catalog had been recognized for some 3,000 years. These occur only in the Northern Hemisphere and do not include stars seen in the southern sky.

Tycho Brahe added two new constellations in the seventeenth century. Since then new ones have been described bringing the present total to eighty-eight. Of these, seventy are visible in the Northern Hemisphere.

The constellations themselves have no astronomical significance except as a means of identifying the location of an individual star, galaxy, nebula, or nova. Traditionally, a celestial object is said to be "in" some constellation. This does not necessarily mean that the object is a part of the constellation. It merely indicates that the object has this constellation as a background or can be seen through the constellation.

Stars and other celestial objects are frequently named for the constellation in which they are found. In the past, usually the brightest star in the constellation was designated by the Greek letter alpha followed by the genitive form of the name of the constellation. Other stars were then assigned Greek letters in decreasing order of brightness, although there were exceptions to this rule. In 1729, a system was adopted in which the stars in a constellation were numbered consecutively from west to east across the constellation, regardless of brightness. As more powerful instruments were developed this, too, became inadequate. Stars are now cataloged by number and position.

The best way to learn the stars is to have someone point them out in relation to the constellation in which they are found. Others may then be found with the aid of star charts.

5.12 SUMMARY

Despite the great distances between the stars and the earth, diligent effort on the part of astronomers has resulted in the accumulation of much data on the distance, temperature, luminosity, size, mass, and color of stars. The data have made possible the classification of stars into types based on remarkable similarities. This system of classification has been most useful in the development of the Hertzsprung-Russell diagram, which relates the class of star to absolute magnitude. The diagram has enabled astronomers to draw some conclusions about the relationship between the mass of a star and its brightness. It has shown, for example, that the brighter stars also have greater mass. The age of stars has been inferred, since it had been theoretically calculated that brighter stars used up their nuclear fuel at a much faster rate relative to their mass than did dimmer stars.

Certain stages of a star's life cycle have also been deduced from the H-R diagram. It is believed that a star originally forms from a contracting cloud of dust and gas and takes a position on the main sequence dependent upon its size and temperature. The major portion of a star's life is spent in this phase. Hydrogen is converted to helium, and eventually the star becomes unstable and begins to expand to become a red giant. What takes place in the state following this is not well understood, but it is surmised that the star shrinks, possibly becoming a variable star, and that formation of some of the heavier elements takes place. Following this, the star may explode or otherwise lose mass, and eventually become a white dwarf.

QUESTIONS

1. Describe the principle of parallax.
2. In what way may the Cepheid variables be used to determine distance to a star?

Stars

3. If star A has an absolute magnitude of 4 and star B has an absolute magnitude of 9, what is their relative brightness with respect to each other?
4. What is meant by a parsec?
5. What is meant by proper motion and radial motion of a star and how are these measured?
6. What is the basis upon which stars are classified into spectral classes?
7. Describe the life cycle of a star in terms of the H-R diagram.
8. Briefly describe the different types of variable stars.

6 Galaxies

Discovery of the Milky Way as a galactic structure.
Dimensions of the Milky Way galaxy.
Motion of stars within the galaxy.
Structure and mass of the galaxy.
Types of galaxies.
Evolution of a galaxy.
How distances to external galaxies are measured.
Motion among the external galaxies.
Distribution of the galaxies in space.
Radio galaxies and quasars.

The sun, all the countable naked-eye stars (about 5,000), and the billions of stars forming the Milky Way make up the aggregate of stars which, together with numerous gaseous nebulae and dust clouds form a flat spiral structure called a galaxy (fig. 6.1). *Gala* is a Greek word for "milk" and was first applied by the Greeks to the irregular band of faint light caused by billions of distant stars extending through the constellations Sagittarius, Cygnus, Cassiopeia, Auriga and Canis Major. The faint light of the Milky Way is visible only on clear, moonless nights far from the artificial illumination found near a city.

6.1 THE GALAXY DISCOVERED

Galileo was the first to recognize the Milky Way as a band of numerous stars at tremendous distances from the earth. However, it was Thomas Wright, an English former sailor, who around the middle of the eighteenth century first

Galaxies

questioned the arrangement of the stars. He saw what others before him had seen, a myriad of stars overhead with a faint band of light passing through the dark sky, representing what appeared to be a much denser population of stars. Wright raised the question of whether the density was really greater in the Milky Way, or whether the stars were just as far apart there as in other parts of sky. He felt that observers were looking broadside through a lens-shaped structure which acted to give the illusion of greater density. This was a revolutionary idea. It seemed strange that in the hundreds of years of viewing by thousands of observers, the same idea had not previously occurred to someone. Wright was able to describe the structure of the star system as a lens-shaped system rather than a sphere, and his reasoning was difficult to refute.

Immanuel Kant, along with his other accomplishments, found time in 1753 to examine Wright's discovery. He made several unusual observations which bordered on the prophetic. He compared the form of the galaxy as described by Wright with that of the solar system. Both were flat, circular-shaped structures. Since the stars had been shown to move, Kant speculated about the possibility that the stars in the galaxy revolved around a center, and that, therefore, the entire galactic structure turned like some gigantic wheel. Kant also felt that there was no reason why the universe should be limited to only one such galaxy. A number of faint, oval-shaped nebulosities had already been observed through telescopes, and Kant thought these might represent distant galaxies similar to the Milky Way. With this reasoning, Kant greatly extended man's concept of the universe and perhaps prompted Johann Heinrich Lambert, in 1761, to stretch this concept to infinity. The question of whether the universe is finite or infinite has still not been resolved.

William Herschel hoped to prove Wright's theory on the structure of the galaxy. His approach was to count stars systematically to determine the density of number in selected positions of the sky. He sampled 3,400 spots, using a 12-inch reflecting telescope, and concluded that the galaxy was irregularly disk-shaped rather than lens-shaped and that the sun was off to one side from the galactic center.

6.2 SHAPE, SIZE, AND POPULATION OF THE MILKY WAY

The modern view of the Milky Way is about as shown in figure 6.1. Much of this was determined by Harlow Shapley who, in 1917, mapped the position of a number of globular clusters associated with the Milky Way. From their distribution, he was able to infer to some extent the shape and size of the galaxy. This flat disk-shaped structure with the bulge at the center is about 100,000 light-years in diameter and 10,000 light-years thick through the bulge. The sun is close to the equatorial plane of the galaxy and much farther off-center than was envisioned by Herschel; it is about 30,000 light-years from the center, or about two-thirds of the distance from the galactic center to the edge.

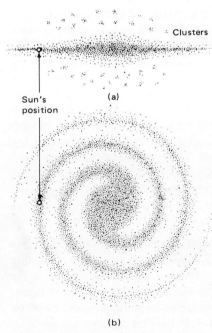

Figure 6.1 (a) Cross section and (b) plane view of the Milky Way.

The total number of stars may be estimated by star counts in representative portions of the galaxy. Herschel first attempted to do this. He assumed that the brightest stars were also the closest and that the stars were uniformly distributed in space. The actual count did not agree with calculated numbers once he got beyond the place he believed was occupied by stars of magnitude 3. He had approached the problem correctly, but he had no way of measuring the true brightness of the stars. At the beginning of the twentieth century, this technique was again applied, and although the true brightness of the stars could then be measured, galactic dust obscured many stars and made the count at great distances unreliable. Present-day counts, allowing corrections for the effects of galactic dust, estimate the number of stars in the Milky Way galaxy at about 100 billion.

6.3 INTRAGALACTIC MOTION

Stellar motion as previously discussed applied only to slight differences in direction and velocity of stars close to the sun. At the same time, these neighboring stars along with the sun move in orbits around the center of the galaxy at a speed of approximately 240 kilometers per second. This latter motion of the sun can be determined only by making observations of very distant objects which do not share in this motion. These objects are distant external globular clusters which do in fact move, but which are not rotating with the galaxy. The motion in this instance is determined in much the same manner as the motion and direction of the sun are determined with respect to local stars. Here, again, in one direction the globular cluster appears to be approaching the sun, and in the other direction to be receding from the sun. Still more evidence of galactic rotation may be obtained by studying the motion of external neighboring galaxies with respect to the sun. Studies of this type reveal that the Large Magellanic Cloud (seen only from the Southern Hemisphere) is receding from the sun at a velocity of about 275 kilometers per second, while the great Andromeda galaxy is approaching the sun with a velocity of about 300 kilometers per second. These velocities are not so much evidence of galactic motion as they are an indication of the motion of the sun within the Milky Way. In fact, these velocities would indicate that local galaxies are moving more or less at random and that their velocities with respect to each other are quite low.

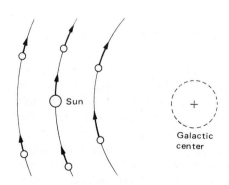

Figure 6.2 Movement of stars around the galactic center.

Does the galaxy rotate as a giant wheel with all stars maintaining the same relative position? This would imply much higher velocities for stars farther from the galactic center. Or do the stars revolve around the center in the manner of the planets around the sun, with the inner stars moving at higher velocities than those farther out? The latter seems to be the answer, based on studies of star motion which show higher velocities on the average for stars closer to the galactic center than the sun and lower velocities for stars farther from the galactic center (fig. 6.2).

Determining the speed and direction of objects in space is a complex task. Reference points from which measurements are made are themselves in motion and this motion in turn is difficult to define. The accuracy of any space measurement is open to argument, but the margin of error is probably not more than 20 to 25 percent. Thus, the velocity of the sun in orbit in the galaxy may be higher or lower than the 240 kilometers per second. This is a reasonable figure, however, and coupled with the distance from the galactic center, also not precisely known, yields an approximate period of revolution for the sun of about 200 million years.

6.4 GALACTIC STRUCTURE

The shape of the Milky Way has been described as a flat disk with a bulging nucleus, but this picture of the structure is not strictly accurate. By studying the many forms of external galaxies, astronomers could see various possibilities and soon suspected that the Milky Way was in fact a spiral galaxy. An examination of external spiral galaxies has revealed that most of the brightest stars (usually also the youngest) and the interstellar material are located in the spiral arms. These conditions appear to exist in the Milky Way with mainly type *O* and type *B* stars and nebulae in the outer reaches of the galaxy. These relatively young stars have been designated as *population I* stars with the sun falling in this category as an older member of the group. Older stars, found in the globular clusters and in the central portion of the galaxy, are considered *population II* stars.

Because of the sun's position in one of these arms, where there is extensive interstellar dust and gas clouds, it is difficult to see any great part of even the neighboring arms. Only in the last decade has a study of distant nebulae enabled astronomers to partially map the spiral arms and find that the sun is located on the inner edge of an arm known as the Orion arm, named for the Orion nebula (fig. 6.3). The arms are about 6,000 to 7,000 light-years apart and extend about 2,500 light-years from the inner to the outer edge. The central bulge of the galaxy is about 4,500 light-years in diameter. The innermost arm of the disk portion of the galaxy is about 15,000 light-years from the galactic center. The second arm is about 21,000 light-years from the center, and the third arm (the Orion arm) is about 27,000 light-years from the center. A fourth arm, the Perseus arm, at 35,000 light-years, and a faint fifth arm about 45,000 light-years from the galactic center complete the structure. More recent radio data obtained in the Netherlands and Australia indicate that the arms show breaks and branching in actual structure.

Surrounding the nucleus and the disk of the galaxy is a spherical halo made up of globular clusters, a few isolated pulsating stars, and a tenuous cloud of hydrogen gas. The objects in the halo do not participate in the galactic motion but have movements of their own. The orbits of these objects around the galactic center are thought to be elliptical in some cases, with one of the focal

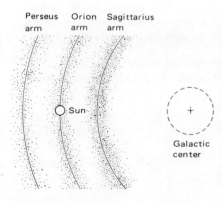

Figure 6.3 Spiral arm structure in the vicinity of the sun.

points of the ellipse coinciding with the center of the galaxy. The radius of the halo appears to be out 60,000 light-years from the galactic center.

6.5 GALACTIC MASS

How much matter is present in the galaxy? Some concept of the number of stars can be obtained by determining the amount of matter in the galaxy. In calculating the mass of a double-star system, Newton's modification of Kepler's third law proved to be invaluable. The same technique may be used here by measuring the gravitational effect of the galaxy as a whole upon the sun. This gravitational force manifests itself in the relationship between the distance from the galactic center and the speed with which the sun revolves around the center. An assumption must be made that gravitational force acts as though the galactic mass were concentrated at the center. However, the value for mass obtained by using Kepler's third law compares favorably with that for the mass adjusted by many factors to account for the fact that the total mass is not at the center. This value is approximately 1.4×10^{11}, or 140 billion times the mass of the sun. The problem involved in making such a computation is that the distance from the galactic center is only an approximation, as is the period of rotation of the sun around this center. These factors, along with the assumption that the mass is concentrated at the center, make the value for the galactic mass only a rough estimate. However, all available information places the value between 10^{11} and 2×10^{11} times the mass of the sun, which is satisfactory for our purposes.

6.6 TYPES OF GALAXIES

Beyond the Milky Way lie countless other galaxies so distant that the light from even the nearer ones started on its journey toward earth long before the dawn of man's recorded history. With characteristic thoroughness, scientists have classified these many galaxies into a few categories for convenience. Basically there are three types which will be described briefly (fig. 6.4).

1. *Spiral galaxies (A)* are by far the most common type of the distinctly visible galaxies, making up slightly more than 70 percent of the total number studied.

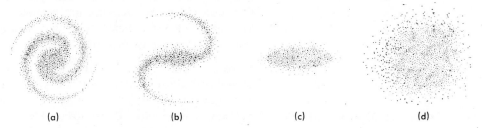

Figure 6.4 Types of galaxies: (a) normal spiral, (b) barred spiral, (c) elliptical, (d) irregular.

The Milky Way is a typical one, wherein the arms attached to a central nucleus radiate outward, giving a pinwheel effect. A *barred spiral* differs in that it is not nearly so compact as the normal spiral, and the arms emanate from a "bar" which passes through the center of the galaxy *(B)*. The spiral galaxies are usually quite large, ranging from 20,000 to more than 125,000 light-years in diameter. The Milky Way and Andromeda are examples of the larger spiral galaxies.

2. *Elliptical galaxies* are generally smaller than spirals, and while much less common among the thousand or so conspicuous galaxies, they are thought to be the most numerous type of galaxy in the universe. Their comparatively small diameter (as little as 5,000 light-years) and, therefore, their relatively low luminosity make them difficult to detect at great distances. While most elliptical galaxies are small, there are some exceptions. Several elliptical galaxies have been discovered which are much larger than any of the spirals. These may be as much or more than 200,000 light-years in diameter, being among the largest such structures in space.

The shape of elliptical galaxies range from an almost spherical form to the flat ellipse shown in figure 6.4C. The population of stars is concentrated in the center and becomes sparse as the rim of the galaxy is reached. Elliptical galaxies are made up entirely of population II stars and therefore resemble the nuclei of the spiral galaxies.

3. *Irregular galaxies* (fig. 6.4D) make up about 5 percent of the total galactic population and are, as the name implies, lacking in symmetry. The Magellanic Clouds, visible from the earth's Southern Hemisphere near the south celestial pole, are typical of irregular galaxies. These are the external galaxies closest to the Milky Way, being approximately 150,000 light-years distant. Spectrographic analysis of these galaxies reveals interstellar gas and dust and type *O* and type *B* stars spread throughout the entire structure. This would indicate a relatively young age for these galaxies, a fact inconsistent with other evidence. Some globular clusters and certain individual stars that are old population I or even population II stars are present in the Magellanic Clouds.

6.7 GALACTIC EVOLUTION

Studying what is known of the life cycle of stars and the motion and structure of galaxies offers some clues as to the possible evolution of galaxies. The age of the sun is about 5 to 6 billion years and may be considered the minimum possible age of the galaxy. If the sun had been formed during the early stages of galactic development, it would be made up primarily of hydrogen forming into helium. However, since it is known that the sun contains heavy elements, it must be a second- or third-generation star formed at least partly from material ejected from earlier stars by the nova or supernova mechanism. Thus, it is permissible to surmise that the galaxy is much older, possibly 10 billion years old.

Stars begin their life cycle as a cloud of interstellar material which contracts and condenses, eventually generating sufficient energy to initiate a thermonuclear reaction. In the initial stages, it is presumed that the protogalaxy was a huge tenuous mass of hydrogen and that no stars had as yet formed. As the great mass of turbulent gas rotated, internal gravitational forces caused the mass to begin to contract. As contraction took place and rotational speed increased, the density of the gases increased. Turbulence within the mass and the force of gravity made some gas eddies sufficiently dense to hold together and evolve into first-generation stars. These first stars may have formed in isolated groups, becoming the globular clusters which orbit around the galactic center. Others may have formed as individual stars with eccentric orbits about the galactic center.

At this early stage of development, the galaxy had not yet taken on a disk-like structure, but as the gas condensed further and the speed of rotation increased, the system began to flatten (fig. 6.5). Only a small amount of the primordial

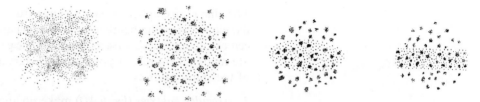

Figure 6.5 A possible means of galactic evolution. Solidly dotted areas indicate regions of contracting gas.

hydrogen was utilized to form the first-generation stars, but the more massive ones quickly converted their fuel to helium and the heavier elements were formed. The hydrogen continued to contract and its density continued to increase, thereby improving conditions for star development. The number of stars which took shape increased, and some of these constituted second-generation stars, since the dust and gas from which they were formed included some of the heavier elements formed in the earlier stars. The disk now formed no longer contracted toward the galactic nucleus but further flattened, and the older population I stars, including the sun, were born. This was probably the period during which the greatest amount of star development was taking place. These stars contained greater quantities of heavier elements than those previously formed.

The sun is now believed to be 5 to 6 billion years old, and many stars have been formed since then, although the rate is at a reduced pace. The density of gas available now is quite low, estimated at about 1 to 2 percent of the total galactic mass. Most of this gas is confined to the outer arms of the galaxy and mainly near the equatorial plane of the disk. The stars being formed at present are, at the very least, third-generation stars with very high concentrations of heavy elements.

What will be the eventual fate of the galaxy? If the evolutionary path of a star and the development of the galaxy are as theorized, then the ultimate end of the galaxy can be predicted. A star goes through a life cycle which terminates in the formation of a white dwarf. Only a small percentage of the material that originally went into the star is returned to interstellar space for the development of future stars. Thus, eventually all galactic material will be tied up in white dwarfs which, as they lose their energy, become black dwarfs or completely lifeless bodies. Ultimately there will be insufficient gas and dust in the interstellar spaces to form new stars, and those that remain will gradually die out to become dense, dark bodies which continue to orbit the galactic center.

In the last days of the galactic life cycle, the night sky will be almost completely dark except for a scattering of dim red stars and a few very dim specks of light indicating the location of the neighboring galaxies. The star around which a planet orbits will be a dim and extremely dense white dwarf radiating its last remaining bit of energy before extinction. The galaxy then will become a vast, invisible structure moving undetected through endless space, but maintaining its internal structure and its position with respect to other galaxies.

The differences in structure and to some extent apparent differences in age have prompted some astronomers to conclude that galaxies go through a sequence of stages during their evolutionary development. Edwin Hubble, during the 1920s, suggested such a sequence as the one illustrated in figure 6.6. He envisioned the elliptical galaxies starting as spheres and gradually flattening into elliptical disks. Subsequently, arms would form, transforming the elliptical galaxies into normal and barred spirals, each developing along a different path.

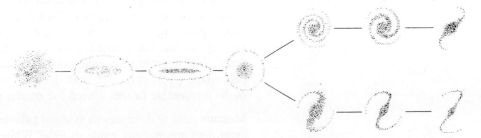

Figure 6.6 Hubble's evolutionary sequence for galactic development.

A more modern version, presented by Harlow Shapley, suggests that the galactic life cycle begins with an irregular galaxy made up of young stars and then branches into the normal and barred spirals (fig. 6.7). Continued contraction of the spiral structure eventually eliminates the arms, whereupon the galaxy becomes elliptical with no arms and only population II stars.

This sequence of events in the evolution of a galaxy, while attractive, does have flaws which cannot be overlooked. Certain double galaxies, it may be assumed,

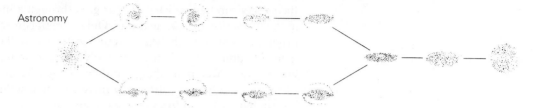

Figure 6.7 Shapley's evolutionary sequence for galactic development.

have formed from the same cloud of gas and should, therefore, be the same age, much as the stars are in a star cluster. If this were so and Shapley's theory were correct, then such galaxies would be of the same type, but frequently they are not. Furthermore, could irregular galaxies develop into spiral galaxies with an attendant increase in mass? Then, too, spiral galaxies are extremely flat when seen edge-on and elliptical galaxies are never that flat. No mechanism has been discovered that would explain why a spiral galaxy would round out in this manner as it progressed to the elliptical stage. It has been suggested that it is no more reasonable to expect a sequence of this type for the cycle of a galaxy than for the evolution of a star. The chances are that a galaxy forms into one type or another, perhaps on the basis of its mass and angular momentum.

6.8 GALACTIC DISTANCES

Distances to objects within the solar system have long been known with a reasonable degree of accuracy. This is now of particular importance, since man is planning to visit various bodies in the solar system. Distances to stars are measurable, but a need for accuracy is not nearly so critical. An error of a few million miles in the tremendous distances involved represents only a small percentage error, and since man does not contemplate so extensive a journey in the foreseeable future, a need for greater accuracy is not required.

Measurement of distances to external galaxies is a relatively recent accomplishment, first successfully made in 1924. While the existence of external galaxies or "island universes" was suggested by Kant and subscribed to by many astronomers, the problem was not definitely solved until Edwin Hubble photographed the Andromeda nebula through the 100-inch telescope in 1924. Individual dots had been seen in Andromeda as early as 1889. However, it was thought that this indicated Andromeda was a part of the Milky Way, for otherwise the dots would have to be giant star clusters or small galaxies in order to be seen at such great distances. Hubble's photographs revealed sharp pinpoints of light that were obviously stars of every type, some of which fluctuated in brightness and were identified as Cepheids. With such information and with the knowledge of the relationship of luminosity to period, Hubble calculated the distance from the Milky Way to Andromeda to be

900,000 light-years. This clearly placed Andromeda in the realm of an external galaxy beyond the boundaries of the Milky Way. Hubble also made a study of the frequent novas that occurred—130 in a few years—and this confirmed the identity of Andromeda as a galaxy. He used the novas as a means of measuring distances to about 100 other galaxies in which he was able to detect individual stars.

Distances to galaxies are even more difficult to measure than was originally thought because of the complicating factor of interstellar dust and gas. This caused Hubble to err in his first determination, but the techniques he developed were of great value, since they provided the means of measuring such distances. He made the same assumption that is made by all other physical scientists, namely, that all the laws of nature which operate on earth or in the stellar vicinity of the earth apply to all other areas of the universe both in space and in time. Hubble assumed that the luminosity-period relationship that existed for Cepheids in the Milky Way applied to the Cepheids wherever they could be seen. This method has proved to be the most reliable one yet found for measuring intergalactic distances. The luminosity of a Cepheid is related to the period of pulsation, and this relationship is well-established. The length of the period can be measured, the absolute magnitude can be determined, and finally from this information the distance can be calculated. Unfortunately, the distance to which this method can be applied is limited to about 20 million light-years. Within this radius only about 30 galaxies with resolvable Cepheids are found, although up to 150 galaxies are visible.

Other methods were developed as equipment and techniques improved and more stars were resolved. Particularly bright stars (red giants and blue giants) were useful in checking the results obtained with Cepheids. The brightness of novas and supernovas has also been of value to a limited extent, but as yet there are insufficient data on their characteristics to make them a reliable means of measuring distances.

The use of resolvable stars as a means of measuring intergalactic distances is quite limited and does not yield data for the very faint galaxies. The intrinsic brightness of the galaxies themselves may be used, but it is difficult to know whether an isolated galaxy is a nearby dim one or a very distant bright one. Only when it is compared with neighboring galaxies can some concept of its brightness be obtained. At present this type of data can be used only in a theoretical, statistical way to give some clue as to the distance of a group of remote galaxies. Much more data must be available to make this a reliable method.

Certain problems plagued the astronomers after Hubble's first values were presented in the 1920s. The Milky Way, for example, appeared to be two to three times larger than any other galaxy in the visible universe. Why this was so was a mystery. Furthermore, novas in the Milky Way appeared to be of a different type from those in neighboring galaxies, and the globular clusters in the Milky Way were brighter than those in the Andromeda galaxy. The influence of interstellar dust and gas on light was gradually recognized, and distances both within the Milky Way and from the Milky Way to external galaxies were corrected. It was not until 1952, with the aid of the 200-inch telescope, that an error in the values calculated for the magnitude and distance of the Cepheids was discovered. An adjustment equivalent to a factor of 4 in luminosity had to be made. This meant that an object that was in reality four times brighter than originally thought must be not only twice as far away but twice as large in order to have the observed apparent magnitude. Thus, the Andromeda galaxy is approximately 2 million light-years from earth and twice as large as previously judged, making it, in fact, slightly larger than the Milky Way.

One other indirect method of measuring distance makes use of the *red-shift phenomenon*. This is a shift of the spectrum to the longer wave lengths of light from the remote galaxies, assumed to be produced by the Doppler shift.*
V. M. Slipher first used this technique in the early part of the twentieth century to measure the velocities of more than forty "nebulae." He was unaware of the true nature of these objects but was able to measure velocities of up to 100 miles per second for these nebulae, most of which were moving away from the earth. During the 1920s, these objects were identified as galaxies, and some evidence indicated that there was a relationship between the velocity and the distance of the galaxy from the earth. Hubble, in the course of his other studies and in collaboration with M. Humason, was able to ascertain by 1930 that the speed with which galaxies receded was in direct proportion to their distance from the earth. This phenomenon became known as the *red-shift law*. Since 1949, work by Humason with the 200-inch telescope has confirmed the value of this law. He was able to establish a velocity of recession of 61,000 kilometers per second at a distance of 2.6 billion light-years for one of the remote Hydra clusters of galaxies. By 1963, the greatest velocity measured was for a galaxy in the handle of Ursa Major. The red shift here implied a velocity of 138,000 kilometers per second or about 45 percent of the speed of light. In 1965, a form of quasar (see section 6.11), which emits no radio waves and which is called a "blue stellar object," was detected by A. Sandage and M. Schmidt. This object was found to be moving away from the earth at a velocity of 210,000 kilometers per second. At the same time, quasar 3C-9, the most distant object from earth yet discovered, had been found to be receding from the earth at approximately 240,000 kilometers per second or at about 80 percent of the speed of light indicating a distance of approximately 10 billion light years from the earth.

Accurate distances cannot as yet be determined by using the red shift. However, some indication of distance may be obtained from the following relationship:

$$r = \frac{V}{H}$$

---

*Doppler shift: An apparent change in frequency due to the relative motion of the source and the observer. Light source moving toward the observer has higher frequency (shorter wave length) than light from source moving away from the observer.

where r is the distance to the remote galaxy, V is the velocity of recession, and H is a value known as Hubble's constant. This constant, although perhaps not strictly accurate, expresses the increase in the rate of recession of distant galaxies per unit of distance, and for purposes here, may be given as an increase of 24 kilometers per second for every million light-years from the earth. This value was obtained by plotting the velocities of distant galaxies against the distances, where the distances could be established by other methods.

Some astronomers doubt the validity of this relationship (speed of recession to distance) and feel that the red shift is the result of some effect on light moving through great distances in space. They attribute the cause of at least a part of the red shift to "a tiring of light as it moves through space" and argue that for this reason the total red shift should be adjusted downward in order to obtain a true velocity of recession. Since no experimental evidence has as yet been presented to support this theory, speed of recession as measured by red shift is accepted.

6.9 GALACTIC MOTION

The earth was taken as the reference point for measuring motion within the solar system, despite the fact that the earth itself was moving. Similarly, the sun was used as a convenient reference point for measuring stellar motion. In the same way, the Milky Way may be used as a reference point for measuring galactic motion. To further complicate the calculations, allowance must be made for the fact that in all three cases observations are made from the moving earth.

Motion within the Milky Way results from the movement of individual stars revolving around a common center. As in the solar system, those bodies closer to the center revolve at a more rapid pace than those toward the rim of the galaxy. The resulting motion as seen from a neighboring galaxy would resemble a gigantic pinwheel slowly spinning in space. The same motion has been detected in other galaxies and appears to be common to all.

The detection of motion of external galaxies through use of the red-shift phenomenon has resulted in some interesting ramifications. It appears from these studies that all but a few neighboring galaxies are moving away from the Milky Way. Regardless of the direction in which observations are made the result is the same—all galaxies appear to be receding, and the greater the distances, the greater the velocity of recession. This movement gives the impression that the Milky Way is in the center of the universe, but an examination of figure 6.8 will quickly reveal that an observer at any point in the universe would see all galaxies receding. An observer on galaxy A (fig. 6.9) would see galaxy B moving away from himself at a rate of 50 units. The point that should be made here is that there is no way of knowing whether this apparent motion is the result of movement by galaxy A, galaxy B, or the net result of their com-

Astronomy

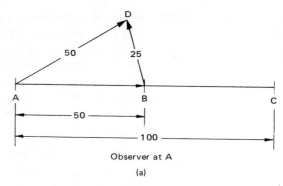

Figure 6.8 Motion of galaxies with respect to each other in an expanding universe as seen from point A. The units are arbitrary.

bined motion. If galaxy C were moving away from galaxy A at 100 units, an observer at B would see A receding at 50 units and C receding at 50 units. Galaxy D would be receding from galaxy A at 50 units, being the same distance from A that B is. Galaxy D, however, is closer to galaxy B, and therefore it would be receding at a slower rate. The diagram depicts this motion in two dimensions only, but by picturing this in three dimensions, it is easy to see the nature of an expanding universe.

The element of time is an interesting aspect of galactic motion, particularly for the very distant galaxies. Light has a finite velocity which means that an expression of distance in light-years (or parsecs) implies time. Light coming from the galaxy in Hydra traveling at a velocity of 300,000 kilometers per second over a distance of 2.6 billion light-years requires 2.6 billion years to make the journey. The light seen at present is a record of conditions that existed 2.6 billion years ago. It must also be recalled that this galaxy was receding at a velocity of 61,000 kilometers per second. On an assumption that it has continued to do so during the past 2.6 billion years, the galaxy would now be more than 3 billion light-years from the earth, and light now leaving the stars within this very distant galaxy would not reach the earth for another 3 billion years.

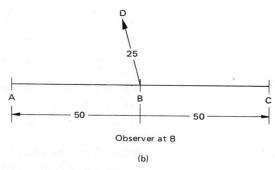

Figure 6.9 Motion of galaxies as seen from point B.

6.10 DISTRIBUTION OF GALAXIES

Galaxies

The distribution of galaxies, first studied by Hubble using the 100-inch telescope, gives some indication of the structure of the universe. Hubble plotted the density of galaxies and found them to be more or less uniformly distributed over the sky except for an obscured region where viewing is blocked by the presence of the Milky Way. It was assumed that a similar distribution occurs in this region of the sky; this assumption has now been proved to be correct by the observation of galaxies through "windows" in the less dense portions of the obscured regions. Hubble was also able to show that doubling the observable distance by longer exposures through the telescope increased the number of galaxies seen in space by about 8 times. This corresponds to the fact that doubling the radius of a sphere increases the volume of the sphere by 8 times.

Detailed study has shown that galaxies are not independent "islands" but are associated in clusters or groups ranging from a few to thousands of galaxies acting in unison. The Milky Way is one of seventeen galaxies that make up what is known as the Local Group. The Local Group is spread over a region of about 3 million light-years, as shown in figure 6.10. The positions are plotted on an arbitrary plane with the Milky Way set in the central position. From this point the galaxies in the Local Group are within a radius of slightly more than 2 million light-years from the Milky Way. In addition, certain galaxies that are thought to be small and dim because of great distance may be dwarflike and actually be within the Local Group.

Many galactic clusters outside the Local Group have been observed. They have provided valuable data in the study of luminosities, types, and sizes, because members of a particular cluster are substantially at the same distance and true relationships may be observed. Some clusters are quite rich in galaxies. The Coma cluster, for example, has about five hundred galaxies in a region that is less than 50 percent larger than that occupied by the Local Group. Dense clusters of this type are of interest in that they contain few spiral galaxies. Many of the galaxies are described as armless spirals. The lack of arms has been attributed to collisions between galaxies. Such an event is not so cataclysmic as it sounds. Stars and their attendant planets occupy only a tiny fraction of the total volume of a galaxy, so the chance of collisions by individual stars in colliding galaxies is extremely remote. Gases within these galaxies are, however, swept out, and since the arms of spirals are rich in dust, these arms are lost. Nor is a collision of this magnitude an instantaneous event. As long as 100,000 years may elapse between the onset of the collision and the separation of the galaxies.

The galactic cluster does not appear to be the largest organized entity in the universe. There are several thousand galaxies within a radius of about 70 million light-years, all organized into clusters, of which the Virgo cluster is the largest. G. de Vancouleurs has found that beyond 70 million light-years

106   Astronomy

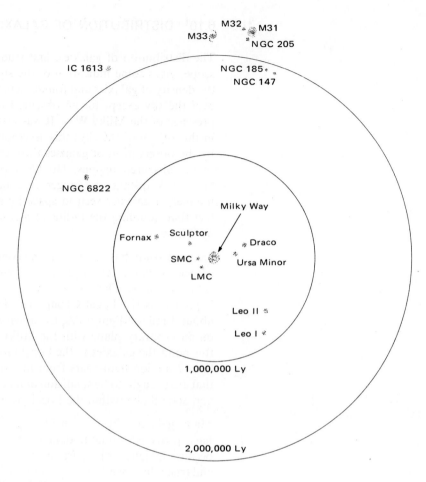

Figure 6.10   Distribution of galaxies in the Local Group.

there is a scarcity of galaxies for a considerable distance and that this local group of clusters constitutes a supercluster, or supergalaxy, 100 million light-years in diameter. He has suggested that the supergalaxy is organized into a flattened disk with the Virgo cluster at its center. The Milky Way lies near the plane of this structure about 40 million light-years from its center. Evidence also points to the possibility that the supergalaxy is rotating. Another example of a supergalaxy appears in Hercules, where several clusters appear to be overlapping.

6.11   RADIO GALAXIES AND QUASARS

Galaxies presumably emit energy over the entire spectrum, although it is energy in the visible portion that first calls them to our attention. When radio waves from space were first detected in the 1930s, it appeared that they origi-

Galaxies

nated within the Milky Way, but soon it became evident that external sources were also responsible for the invisible radiation. Several radio sources were detected in regions where no apparent objects seemed to be located. In 1954, Walter Baade, with the aid of the 200-inch telescope, was able to identify visually a radio source previously discovered. The radio source coincided with the position of a pair of colliding galaxies in the constellation Cygnus. This object, now known as Cygnus A, located 700 million light-years from earth, is an energetic emitter of radio energy.

More than one hundred discrete radio sources have been identified as coming from visible galaxies. These galaxies have been classified into two general groups, namely, the *normal galaxies* and the *peculiar galaxies*. Although there is no clear-cut distinction between them, the categories have been found useful. The normal galaxies include most of the spiral and irregular galaxies, both of which are weak radio emitters. The Andromeda galaxy is the first of the external galaxies from which radio emissions were detected. This is a normal galaxy and was so categorized on the basis of data obtained with the aid of the 220-foot diameter Jodrell Bank radio telescope. Cygnus A is an example of a peculiar galaxy having strong radio emissions apparently resulting from the collision of two galaxies. Cygnus A has also been thought of as a galaxy whose nucleus is splitting into two galaxies. As yet no satisfactory hypothesis has been presented which explains to everyone's complete satisfaction the source of so much energy.

Quasi-stellar objects, or *quasars*, as they are commonly known, are another group of radio emitters with some unusual characteristics. First discovered in 1960, the quasars were found to exhibit large red shifts, indicating that they are moving away from the earth at high velocity; they are perhaps the most remote objects now known in the universe. The quasar 3C-9, previously mentioned (section 6.8), is in this category. Distance alone is not the unusual factor; the fact that the objects emit tremendous amounts of energy, yet appear to be relatively small, also presents some problems. To be visible at such an enormous distance (10 billion light years) requires that a luminous object emit as much energy as two galaxies. Yet some of the quasars fluctuate in brightness over a period of several weeks to a month. Since nothing can travel faster than light, any object that changes in brightness cannot be larger than the distance light would travel during the period of fluctuation. This would indicate that quasars are relatively small, perhaps less than one light-year in diameter, compared to the Milky Way which is 100,000 light-years in diameter. How can so small an object emit so much energy as to permit it to be visible at the distance indicated by its red shift?

One suggestion has been that quasars are not as distant as indicated by the red shift but are objects relatively close to the earth. This explanation, however, would cast doubt on the validity of the red-shift law. No theory or explanation has as yet resolved the enigma that is presented by the quasars. The situation is best summed up by Maarten Schmidt, "If you get the impression of uncertainty about the state of knowledge of quasars, you are right. That is what the situation is."

6.12 SUMMARY

The twentieth century has been a time of tremendous discovery in astronomy. Not only has the true nature of the Milky Way been revealed, but the existence of external galaxies has been verified. With the techniques now available, it is possible to estimate the distances and the masses of objects so far away that the light from these objects left its point of origin before life developed on earth. Much is known now about the structure of galaxies, and theories have been formulated on the possible mode of their evolution. Some clues based on the probable arrangement of the visible universe seem to indicate that the galaxy is not the ultimate structure. Evidence is available that hints at a "galaxy of galaxies," or a supergalactic structure flattened into a disk and possibly rotating. Is it conceivable that further organization exists—a galaxy of supergalaxies? No data are available to confirm or deny this premise. Future studies will be directed toward securing more accurate and comprehensive data, proving or disproving existing theories, and theorizing further about the structure and the origin of the universe.

QUESTIONS

1. What is the modern view of the Milky Way and how many stars are thought to make up this galaxy?
2. Describe the motion of stars closer than the sun to the galactic center; at the same distance as the sun from the galactic center, and at a greater distance than the sun from the galactic center.
3. What is the approximate velocity and period of the sun around the galactic center?
4. Briefly describe the three types of galaxies.
5. What is thought to be the manner in which a galaxy evolves?
6. What is Shapley's suggested evolutionary sequence of galaxies?
7. What were some of the problems raised by the first values obtained in determining distance to the Andromeda galaxy?
8. Describe an indirect method to measure distance to remote galaxies. Why is such a method necessary?

7 Cosmology—Life in the Universe

Ancient theories on the structure and formation of the universe.
Views on the evolutionary universe.
Views on the steady state universe.
Is there life in the universe?
Essential requirements for life as we know it.
Life in the solar system.
Life in other systems.
Contact with life in other systems. What form will life in these systems take?
Influence of astronomy on society.

Consideration of the arrangement of stars, galaxies, and supergalaxies within the visible universe leads to the question of how it all began. This text is not concerned with the Creation itself but rather with events which have occurred since that first moment. It is important to avoid the confusion that sometimes occurs in attempts to reconcile theology and science in a discussion of the origin of the universe. It should be clear that science attempts to answer questions such as, What has been occurring? and How did it take place? Science attempts to trace the sequence of events from that very first moment. It is relatively easy to interpret events in the recent past. But in order to understand events in distant time and space, the scientist must extrapolate, since data are extremely scarce. The theologian, on the other hand, attempts to explain why the universe was created. He is involved in the problem of absolute beginnings and the philosophic reason for Creation.

7.1. HISTORY OF COSMOLOGY

Cosmology, the science that deals with the evolution and organization of the universe, is of relatively recent origin, yet its roots go back to the ancient astronomers. More than two thousand years ago Chinese Taoist astronomers attempting to describe the universe developed the *hsuan yeh* or "empty infinite space" theory. Previous Chinese theories had put limits on the universe, but according to hsuan yeh, space was empty and had no shape. It was occupied by the earth and other visible celestial bodies which were propelled by "hard winds" through an infinite space. Over the centuries, the basic concept of empty infinite space was incorporated with other theories and eventually abandoned.

The Greek astronomer Aristarchus (ca. 280 B.C.) also suggested that space was not bounded by the sky containing the stars. He explained the lack of parallax of the stars by suggesting that they were at an infinite distance from earth—a distance too great for parallax to be observed. He thus subscribed to what was to become the modern view of an infinite universe.

During the Renaissance a certain amount of speculation existed about the infinite nature of the universe. For example, in the fifteenth century, Nicolaus Cusanus, Bishop of Brixen, felt that the heavens were of the same nature as the earth and were inhabited by earthlike creatures; the universe was infinite, and regardless of what position in the universe one occupied, it would appear as though that were the center and all else was in motion around it.

Perhaps the first true cosmologist was Giordano Bruno (1548–1600) who was neither an astronomer nor a scientist but rather a philosopher. Wherever he went he invoked wild enthusiasm or bitter hatred by preaching the Copernican doctrine. Where Copernicus had replaced the earth by the sun as the center of the universe, Bruno went a step further, implying that the sun was a star among many stars in a boundless universe. He had no evidence to support this contention, and it would be several hundred years before sufficient data became available to prove him right. He described a universe filled with countless suns and planets and proposed the possibility that these planets supported life. Bruno was far ahead of his time, and his ideas were not appreciated.* Centuries passed before the full impact of his reasoning was felt and found valid.

Thomas Digges, a contemporary of Bruno's, reasoned along the same lines as Bruno. An early English translator and advocate of Copernicus's works, Digges no longer felt bound by the sphere of fixed stars surrounding a central earth. He viewed the cosmos as a collection of stars all placed in an infinite universe at varying distances from a central sun.

The concept of infinite space was strengthened in the generations that followed Copernicus by the slow but continual expansion of knowledge. Kant, Wright,

---

*In fact, he was burned at the stake as a heretic for his ideas.

and Herschel first visualized the structure of the galaxy. On the basis of new information, several important cosmological theories have been formulated. These differ primarily in the interpretation given to the recession of distant galaxies as measured by the red-shift law, which would indicate an expanding universe.

7.2 THE EVOLUTIONARY UNIVERSE

George Lemaître, a Belgian astronomer, is credited with first enunciating the evolutionary theory. His proposal, published in 1931, considered that the universe originated as an extremely dense (100 million tons per cubic centimeter) hot mass which he called the *Atome Primitif*, which translated means "primeval atom," now commonly called "primeval nucleus." This mass contained all the matter in the universe and had a diameter of about 200 million miles or about the diameter of the earth's orbit. This volume was obtained from what was thought to be the average density of the universe and from Einstein's estimate of the diameter of the universe. This mass exploded, according to Lemaître, and the particles which he called "atom stars" were scattered at high velocity in all directions. The matter continued to expand, thinning out, cooling, and emitting radioactive particles during the process. Lemaître suggested that during this period the heavier elements were formed. He believed that elements such as uranium with long half-lives have retained their radioactive character.

Lemaître believed that the universe, at that very early moment, was restricted to the boundaries of the primeval nucleus and that nothing existed beyond this. When the initial explosion took place, the universe began to expand at a rate equal to the velocity of the fragments from the primeval nucleus. With expansion, there was a decrease in density of the matter scattered in space. The particles, however, maintained their mass and therefore exerted a gravitational force. This gravitational force gradually slowed down the rate of expansion until after a few billion years expansion practically ceased, and the universe was in a state of relative equilibrium. The universe became nearly static, a condition already described by Einstein, where the forces of expansion and gravity are in balance. It was during this phase that the galaxies began to form by the condensation of matter. During this stage there was considerable instability. After several billion years, according to Lemaître, the forces of expansion overcame the gravitational forces and the universe resumed its expansion.

Lemaître's theory faced many problems after it was first presented. For example, the age of the universe as measured by the recession of the galaxies was established at slightly less than 2 billion years. This figure was contradicted by the discovery of rocks on the earth that were in excess of 3 billion years old. The problem was resolved with Baade's discovery in 1952 that the magnitude of Cepheids had been underestimated and that distances to the galaxies were more than twice what they had been thought to be. This raised the possible age of the universe to 5 billion years and disposed of the apparent contradiction. Discoveries in the past few years have further extended this age to almost 13 billion years.

In 1946, George Gamow proposed a modified form of the evolutionary theory, and this form is one of the two theories currently being given serious consideration. Gamow's theory differed in several respects from Lemaître's theory.

In Gamow's universe, the very dense state of matter, or YLEM, as Gamow refers to it, was essentially gaseous with radiant energy predominant. At this initial stage, before expansion began, the matter was composed of protons, neutrons, and electrons. Five minutes after expansion began, the mass was cool enough so hydrogen and the heavier elements could be formed from the protons, neutrons, and electrons. This process continued for about 30 minutes until the temperature became too cool for these reactions to continue. Within an hour after expansion started, the universe had cooled to 250 million degrees, and in 200,000 years to the temperature of the sun's surface. By the time the universe was 250 million years old, the temperature had dropped to 100 degrees below the freezing point for water. This particular point is important in Gamow's theory. It was at this time that the gravitational effect of matter began to dominate the effect of radiant energy, because in the expanding process, the density of radiant energy decreased faster than did the density of matter. The radiant energy decreases in density as the fourth power of the distance of expansion. As the radius of the system doubles, the density of the radiant energy drops to one sixteenth. The density of matter, on the other hand, declines as the third power. A doubling of the radius means an eightfold increase in volume, or eightfold decrease in density of matter. Gas clouds separated from one another as the universe continued to expand, and these chaotic gas clouds condensed and formed the galaxies.

Gamow's theory also differs from Lemaître's in the effect of the initial explosion of matter. According to Lemaître, the universe came to a resting state, or state of equilibrium, and expansion continued only after the galaxies were formed. In Gamow's theory, the galaxies were formed at a given period without a discontinuance of expansion. Expansion has continued unabated from the very moment of that initial explosion.

The fact that the expansion does take place and did start at some point in the distant past is a feature both the Lemaître and Gamow theories have in common. There is also the implication of an "all-in-one" creation at some remote moment.

Gamow predicted that it would now be possible to detect weak radio radiation from the initial explosion if such an event had taken place. Extraterrestrial radiation of radio wavelengths was discovered in 1965 and confirmed in 1968. This radiation appears to be coming from all parts of the universe and is considered support for the evolutionary theory.

One other view of the evolutionary universe was suggested by Allan Sandage in 1965. He proposed that the universe alternately expands and contracts—

that is, it pulsates in a continuous cycle. The universe contracts to a maximum density, then explodes and expands to a point of minimum density, completing a whole cycle in 82 billion years. The universe is now in the expanding phase, according to this concept.

7.3 THE STEADY-STATE THEORY

In 1948, Thomas Gold, Hermann Bondi, and Fred Hoyle of England proposed another cosmological theory. They suggested that creation is a continuing process and that hydrogen is continually being formed. The main point of this theory is that the universe is essentially unchanging. The features of a billion years ago are much the same as they are now and will be a billion years in the future. The overall population of galaxies does not change, although the individual components do. This idea was based on a "cosmological principle" first suggested by Bondi and Gold; namely, that the universe is unchanging in space and in time. From whatever point in space or time an observer views the universe, it shows the same general features. This implies a steady state, a state of continuous being without change (fig. 7.1). But the universe is in motion and galaxies are receding in all directions.

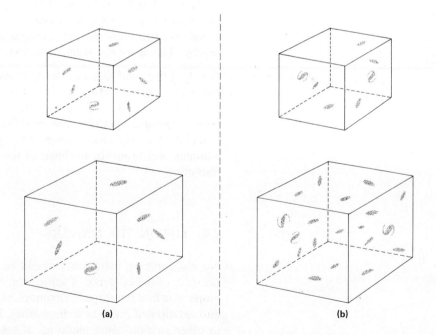

Figure 7.1 In the expanding universe (a) the same number of galaxies would, over an extended period of time, move away from each other, resulting in the same number of galaxies occupying a greater volume of space. Under steady-state conditions (b) the galaxies would move away from each other but new ones would be formed thus maintaining a constant galactic population density for any given volume of the universe.

How is it possible to maintain a continuous steady state in which any portion of the universe will maintain the same population of galaxies over extended periods of time? This is where the continuous creation of hydrogen plays an important role in the theory. As galaxies move out of range, new galaxies are formed from the spontaneously created hydrogen to replace them. The assumption is made that hydrogen is formed at a rate sufficient to replace and maintain the density of the galactic population. Individual galaxies are born, evolve through their life cycle, and gradually fade out as the stars become black dwarfs and die. If such be the case, there are, in any locale in space, old and young galaxies in about the same proportions.

Hoyle best describes this concept by comparing the universe to a settlement of humans. From a distance, it is possible to distinguish the old, middle-aged, and young who inhabit the town and make up the total population. Assuming no change in population, the old die and are replaced by the birth of new individuals, so that if the settlement were viewed one hundred years later, it would appear unchanged except for the difference in individuals. The universe, according to Hoyle, has existed in this manner for an infinitely long time, is infinitely large, and has neither end nor beginning.

More recently, the discovery of quasars has cast some doubt on the validity of the steady-state theory. Quasars are generally found at very great distances (if the red-shift law is believed) and represent events which took place several billion years ago. If the steady-state concept were correct, quasars should be found uniformly distributed throughout space—some very distant and some nearby. This does not seem to be the case.

Which theory will win out depends on the results of continued observation and measurement. Many technical problems must be overcome. The interference presented by the atmosphere has been eliminated in part by the use of radio telescopes. Perhaps the establishment of satellite observatories or an observatory on the moon, where atmospheric disturbances and distortions are minimal, will bring the problem of the origin of the universe closer to a final resolution.

7.4 LIFE IN THE UNIVERSE

No discourse on astronomy would be complete without a discussion of the prospect of the existence of some form of life in outer space. When the spectrograph was first developed, astronomers viewed its possibilities enthusiastically and anticipated many new discoveries. It was soon found that other stars even in other galaxies were made up of essentially the same material as the sun, and that the construction of the universe was amazingly uniform. No elements were discovered in very distant galaxies that did not occur in the solar system. Temperatures and pressures of stars, while varying from those of the sun, obeyed the same physical laws that controlled the activity of the sun. Under these circumstances, it would not be surprising if life occurred in other parts of the universe. In fact, it would be surprising if it did not.

Life is generally described in terms of those forms found on earth. It is based on carbon, a very versatile element which has the capacity to combine with other elements, notably oxygen and hydrogen, in a tremendous array of molecules. Water is essential as both a solvent and a medium in which the many complex reactions within the organism can take place. These elements, along with the other elements necessary for life, are quite commonplace in the universe. This does not mean that life is dependent on the presence of these specific elements. It is possible that some life form may have evolved in some distant place based on chlorine or silicon instead of carbon, since these elements are almost as versatile. However, there is no evidence to support such an idea, so it is necessary to limit the discussion to the requirements of a carbon-based form of life.

Life, as it exists on the earth today, is adapted to a variety of environments and conditions. From this it is possible to establish some environmental conditions that must be met in order for life as we know it to exist. Aside from the presence of the essential elements (of which there is no question, since these elements are universally present), conditions conducive to the development and continuance of life include a moderate temperature ranging from about $-7°$ C to $45°$ C. Chemical activity is very sluggish at the lower end of the scale, since pure water freezes at $0°$ C. The presence of salts in water will lower the freezing point and, therefore, permit some activity down to a temperature of about $-7°$ C. At the upper end of the temperature range, certain molecules important to life break down when temperatures reach much above $45°$ C. Most life forms can tolerate temperatures outside these limits but only for limited periods of time. Man has been able to survive greater extremes because of his ability to modify his environment to suit his needs.

On the basis of temperature, it is possible to establish the so-called *life zone* within the solar system. This is a zone ranging from Venus to Mars, wherein temperatures on planets may be sufficiently tolerable to sustain life. In terms of the physical characteristics of the planets, it is possible to eliminate Mercury as being too hot (bright side) or too cold (dark side) for life to survive, much less develop. Only Venus and Mars are within the limits of a zone where life is possible, although some astronomers feel it may be possible on Jupiter. Venus, however, has a cloud layer which may entrap heat close to the surface, making life insupportable. Only Mars, on the outer edge of the life zone, appears to have the prerequisites for some form of life. Temperatures are somewhat more rigorous than on earth, but they are within tolerable limits for life. Mars has much rarer atmosphere than earth's and lacks sufficient oxygen for animal life, but it is satisfactory for some forms of plant life. The Martian atmosphere contains up to ten times the carbon dioxide found in the atmosphere of the earth, a compound necessary for plant life—at least on earth. Life on Mars, if in fact it has developed, will be adapted to conditions of low moisture, low temperature, and low atmospheric pressure. Lichens, which grow under rigorous conditions, have been tested in a simulated Martian environment and survived. While lichens may not be found, plant life similar to this may possibly exist on Mars, although recent data does not support this possibility.

One other celestial body, the moon, is found within the limits of the life zone. The moon lacks an atmosphere and, for this reason, temperatures may be too extreme for life to have developed. Thus, great care is taken to sterilize lunar probes so that no earthlike organisms will contaminate the lunar surface. Thus far no life has been detected in the samples returned from the moon, and it now appears a certainty that the moon is devoid of life.

Beyond Mars, the planets are too far from the sun to receive sufficient heat for life to evolve. Temperatures of the visible portions of the planets (their upper atmospheres) are far below freezing, being at most $-130°C$ on Jupiter and lower still on the planets beyond Jupiter. It is now thought possible that temperatures are sufficiently high beneath Jupiter's cloud layer to support life, but this has yet to be confirmed.

The atmospheres of the outer planets (Jupiter and beyond) are made up primarily of methane, ammonia, and hydrogen and are extremely toxic to existing life forms on earth, although it is suspected that the atmosphere of the earth was once composed of these gases and that life may have first evolved under these conditions.

From all that can be observed, it appears that life within the solar system is confined to the earth, or at least this may be said for advanced forms. The possibility of some form of primitive life on Mars exists but it is slim. Extreme temperatures and hostile atmospheres have in all probability inhibited the development of life on other planets or satellites in the solar system.

Does life exist beyond the solar system? This question cannot be answered positively because of the tremendous distances to even the nearest stars. Even with the finest equipment presently available, it is not possible to see planets associated with these stars, and it is unlikely that these stars will be visited in the foreseeable future. All that can be done at present is to speculate on the possibilities of life based on the existing knowledge of conditions associated with the various types of stars. The tremendous number of stars in the universe, more than 10^{20}, permits ample opportunities for the duplication of conditions similar to those which exist in our solar system. The fact that the chemistry of the universe is no different from that within the solar system, and the assumption that the same physical laws apply in all parts of the universe, gives cause for believing that this duplication may occur many times.

In 1930, Sir James Jeans described the occurrence of life on earth as an accident of nature. According to Jeans, space was extremely cold and the stars too hot to support life. The fact that such an "accident" occurred was due to the extreme vastness of space and the long period of time during which it existed. Sir James quoted Huxley, who said that if six monkeys were set to typing continuously for millions of years, they would in that time write, purely by chance, all the books in the British Museum. Although there has been a change in thinking on how often such an "accident" may occur, the present feeling is that planetary systems are quite commonplace and that life on at least a small percentage of these planets is possible.

Although a planet may be associated with a star and be within the life zone, there is no assurance that intelligent life would exist there. Life forms are assumed to progress up the evolutionary path from lower to higher forms over an extended period of time, perhaps 2 to 3 billion years. If the current theories of star evolution are correct, then some stars do not have a sufficiently long life cycle to permit life to develop. Type O and type B stars (table 5.2) are bright, hot stars which have a short life cycle—too short to permit life to evolve to any fair degree on planets which may be associated with them. This is true also of type A and possibly early type F stars. Although their life cycles are too short, these stars are adequately luminous, so that even a planet with a slightly elliptical orbit would fall within the life zone. By the same token, stars of type M and those lower on the main sequence have, because of their small size, such a limited life zone that only a planet with a perfectly circular orbit would be capable of supporting life. It is rather unlikely that planets achieve such an orbit, thus eliminating these stars as likely prospects. This narrows down the field of main-sequence stars capable of having a large enough life zone and a long enough life cycle to late types F and G, and early K. The sun, it will be recalled, is a type $G2$ star.

Of the three dozen stars located within a radius of 5 parsecs of the sun, only a few are of these types. Epsilon Eridani is a type $K2$, located 3.3 parsecs from the sun, and Tau Ceti, a $G8$ type, is 3.64 parsecs from the sun. The balance of the sun's neighbors are mostly M-type stars or are multiple star systems where planets are either unlikely to occur or have orbits too elliptical to remain within the life zone.

Approximately one-third of the stars in the neighborhood of the sun are members of multiple star systems. G. P. Kuiper has expressed the opinion that planetary systems are merely variations of the multiple star systems. If the angular momentum of the original dust and gas cloud is high, the cloud probably splits into two more or less equal masses to form a binary star. A smaller angular momentum results in one large and one small star, while a still smaller angular momentum may result in the star and a series of planets made up of 10 percent or less of the total mass. A very low speed of rotation results in a disk around a star, with a density too low to form planets. Studies of this phenomenon indicate that about 1 percent of the stars could have planetary systems. Some astronomers, however, believe that this figure may be as high as 10 percent.

What would this mean in terms of actual numbers of planets capable of supporting life? The total number of stars in the Milky Way is estimated to be about 100 billion. If only 1 percent of these have planetary systems, there is still a sizable number of them, something on the order of one billion. Obviously, not all the planets associated with these stars are inhabited, because they would have to be located within the life zone and have sufficient mass to retain an atmosphere. Within the solar system three planets are in the life zone and seven have atmospheres. If this is typical of other systems, then many planets exist and are capable of sustaining some form of life.

Although life may exist in other parts of the universe, it is not necessarily as advanced as that found on earth. At the same time, if the star is a billion years or so older than the sun, a much more advanced type of civilization may exist. If this is the case, is it possible that the earth has been visited by beings from outer space at some time in the past? Such an idea would have been thought impossible a generation ago, but since then it has been suggested that life on earth resulted from the dumping of refuse from a visiting space vehicle. While such a contingency is considered extremely unlikely, it does raise the question of whether a voyage of this magnitude is possible. Epsilon Eridani, a star likely to have planets in the life zone, is 3.3 parsecs (10.8 light-years) from the earth. Even if speeds of 1/100 that of light could be attained, the round trip would take more than 200 years. If man were to make such a voyage, it would stretch over several generations, and, in fact, some would live their entire life span in space. Such a voyage from earth is impossible in the foreseeable future with our present technology, but this does not mean it would be forever impossible.

For the present, man is satisfied that initiating physical contact with a distant civilization is impossible, but some astronomers are attempting to communicate with other beings. Communications is possible with the equipment now available. A project in progress at the National Radio Astronomy Observatory in West Virginia is attempting to detect artificial signals from Tau Ceti which would indicate some intelligent control of the signals. The type of signals that could be expected would be the simplest form of abstraction involving basic numbers. Great patience is required, since the minimum time between sending a message to Tau Ceti and receiving an answer, provided it is returned immediately, is more than 20 years. However, the patience required would be well rewarded, for an achievement of this type would be one of the most important scientific events in human history.

If the presence of some form of intelligence can be established, it would be natural to speculate on the nature of the creatures that had sent the signals. There is no way of knowing what direction evolution may have taken on some other planet, but certain physical characteristics are essential to a creature possessing the intelligence necessary to communicate. The creature must possess a high degree of mobility in order to gather raw materials necessary for communication equipment, and a high degree of manipulative ability to convert these materials into a usable form. This would imply some form of structure having features similar to human hands and legs. Since communication equipment and all equipment necessary for its functioning is complex, the creature would require vision and hearing. Many biologists believe that for the purpose of balance such alien organisms would possess a degree of symmetry such as is found in life on the earth. This would mean that intelligent forms may not be entirely dissimilar to human beings. Some scientists, however, feel that we are unique in the universe and that intelligent creatures on other planets are so unlike the humans on earth that they could not be classified on the same basis.

7.5 ASTRONOMY AND SOCIETY

Of what value is the study of astronomy? In Chapter 2 we saw that the measurement of time and determination of direction—both vitally important to us—resulted from the study of astronomy by ancient people. While there are more accurate methods for measuring short intervals of time, the rotation of the earth still serves to measure day length, and the earth's period of revolution defines the calendar year. Seasons may be accurately defined from the relative position of the sun, and the month is loosely interpreted as the period of the moon's orbit. Our activities are deeply dependent upon a time-telling system, and our activities are also dependent upon our ability to move from one point on the earth to another. Again astronomy provides a means of doing so by using relatively fixed points (stars) in the sky whereby position and direction on earth can be determined.

But what direct benefit do we derive from knowledge of such distant objects as the stars and galaxies? This question is difficult to answer, unless it is pointed out that man is not a creature that lives only for the perpetuation of the species. Man is a complex being who needs more than bread to fulfill his many needs.

There are those people, the astronomers, who derive great satisfaction from unraveling the mysteries of the universe and solving the problems related to the laws of nature. They seek reasons for beginnings and mechanisms whereby these beginnings were achieved. Many on an amateur basis enjoy the observation and study of the numerous celestial objects, and enrich their lives by participating in various ways in the search for new knowledge. For example, most comets are discovered and tracked by amateur astronomers. Others derive pleasure from viewing the heavens for esthetic reasons, and while they have little knowledge of the universe they are awed by its magnificence and scope. These are intangible benefits, benefits which can be enjoyed only after all our more basic needs have been satisfied.

Can astronomy gratify the basic desires of man and his society? In certain instances religious needs have been satisfied through study of astronomy. However, most of the benefits from astronomy have been indirect—but then, this is the nature of science. For example, electricity, now a servant of man, could not have provided its many benefits unless the scientist first learned what electricity was and how it was generated. Then the engineer could develop ways to supply and utilize electricity. So it is with astronomy. Until the 1960s astronomy was an earthbound science indulged in by few. But because of our desire to know what is beyond the immediate sphere, our technology has made it possible to boost the astronomer and his instruments above the blurring effects of the atmosphere.

The effort to explore space has greatly accelerated the need for workers in many industries only indirectly related to astronomy, and it is work that provides for the needs of many individuals. New techniques and processes—many not previously available—have proved useful in various ways unrelated to the space exploration program that prompted their development. A new philosophy in dealing with problems came into being, and hopefully this new approach will be useful in solving old issues. Team efforts, computer data analysis, and the examination of all possible avenues of approach regardless of how trivial may help in solving some of society's most pressing problems.

7.6 SUMMARY

The science of astronomy has developed from a description of a limited universe which placed the earth at the center, to the infinite universe where the earth revolves around one of many billions of stars located near the outer edge of one of billions of galaxies. From the very beginning, man has sought an explanation and proposed many theories. Even in ancient times, many of these theories described the universe in terms of infinite space. This thinking has culminated in the current concepts of the evolutionary universe and the steady-state universe. Both theories have arisen from essentially the same observations —observations that have been interpreted in different ways. Additional data and tests of existing data will aid in determining the validity of one or the other theory or lead to the formulation of new theories.

Life forms on other planets have also been the subject of study in recent years. Until recently, life in the universe was thought to be unique to the earth. Now that our environmental needs and tolerances are better understood, and more is known about the nature of stars and planetary systems, it becomes evident that there is a good statistical possibility of some form of life on distant planets. Direct evidence to support such ideas is not available, and communication with other life-forms has not been established. With present technology, there is little possibility of visting some remote planet outside the solar system and discovering a new form of life. But humanity's capacity for solving problems is boundless, and ultimately our exploration of space will be bounded only by the time barrier.

QUESTIONS

1. What are the basic features of the evolutionary or expanding universe theory?
2. What are the basic features of the steady-state theory?

Cosmology—Life in the Universe

3. What features do the evolutionary and steady-state theories have in common?
4. What are the conditions upon which life as we know it are based?
5. Briefly describe what is thought to be the chances of life occurring on each of the planets in the solar system; beyond the solar system.
6. Around what type of stars is it thought life may occur and why?
7. What is the prospect for contacting an alien form of intelligence?

GEOLOGY

8 Introduction to Geology

Early concepts and principles in geology.
Ideas on the early form of the Earth.
Modern view of the Earth's structure, oceans, and atmosphere.
Continental drift—evidence for it and possible mechanism.

Geology is a science dealing with phenomena whose influence on society is not always readily apparent. It is the study of changes on and in the earth and of the materials of which the earth is composed. It is a descriptive and analytical science dealing with the history, development, and nature of the earth. Geology is a very broad subject—too broad to cover entirely in the short space of this text. The discussion will therefore be limited to some selected topics.

The study of geology attempts to provide, among other things, reasons for the occurrence of the various landforms by developing an understanding of the processes involved in shaping the land. The development of civilization has been controlled to a certain extent by landforms. We can readily see that the centers of early societies did not occur on mountain peaks or other inhospitable terrain but rather in the broad river valleys sculptured by the erosional action of water (fig. 8.1). The river provided the water for domestic purposes as well as irrigation water for the crops of the predominantly agricultural communities. The river also provided a means of transportation and was, for this reason, important in the development of commerce.

The initial use of the term *geologia* was suggested in 1473 by Richard de Bury, Bishop of Durham. He designated the study of law as "geologia," or "earth science," to distinguish law from theology, which was the study of divine things. Not until the seventeenth century was *geology* used in the modern

Geology

sense of studying the properties of the earth. Thus, as an organized science, geology is relatively new compared to other sciences such as astronomy or physics. But despite this, geology has already been subdivided into a number of subdisciplines such as geochemistry, geophysics, mineralogy, and paleontology.

8.1 SOME EARLY PRINCIPLES

Ancient man did not consciously separate the sciences into different categories. To him astronomy and geology were one. Conditions in the heavens influenced conditions on the earth. Despite this, man was aware of certain geologic conditions which influenced his life. Even before written history, information of a geologic nature was passed from generation to generation by word of mouth. In the Stone Age, man was interested in rocks as raw materials for weapons and tools. Certain rock such as flint and obsidian served his purpose better than other forms of rock which were frequently too soft for such use.

The advent of the Bronze Age found man's interest in the minerals of the earth increasing. Mining became an established industry requiring at least a rudimentary knowledge of mineralogy. Gold and gems were important trade mediums and, being of value, were actively sought. Materials such as tar were used in building, and the quarrying of rock for building purposes was a well-established activity when the first pyramids were constructed in Egypt.

Figure 8.1 Societies more readily develop in the river valleys (above) than in the hostile mountain regions (at right).

Although information on ancient Chinese culture is scarce, it is known that the Chinese were proficient in the manufacture of pottery, which required a knowledge of clays. Thus, it can be seen that the practical application of geologic principles was practiced by man long before the establishment of the science of geology.

However, very few significant scientific principles were contributed to this science by ancient peoples. Even the Greeks, intellectuals that they were, offered little in the way of organized knowledge on this subject, although they did speculate on the cause of certain natural phenomena now considered to be

Figure 8.1 (Continued)

within the realm of geology. For example, Aristotle (384–322 B.C.) is reputed to have presented an interesting (although incorrect) theory on the manner in which minerals were brought into being. This was set forth in a book bearing the title *Meteorologica*. (Later evidence has indicated that the work may have been by an Arabian author several centuries after Aristotle's death.)

In this volume, the earth was described as being composed of the elements fire, air, water, and earth. Aristotle claimed that by changing the relative proportions of these elements, it was possible to transform one material into another. He cited as an example the heating of water which would result in the production of air (steam), and if the heating were continued until the water has completely evaporated, some earth would be formed and remain as a residue. Aristotle also described the "two exhalations" given off by the earth. The first was a "moist exhalation" which rises as vapor from the surface of bodies of water, and the second was a "dry exhalation" which he described as being akin to smoke. It is from the condensation of great clouds of "dry exhalations" that stones were formed, according to Aristotle, falling to earth in the form of thunderbolts. Also, according to Aristotle, these two exhalations transformed into rocks, minerals, and metals within the interior of the earth as a result of the influence of the sun and planets. Those materials containing a higher proportion of dry exhalations formed the rocks and minerals which did not readily melt, while moist exhalations formed into the metals such as gold, silver, iron, and copper.

Perhaps the greatest work of classical times was the *Natural History* written by Pliny the Elder and dedicated to the Roman emperor Vespasian in 77 A.D. This was a work of thirty-seven volumes covering some twenty thousand subjects—the most complete encyclopedia of existing knowledge of that time. The last five volumes are of special historical interest to geologists, for it is in these that Pliny deals with subjects of geological significance, although much of it is misinformation and of little value. He discusses, for example, the magical properties of many materials, such as the mysterious reaction between a diamond and a lodestone wherein the lodestone loses its magnetic qualities when a diamond is brought near it. Despite the inclusion of such erroneous materials of this type, the *Natural History* was drawn upon as factual by many authors as late as the seventeenth century.

The sixteenth and seventeenth centuries brought about a change in the approach used by scientists, especially those concerned with the geologic processes. Less speculation and more field observation was practiced, and some important geologic principles gradually evolved. Nicolaus Steno (1638–1686), a court physician to the Grand Duke of Florence, used field observation in his study on the origin of mountains. He made detailed studies of the hills around Tuscany and made an assumption never before proposed. Steno proposed that the sedimentary rock strata found in Tuscany were originally deposited in water in a basin (fig. 8.2). The bottom layer conformed to the shape of the basin and the upper layers were parallel to the horizon. Any subsequent sediments would have as its base the previous top horizontal layer. He also saw

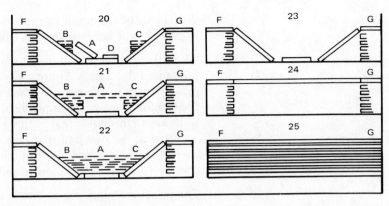

Figure 8.2 From Steno: Prodromus representing his view of six successive stages in the geological history of Tuscany.

continuity of sediments from one mountain to another, and, in this way, he could trace the extent of the sedimentary deposits for some distance.

Steno derived several important geologic principles from his observations. The first is the *law of original horizontality*, which states that water-deposited sediments are laid down parallel to the surface on which they are accumulating. If this principle is accepted, then one must conclude that sedimentary deposits not now horizontal must have had their position altered by some force operating on the earth's crust. There are a few exceptions to this, as in the case of deposition on a river delta where sediments are deposited on a slight slope, or flow-banding in volcanic rock which resembles layered strata. The second principle is the *law of original continuity*, which proposes that the water-laid deposit will be continuous in all directions until it borders against the edge of the basin within which the deposition is taking place. A third principle is the *law of superposition*, which states that those rock strata above are younger than the underlying strata. While these now appear to be truisms, they have nevertheless proved very important in interpreting the geologic evolution of the landscape. Unfortunately Steno's influence was not great, and his ideas were not adopted until their rediscovery in the nineteenth century.

By the beginning of the eighteenth century, geology as a science had begun to take form. For pragmatic reasons, this new science had its roots in mineralogy, because mining had been an important industry for centuries, and the increased demand for more and more metals required better technology and understanding of the occurrence of ore bodies. Despite this rather narrow beginning, a broadening in scope led to the development of several theories or schools of thought in geology. One school of thought, the Neptunists, considered water the sole agent responsible in building up the successive layers of the earth's crust; another, the Plutonists, felt that fire and water played a major role in the development of geologic succession. Other geologic theories were advanced during the eighteenth century, ranging from catastrophism, in which all events causing large-scale change on the earth's surface were thought to be the result of sudden and violent action, to uniformitarianism, wherein it was believed that all changes were gradual and continuous.

Of all these theories and ideas, only the principle of uniformitarianism, first formalized by James Hutton (1726–1797), was of real significance. This principle was to become the keystone of modern geology; the others fell into oblivion. In essence, the principle of uniformitarianism states that "the present is the key to the past"; that is, that all geologic phenomena now occurring have occurred in the past and that past events may be interpreted on the basis of current geologic action. Hutton felt that the forces now acting on the earth's surface have acted continuously and uniformly throughout the earth's history.

The principles suggested by Steno and Hutton have served as a basis for the development of geology into a modern science. Many men have contributed to the gradual evolution of geology, extending not only the knowledge but the scope of this science. Although early geologists were often mistaken in their concepts, they at least laid a foundation for learning the true nature of things. Not all problems have been solved, and the search for answers continually reveals the existence of new areas for study. One of the more elusive problems, because of the great lapse in time, is the nature of the earth in the very early stages of its formation. It is generally conceded that the earth was formed along with the solar system from a gas and dust cloud over 4.5 billion years ago. One of the questions asked by the geologists is, How did the earth appear after it was first formed?

8.2 THE NEWLY FORMED EARTH

The picture of the newly formed earth is not a clear one. Some scientists theorize that the earth was in a molten condition from the very beginning and could be pictured as a giant cometlike object streaming gases behind it as it hurtled through space. Others feel that the earth was comparatively cool and gradually heated to incandescence during the first few hundred million years of its existence before cooling to its present state. Whatever its early form, the consensus is that during some part of its early development, the earth was molten or at least very hot, and it was during this period that the heavier components migrated to the center and the less dense rock accumulated on the surface and formed the continents. Large quantities of gas and steam released from the interior were added to the gases already in the atmosphere.

There is doubt as to the condition of the oceans during that first half billion years. One opinion expressed is that as little as 5 percent of the present seas were formed during that early part of the earth's history. It is felt that the balance of the water was gradually released by volcanic activity and filled the ocean basins during the next 4 billion years.

A different view holds that the ocean basins were filled during the first billion years and that only small amounts of water have been added since then. The amount of water in stony meteorites has been measured and found to contain about 0.5 percent water. If the rock in the earth, excluding the core, contains the same amount of water, then it is estimated only about 5 percent of the total existing water has escaped from the interior of the earth.

The atmosphere is believed to be quite different now than it was at the beginning of the earth's history, but again there is a diversity of opinions as to its original composition and formation. One theory has it that if the earth were molten in its early stages, then most of the gases were to be found in a hot, primitive atmosphere. The alternative theory suggests a gradual release of gases from the earth's interior over the entire earth's history. Both these hypotheses parallel those on ocean formation: It is thought the hydrosphere (ocean) and atmosphere (air) were formed in a similar manner and from the same source, the interior of the earth.

The exact composition of the early atmosphere is not known, but there is little doubt that it differed from the present atmosphere. It is generally believed that most of the original gases were replaced by a secondary atmosphere composed principally of methane (CH_4), ammonia (NH_3), sulfur oxides, hydrogen sulfide (H_2S), nitrogen oxides, carbon dioxide (CO_2), and water. Oxygen, so necessary for life, was largely absent at this stage. The initial formation of atmospheric oxygen probably came from the photodissociation of water vapor and to a lesser extent carbon dioxide (see section 13.2).

Hydrogen, because of its low atomic weight, escaped from the earth's atmosphere during the early stages of the earth's development. Methane and ammonia are not stable in the absence of free hydrogen and these also broke down, releasing more hydrogen and eventually evolving into an atmosphere of nitrogen and carbon dioxide. This is estimated to have occurred by the time the earth was 100 million years old.

The amount of nitrogen in the atmosphere has changed but little since that early time. Nitrogen is not very reactive with other elements, and only a relatively small percentage of the total available has been utilized in plant and animal protein. Carbon dioxide, on the other hand, is thought to have been rather abundant in the primitive atmosphere, but great amounts have since been withdrawn by chemical reaction with other elements to become the deposits of limestone and dolomite rock.

Gases of volcanic origin contain little or no oxygen, except that which can be accounted for as contamination. Oxygen was probably produced in the primitive atmosphere by photodissociation, especially in the upper atmosphere. The hydrogen released by the process was lost in space. However, it is felt that this oxygen did not accumulate as such in the atmosphere but rather became tied up, largely in the oxidation of certain minerals on the earth's surface.

Biologists have shown an interest in the composition of the early atmosphere because of the relationship of plants and animals to the presence of certain types of atmospheric gases. Amino acids have been formed by passing an electric charge through a mixture of methane, ammonia, hydrogen, and water vapor, indicating that complicated organic compounds could have been synthesized without the presence of oxygen. In fact, many biologists feel that primitive life developed in just such an oxygen-deficient environment and possibly could not have formed if oxygen had been present. Green plants utilize carbon dioxide and water in the photosynthetic process and release oxygen as a by-product. This reaction may have contributed significantly to atmospheric oxygen after plants became prominently established, and could be principally responsible for the oxygen content of the present atmosphere.

8.3 THE PRESENT EARTH

The earth's surface is made up of the land masses, which cover 29 percent of the surface, and the oceans, which comprise the remaining 71 percent of the surface. The land mass has an average elevation of about 0.8 kilometers above sea level, while the oceans average 4 kilometers deep. Separating the exposed portion of the continental mass from the ocean basin is the continental shelf, a shallow zone extending from the tide line to a point where the water is about 200 meters deep. Along some coastal areas, the width of the shelf extends to more than 160 kilometers from shore, while elsewhere the width may be negligible. Beyond the continental shelf is the continental slope which drops more abruptly to the ocean basin (fig. 8.3). The continental slope is considered by some scientists to be the zone where ocean basins, composed primarily of basaltic rock, join the continental masses, composed mostly of granitic rock.

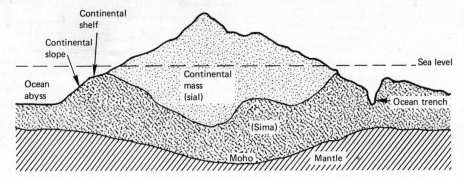

Figure 8.3 Hypothetical profile of the crust showing continental mass and some features of the ocean basin.

The earth's interior is not a homogeneous mass, but appears to be made up of layers of lighter material on the surface and progressively heavier materials toward the center. The evidence supporting this theory is the fact that the density of surface material is 2.65 grams per cubic centimeter, while the average density of the earth as a whole is 5.52 grams per cubic centimeter. The conclusion is that the density of rock toward the center of the earth is somewhat greater than that on the surface in order to achieve an average of 5.52 grams per cubic centimeter. Additional information comes from the varying speeds with which earthquake waves travel through the earth's interior and the manner in which these waves are refracted. This evidence indicates that the earth is made up of concentric spheres of progressively heavier material surrounding a denser core.

The outer layer or *crust*, is a relatively thin shell ranging in thickness from about 8 kilometers beneath the oceans to about 40 kilometers or more in the continents. Why do the continental masses project so unevenly above sea level? In the midnineteenth century, two explanations were offered. One was made by a British churchman, J. H. Pratt, who felt that the continental masses were composed of materials of slightly lower density than the rock beneath (fig. 8.4). The lighter blocks of rock were supported to a higher level to form mountains and high plateaus, while the material of slightly greater density settled lower into the underlying basaltic rock to form low-lying plains. At about the same time, another suggestion was made by G. B. Airy, Astronomer Royal of Great

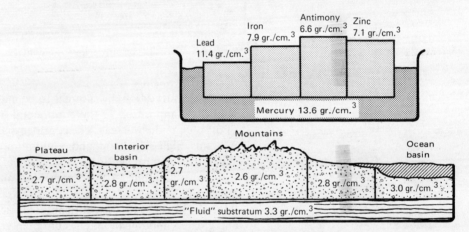

Figure 8.4 Pratt's view of the cause for continents projecting above the general level of the crust.

Britain. He modified Pratt's view by stating that the granitic rock was all of the same material and density but of different thickness. Airy expressed the idea that the mountain block floated higher but also sank deeper and that the height of the mountains was compensated by a root which projected into the basaltic rock beneath (fig. 8.5).

The basaltic rock beneath the continents is not in a fluid condition, but it is known that there is considerable plasticity in the rock due to the enormous pressures that exist at those depths. This plasticity permits a degree of shifting or drifting of the continental mass so that gravitational equilibrium is maintained. The term *isostasy* (derived from a Greek word meaning "balance" or "equal standing") was applied to the idea of an equilibrium condition between the continental masses and the underlying rock.

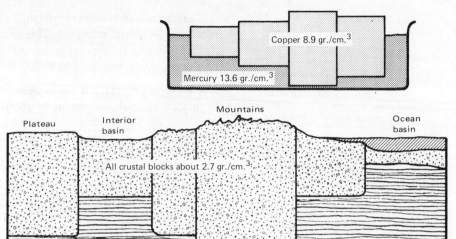

Figure 8.5 Airy's mechanism whereby mountains project above the average level of the crust.

The crust is traditionally thought to be made up of relatively lightweight rock, with the composition of the continental masses varying somewhat from that of the rock under the seas. The continents are mainly granitic rock composed of minerals high in silicon and aluminum. An Austrian geologist, E. Suess, suggested the term *sial* for this material after the chemical symbols for silicon (Si) and aluminum (Al). This granitic material is relatively lightweight and appears to "float" on the more dense basaltic rock which makes up the ocean floor and lies beneath the continental masses. This basaltic rock contains silicon and greater amounts of magnesium than granites do, therefore Suess suggested the term *sima* (see fig. 8.3).

A more or less abrupt change in density occurs in the rock at the base of the crust, marking the beginning of the *mantle*. This was first noted in 1909 by the Croatian geologist, A. Mohorovicic, who discovered a change in the velocity of earthquake waves at those depths now designated as the base of the crust. The zone of change became known as the *Moho* (short for Mohorovicic) *discontinuity*, which refers to velocity discontinuity or change.

Beneath the Moho discontinuity, the mantle made up of increasingly denser rock extends to a depth of about 3,000 kilometers, where another important discontinuity occurs that marks the boundary between the mantle and the *core*. The density of the mantle ranges from 3.3 grams per cubic centimeter just below the Moho discontinuity to an estimated 5.5 grams per cubic centimeter at the boundary between the mantle and the core. Other discontinuities of less importance occur within the mantle.

The core of the earth is a sphere about 6,800 kilometers in diameter, presumably composed of a very dense fluid material, possibly of iron-nickel composition.

Introduction to Geology

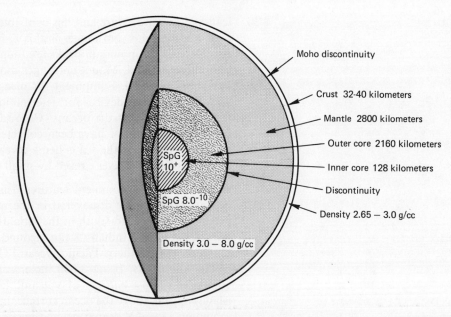

Figure 8.6   Cross-section of the earth showing crust, mantle, and core.

The core is thought to have a density in excess of 10 grams per cubic centimeter, although neither iron nor nickel achieve this density on the surface. It is felt that the higher-than-normal density levels of these materials is the result of the extremely high pressures occurring at that depth.

Although the core has never been reached, indirect evidence indicates its fluid state. Secondary earthquake waves (see Chapter 10) will not pass through a fluid medium, and this type of earthquake wave does not pass through the earth's core. This evidence, together with the extremely high pressures and temperatures that must exist at the earth's center, leads geologists to conclude that at least the outer core is in a fluid state. The inner core, with a radius of 1,350 kilometers from the earth's center, appears to have different properties from the outer core. This has led some geologists to believe that the inner core is solid, but the evidence is not conclusive.

8.4   THE OCEAN BASINS

The oceans cover about 71 percent of the total surface of the earth and constitute the last earthly frontier for exploration. Up until the 1920s, the ocean floor was viewed as a vast, featureless plain with few ridges or valleys. The available information was obtained by time-consuming lead-and-line soundings and bottom samples taken in relatively shallow water.

In the past fifty years, new devices and methods have greatly facilitated the collection of ocean floor data. Echo sounding equipment has replaced the lead-and-line procedure and has contributed immensely to the present picture of the ocean floor. No longer is this hidden portion of the earth considered flat. Instead, the mountain peaks are found to be higher and the valleys deeper than those on land. Mountain ranges more extensive than any found on land have been revealed. Equipment for measuring temperatures and densities of seawater as well as devices for determining current flow at any level are being used in the study of the oceans. Core samples to depths of 30 to 35 meters below the ocean floor have been collected, and the nature of the rock strata of the crust beneath the sea is being revealed by studying the passage through the crust of shock waves caused by small explosive charges.

Among the more prominent features rising from the ocean floor are the mid-oceanic ridges. First discovered in the mid-Atlantic in the 1920s, the ridges extend from north to south in the Atlantic Ocean, around the southern tip of Africa, across the Indian Ocean, around the southern part of Australia, and north to the northern Pacific Ocean. The system is almost continuous for 65,000 kilometers, rising 3,000 meters or more from the ocean floor and is composed almost entirely of volcanic lava and debris. Occasional volcanic islands appear above the ocean surface, indicating high peaks in the mid-ocean mountain range. Volcanic peaks that do not break the surface are scattered over the ocean floor by the thousands and are called *seamounts*. Many of these have truncated tops, forming a flat summit. These *guyots*, as they are called, may be inactive volcanic islands whose tops were eroded flat by the beating action of ocean waves. Some are a considerable distance beneath the surface, indicating that the sea level must have been much lower at one time than it is at present in order for the eroding power of waves to be effective, or that the ocean floor upon which the guyot was built has subsided, or that the guyot has been moved to lower depths by sea-floor spreading (see section 8.5).

Ocean submarine *scarps*, or cliffs, have been detected, some as much as three kilometers high. These are scored with great canyons, indicating the movement of underwater currents with a force and volume greater than that of some of the largest rivers. Detailed mapping of the continental shelf and slope reveal the existence of many such canyon formations. Some of these canyons may be the outlets of prehistoric rivers that drained into the sea when it was at a much lower level, due possibly to the accumulation of water as ice in the extensive glaciers which existed in the past (see Chapter 11). Some of these canyons resemble river valleys cut by large rivers on land that subsequently subsided beneath the ocean. As yet, the origin of sea canyons is not clearly understood and is being studied.

One other feature of interest is the *deep ocean trench*, or simply *deep*, recognized as the deepest point in the ocean. Strangely enough, these trenches occur close to the continents, not in mid-ocean. Trenches are associated with the *island arcs*, which in most cases are chains of islands that lie off the continental margins. The island arcs, such as the islands off the east coast of Asia (including Japan) and the West Indies, are areas of great volcanic and earthquake activity and are thought to be regions of continental growth. The trenches, lying close to the island arcs, are long, narrow troughs plunging down to depths of over 11 kilometers.

Much of the ocean basin, or the *abyssal plain*, is, as was first thought, a flat, featureless plain. Seismic waves reveal that sediments on the plain may be thousands of feet thick, and core samples taken from these sediments reveal interesting facts about the earth's history. Thus far only a few areas have been explored. Continued exploration in the future will bring to light many answers to questions currently being asked about the structure and formation of the earth's crust. (See inside cover.)

8.5 CONTINENTAL DRIFT

By ignoring the oceans (which represent only a very thin layer on the earth's surface), it is possible to view the surface of the crust and ask if the gross surface features always appeared as they do now. Present thinking, after many years of debate, favors the idea that continents once existed 200 million years ago as two large land masses which have since broken up and drifted apart. This is not a new concept. Francis Bacon first suggested in 1620 that the Eastern and Western Hemispheres may at one time have been a single continent. A number of geologists later considered this concept seriously, and in the nineteenth century, Edward Suess offered some data which indicated that certain geologic features in Africa and South America corresponded rather closely. In 1910, Alfred Wegener felt that the continents may all have been one large land mass prior to 200 million years ago and enlarged upon the idea of drifting continents. Currently, the available data seems to favor the existence of two large land masses in the past: *Laurasia* in the north, representing North America, Europe, and part of Asia; and *Gondwanaland* in the Southern Hemisphere, composed of South America, Africa, India, Australia, and the Antarctic (fig. 8.7).

Aside from the somewhat obvious fit of the east coast bulge of South America into the west coast of Africa, certain geologic correlations were found which lent support to what is now known as the *continental drift* theory. Evidence of glacial activity, which occurred 250 to 300 million years ago, was found in the southern part of Africa, India, South America, Australia, and in the Antarctic. The glaciers left a distinct record as the huge masses of rock-filled ice moved over the solid rock, planing it flat and grooving the surface. On top of this was deposited the rock debris contained in the ice when the glaciers melted. Evidence of this activity can be traced from one continent to another, indicating the likelihood of one common land mass in the past.

In late 1969, the remains of a lystrosaurus was found in the Antarctic. The lystrosaurus was a hippopotamus-like reptile about four feet in length whose remains are widely distributed in the Southern Hemisphere. The discovery of such a vertebrate in the Antarctic far removed from the other continents lends considerable support to the single-continent concept. This evidence,

Geology

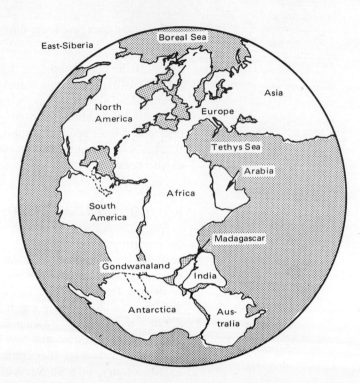

Figure 8.7 Supercontinents as they may have appeared 150–200 million years ago.

plus the occurrence of certain fossils of identical plants in specific layers of rock that can be traced from one continent to another, indicates that they were once joined. It is considered very unlikely that the identical animal or plant species could have evolved on lands separated by large bodies of water or that similar glacial features would have occurred at such widely separated locations.

Dating the age of rock by radioactive-dating techniques has also contributed evidence to support the continental drift theory. A sharp boundary line was discovered in West Africa between rock that was 2 billion years old and rock 600 million years old. The boundary enters the sea near Accra in Ghana and heads generally in a southwesterly direction. It was determined that had South America been joined to Africa, a continuation of the line would occur near São Luis on the northeast coast of Brazil. The line was found where predicted, the appropriately aged rock in the proper juxtaposition, indicating that the continents could have been joined at some time in the geologic past and subsequently drifted apart (fig. 8.8).

Another source of evidence of continental drift has been the study of "fossil" magnetism in rock. This results from iron-bearing rock being magnetized by the earth's magnetic field at the time the rock solidified from a molten condition. By studying rocks of different ages, it is possible to determine the position

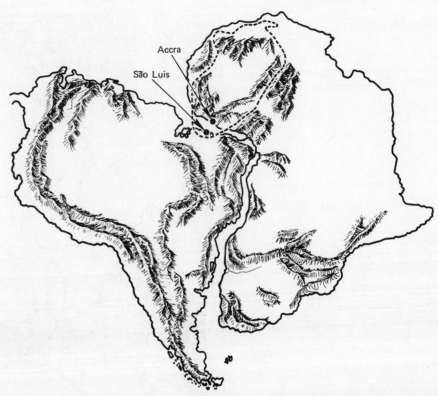

Figure 8.8 Through age-dating of rock formations in Africa and South America it was found that a region near Sao Luis, Brazil, was part of the rock structure near Accra in Ghana.

of the magnetic pole at different periods of the earth's geologic history. From such studies it was found that for any given time period the position of the pole seemed to vary for each continent. The conclusion reached was that after the rock had solidified the continents had drifted from earlier positions to their present locations during the course of millions of years. The study of magnetism in rock was useful in determining the early position and orientation of the continental masses.

The discussion thus far has dealt with the concept of continental drift and the evidence supporting the conclusion that continental drift has occurred. Now it is necessary to discuss the phenomenon called *sea-floor spreading*, which is thought to be related to the cause of continental drift.

Internal forces, generated by heat, are thought to be responsible for breaking up the earth's *lithosphere* into "plates," and also for moving these plates about, thereby causing continental drift. The lithosphere is solid rock that extends to a depth of about 50 kilometers and includes the crust and part of the upper mantle. At present there appear to be six large plates and several smaller ones involved in this motion. The lithosphere grades into an extremely hot layer in the upper mantle called the *asthenosphere*, which is relatively deformable (fig. 8.9).

140

Geology

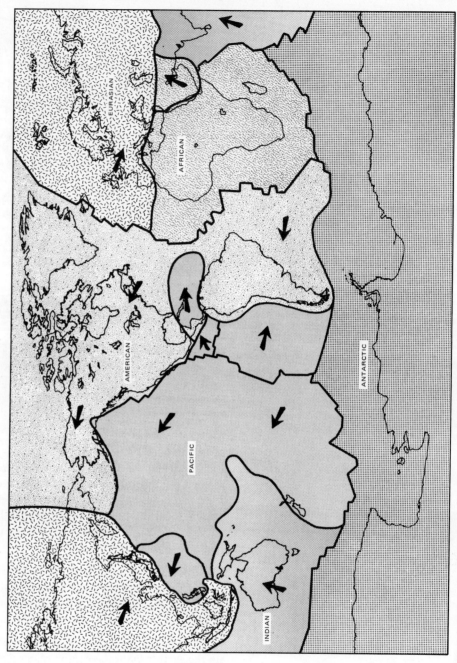

Figure 8.9 Generalized view of six major crustal plates thought to account for continental drift. The African plate is assumed to be stationary. Boundaries for all plates are not known for certain.

The upwelling of molten rock from the asthenosphere beneath the mid-ocean ridges causes adjacent plates to move apart. This results in the formation of new ocean-floor crust on either side of the mid-ocean ridge and explains why the ocean floor is composed of relatively young material, geologically speaking. (No ocean floor samples have been found to be much older than 125 million years.)

As the new crustal rock was formed, it was magnetized by the earth's magnetic field, thus providing a history of sea-floor spreading. This history was based on the discovery that the earth's magnetic field had reversed itself a number of times during the earth's geologic past, forming matched strips of alternately normal and reversed magnetized rock on either side of the mid-ocean ridges. The reversals have occurred on the average of every 400,000 to 500,000 years, although the most recent major reversal occurred 700,000 years ago. Since the age of the reversals are approximately known, it is possible to calculate a rate of sea-floor spreading which varies from one to nine centimeters and more annually.

The process of sea-floor spreading results in the movement of the plates (fig. 8.10). The upwelling of material along mid-ocean ridges and movement of the plates also results in surface material being returned to the deep earth at other locations. The sinking tends to pull the surface crust toward these regions, and a downwarping of the crust results at the sites. These downwarped areas, sometimes close to continents, are thought to be the deep-ocean trenches. In 1969, J. Tuzo Wilson described the whole process as resembling a pot of boiling soup, wherein the surface is comprised of two phases, froth and broth. He likened the continents to the froth which floats on the surface slowly moving about accumulating more of the lightweight (sialic) material. The ocean floors he compares to the broth, where new material wells up from beneath and is again carried beneath the surface at other places.

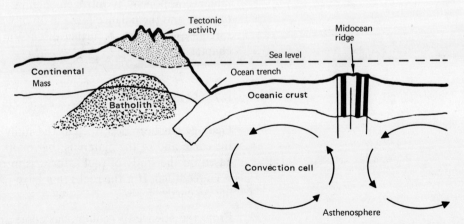

Figure 8.10   Convection currents in the crust along the mid-Atlantic ridge is thought to be the mechanism for sea floor spreading.

The process of continental drift appears to be continuing but does not seem to be occurring at a uniform rate. Following is an approximate timetable of events as the continents slowly drift apart.

Time (Millions of Years Ago)	Event
200 to 300	North America separated from Eurasia.
150	South America separated from Africa.
110	Africa separated from balance of Gondwanaland which included India, Australia, New Zealand, and Antarctic.
80	New Zealand separated from Antarctic.
60	Greenland separated from Norway.
40	Australia separated from Antarctic and the Bay of Biscay opened as the Iberian peninsula rotated south.
30	Gulf of Aden opened up.
5	Gulf of California began to open.

8.6 THE EARTH'S PRESENT ATMOSPHERE

The present atmosphere is an envelope of gas which surrounds the earth to a height of perhaps 2 earth radii. It is composed of a uniform mixture of several gases, the combined weight of which is 14.7 pounds per square inch at sea level (approximately 1 kg/cm^2) or a total of about 6×10^{15} tons over the entire earth. About 50% of the gases are concentrated within a zone 5.5 kilometers above sea level, and 90% are concentrated within the first 27 kilometers. The major component is nitrogen, comprising about 78% by volume; 21% is oxygen; and the balance is made up of water vapor, carbon dioxide, and some rare gases and dust. A further discussion on the atmosphere may be found in chapter 13.

8.7 SUMMARY

Geology has been described as the study of changes in the surface features of the earth and of the materials that make up the earth. Essentially, geology is based on the study of rocks, for it is in rock that the evidence of change has been recorded. It is the rocks that have revealed the record of the evolution of life on earth.

From the science of geology has come an important principle—the principle of uniformitarianism—which views events in the past in the light of what is occurring today. The same processes that produced naturally occurring changes on earth at present have produced similar changes in the past. Uniformitarianism may have much broader application in more than the study of geologic changes, for it may readily be applied to the study of the development of the universe.

Little is known about the earth immediately following its formation, because the rock record from that time has been almost completely altered or destroyed. There is little doubt about the lack of life at that time: the earth is pictured as a barren rock surface with an atmosphere of water vapor and possibly methane and ammonia. With the passage of time, the oceans and atmosphere formed and life gradually evolved to its present level. The process continues unabated, and continual changes may be expected as the natural course of events.

QUESTIONS

1. Define the three geologic principles developed by Steno in the seventeenth century.
2. What is the significance of the "principle of uniformitarianism"?
3. Give the two suggested means by which the atmosphere and oceans are thought to have evolved.
4. How did the secondary atmosphere differ from the present atmosphere? How did the change come about?
5. Make a diagram showing important crustal features of the continental masses and ocean basins.
6. What is the evidence supporting continental drift?
7. Describe the mechanism responsible for the occurrence of continental drift.

9 The Earth's Materials

What are minerals?
Physical properties of minerals useful in their identification.
Classification of minerals.
Value of minerals to society.
Rocks—igneous, sedimentary, metamorphic. Properties and mode of formation.
Weathering—the breakdown of rocks and minerals.
Importance of weathering to society.

The earth's materials—rocks and minerals—have long been regarded by man as having special significance. The rocks and minerals of the earth's crust are the source of raw materials used to make the tools upon which civilization is based. It has been stated that a civilization's advance is dependent upon the materials at its disposal. This may well be so, considering the technical and attendant social progress that has occurred in the past seventy-five years—a period when many new uses have been found for substances which earlier were merely a curiosity or had not yet been discovered. Such materials include germanium, necessary in the microminiaturization of components of communications systems; titanium, for light strong alloys; and uranium, for a source of energy.

9.1 SOME MINERAL MYTHS

Many curious and interesting beliefs have sprung up through the centuries on the nature and origin of rocks, and some of these beliefs persist to the present. The ancient Greeks, for example, thought that the mineral quartz was ice that became permanently crystallized because of long exposure to intense cold. Minerals were thought to "mature" from a base metal into precious metals, and the belief was held that pearls were in reality dewdrops taken up by oysters and transformed into the pearl, preserving the original form and beauty of the dewdrop.

Selenite, a form of gypsum, was believed by ancient people to have originated on the moon. It has a silvery, translucent luster suggestive of moonlight, and it was imagined that the image of the waxing and waning moon could be seen by looking into the smooth surface of a piece of selenite.

Of all the minerals, none drew more attention than the lodestone (magnetite) because of its unusual ability to attract iron. Pliny, the Roman philosopher, discussed this mineral at great length, telling of its magnetic qualities. In his *Natural History*, Pliny erroneously described the properties of two mountains near the river Indus. One of the mountains was lodestone and attracted iron, and the other mountain repelled it. So fabulous became the properties of lodestone, according to this myth, that great disasters were attributed to it. Ships that came too close to the mountains were said to have had all the iron nails pulled out, causing the ships to fall apart. Various writers described these scenes, but they allegedly occurred in remote places and could never be verified.

Much attention has been given to the source of this strange magnetic power of lodestone, but the mystery is still unsolved. Aristotle, writing on the subject, discusses Anaxagorus's opinion that the lodestone is a "living stone." This opinion was still held in the sixteenth century when many minerals were considered to be "living" and thus requiring nutrition. For this reason, it was thought that lodestone attracted iron from which it drew its sustenance.

9.2 MINERALS DEFINED

So important are chemistry and crystallography (the study of crystal structure) to mineralogy that the definition of a mineral must include some mention of these sciences. Thus, a mineral is defined as a naturally occurring inorganic substance having a definite chemical composition within narrow limits and a specific crystal structure.

A "naturally occurring inorganic substance" implies that the mineral occurs and is formed in nature and is not artificially produced. This does not mean that man cannot artificially duplicate nature's work, for many minerals have been duplicated in the laboratory. It simply means that minerals were originally formed under natural conditions, and we can therefore eliminate, for example, manmade plastics, which have a superficial resemblance to minerals but do not occur in nature. The definition also eliminates most organic materials that have their origin in living substances. Inorganic products of animals in a few instances may be considered minerals, such as, for example, fossiliferous limestone formed from the calcareous shells of certain organisms. Substances such as coal and other so-called fossil fuels are occasionally included in lists of economic minerals, although in the strict definition, coal is not a mineral because of its organic origin. But it is at times termed a mineraloid (mineral-like).

A mineral must have a definite chemical composition, although from a practical standpoint, an analysis will occur within a narrow range of the ideal composition. During the formation of a mineral, other substances are unavoidably included, in some instances, in sufficient amounts to change the physical properties of the mineral so that it takes a slightly different form. For example, quartz (silicon dioxide) and amethyst have the same chemical composition, but because of a slight impurity in amethyst, it is classed as a variety of quartz. In some cases, materials of the same chemical composition will form different minerals as a result of the manner in which they originated. Calcite (calcium carbonate) for example, will form from cold solution; but aragonite, which has the same chemical composition as calcite, forms from a hot solution and becomes a different mineral.

Minerals are also categorized on the basis of crystal form; that is, they will take the form represented by one of the six recognized crystal systems. The crystal system in which a mineral occurs is basically the result of the internal arrangement of the individual atoms. The growth of the crystal is largely dependent on the external environment. Large well-formed crystals may grow in cavities or crevices, permitting growth without interference by surrounding rock. Rapid crystallization of the solution from which the minerals form will usually result in very small crystals. If there is interference in growth (the normal condition) the crystals impinge upon each other to form distorted crystals or irregular masses of the mineral material.

For example, well-formed, transparent quartz crystals can form, provided no interference occurs and a ready supply of silicon dioxide solution is available. The same solution will form a milky, amorphous mass of quartz if space is lacking, if cooling is more rapid, or if minute fluid inclusions are present. However, both of these minerals have the same general chemical composition (silicon and oxygen) and internal atomic arrangement.

A crystal is a three-dimensional figure bounded by smooth plane surfaces, or *faces*. The positioning of the crystal faces is described on the basis of their axial relationships. For example, a cube-shaped crystal has three crystallographic axes connecting opposite faces (see *A*, fig. 9.2). The axes are equal in length and are at right angles to each other. Minerals that vary in chemical composition may occur in the same system. For example, the metallic minerals galena (lead sulfide) and pyrite (iron sulfide) occur in the isometric system along with the nonmetallic mineral halite (sodium chloride), even though they are entirely different materials. A few of the minerals such as opal have a random arrangement of the atoms in their internal structure and are therefore designated as amorphous.

Figure 9.1 A quartz crystal mass.

The Earth's Materials

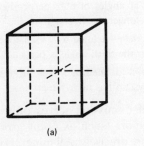
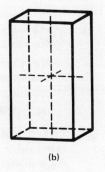
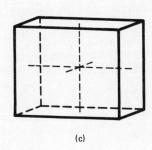
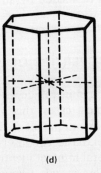
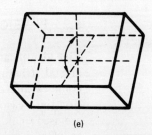
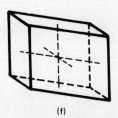

Figure 9.2 The six basic crystal systems: a) isometric; b) tetragonal; c) orthorhombic; d) hexagonal; e) monoclinic; f) triclinic. Sometimes seven systems are defined by adding a trigonal subsystem under hexagonal.

Despite the fact that a variety of crystal forms are possible in minerals, all of them may be included in six main groups or crystal systems. The crystal systems are differentiated on the basis of their angular relationship and the length of the respective crystallographic axes, these axes being imaginary lines running parallel to the edge of the crystal face. For example, an imaginary line running parallel to two edges of the cube would describe one crystallographic axis. The six generally recognized crystal systems are as follows:

Isometric (cubic) system—defined by 3 axes of equal length that intersect at right angles to each other at their respective midpoints; this system has the highest symmetry of all the crystal systems.

Tetragonal system—defined by 3 axes all at right angles to each other, with 2 axes being of equal length, and the third axis either longer or shorter than the other 2.

Orthorhombic system—defined by 3 unequal-length axes intersecting at right angles.

Hexagonal system—defined by 4 axes, 3 of which are of equal length on the same plane and intersect at angles of 60°; perpendicular to the plane of these axes is a fourth axis which is longer or shorter than the other three.

Monoclinic system—defined by 3 axes, all of unequal length, 2 of which intersect at an oblique angle; the third angle is perpendicular to the other two.

Triclinic system—defined by 3 axes, all of unequal length and intersecting at 3 oblique angles; this system has no symmetry.

This very brief discussion serves only to identify the crystal systems, there being other factors (such as point groups and space lattices) used to define the systems more precisely. All minerals belong to one of the six crystal systems, but when closely examined the system may not be immediately obvious or identifiable. For this reason, a number of subdivisions or classes have been designated for each crystal system, thus permitting a more detailed classification of minerals. For example, sphalerite (zinc sulfide) crystals are usually found to form in a tetrahedron (a 4-sided pyramidal figure) belonging in the isometric system.

9.3 PHYSICAL PROPERTIES OF MINERALS

The identification of a mineral depends on the chemical composition and crystal structure. In most cases, this poses a considerable problem, because complex and expensive equipment is necessary to make a detailed analysis. For this reason, other basic or observable physical properties of minerals are used to aid in general mineral identification. These include such properties as luster, hardness, color, streak, fracture, and specific gravity. In most cases, the tests are relatively simple and require a minimum of equipment, time, and knowledge. Most of the common minerals may be readily identified by an understanding of these characteristics. There is no standard set of rules for the order in which the physical characteristics should be determined, but many systems begin with simple things like luster, which immediately separates minerals into two broad categories (metallic and nonmetallic). The characteristics discussed below should be determined on a fresh surface of the mineral and not on a surface that has been exposed to considerable weathering.

Luster is defined as the character of light reflected from the mineral surface. Minerals may be separated into two categories on the basis of luster, namely, *metallic* for those minerals that reflect light to give a metallike appearance, and *nonmetallic* for all others. The nonmetallic category is again subdivided on the basis of descriptive characteristics such as *earthy* for dull surfaces, *adamantine* for brilliant surfaces like a diamond, or as *waxy, pearly, vitreous, greasy, silky*, and *resinous*. To be able to recognize these features requires a little practice. Minerals that have a metallic luster are usually opaque and include the native metals (gold, silver, copper, etc.) and most of the metallic sulfides. Those minerals in the nonmetallic category, which includes by far the largest group of minerals, may be opaque, translucent, or transparent. A submetallic category is also recognized, which represents a transition from metallic to the nonmetallic groups. However, for simplicity's sake, this category is omitted in most manuals for mineral identification.

Hardness is the resistance a mineral offers to abrasion or scratching. The scale now most frequently used was originated in 1822 by F. Mohs, an Austrian mineralogist. He proposed a relative scale divided into 10 subdivisions in which the intervals of hardness are as uniform as possible. He assigned a standard mineral for each level to represent the degree of hardness for that level. Hardness appears to be related to the size of the atoms making up the mineral (the smaller the diameter, the harder the mineral), the type of chemical bonding, and the packing of the atoms (generally the more closely packed, the harder the mineral). Other hardness systems have been suggested, but it is to Mohs's credit that his has survived unchanged since first presented.

The scale of hardness as described by Mohs with a representative mineral for each degree of hardness is given as follows from 1 (the softest) to 10 (the hardest):

1. Talc
2. Gypsum
   (fingernail)
3. Calcite
   (copper penny)
4. Fluorite
5. Apatite
   (knife blade, glass)
6. Orthoclase (Feldspar)
7. Quartz
8. Topaz
9. Corundum
10. Diamond (diamond is considered ten times harder than the next hardest mineral)

Each of the representative minerals will scratch those lower on the scale and will in turn be scratched by those higher on the scale. Thus, if an unknown mineral is scratched by fluorite but, in turn, scratches gypsum, it will have a hardness of between 2 and 4. If an unknown mineral scratches orthoclase but is scratched by quartz, its hardness is between 6 and 7 (fractional values between the standard intervals of hardness are recognized).

A set of small pieces of the standard minerals are usually used as hardness sets for testing the hardness of minerals. However, in actual field work some mineralogists and most rockhounds find the fingernail ($H=2.5$) and the pocket knife ($H=5.5$) adequate tools for determining hardness. Minerals of hardness 1 often have a greasy feel and minerals up to 2.5 can be easily scratched with a fingernail. Over hardness 2.5 the mineral cannot be scratched with the fingernail, but is subject to scratching with a knife or a copper penny ($H=3.5$). Actually, minerals with a hardness up to 3.5 can be cut with a knife blade, up to 4.5 easily scratched, and up to 5.5 scratched with more difficulty. Those minerals with a hardness above 5.5 cannot be scratched with a knife. Minerals with a hardness above 7 are generally very rare.

Color of a mineral is not a very reliable property for identification purposes, because some minerals range through various shades of a color or may occur in several different colors. The color is the result of the absorption of certain wave lengths of light and represents the wavelengths reflected. Dark-colored minerals are those which absorb all wavelengths of light in about equal quantity and none are dominant. White or gray also absorbs all wavelengths of light, but more is reflected than in the case of the dark-colored minerals.

The cause of this phenomenon is quite complex and is related to chemical composition and to the internal atomic structure. Quite frequently, small amounts of impurities will result in a change of color. Thus, the material that produces normally clear quartz may become purple (amethyst) by the inclusion of traces of titanium. Graphite (black) and diamond (clear) are both composed of carbon, but the internal arrangement of the carbon atoms results in a difference in color as well as other physical properties. Generally, the true color of a mineral is its color in powdered form.

Streak is the color of the powdered mineral generally obtained by scratching the mineral on an unglazed porcelain surface. The porcelain has a hardness of 6, which limits its usefulness for this purpose to those minerals softer than the porcelain. Streak is related to the color of the mineral, but it is not necessarily the same. Streak is more reliable than color, since it does not vary as much as the color variations of a given mineral. Minerals and especially nonmetallic minerals with a hardness greater than 6 will not yield a streak.

Cleavage: In the discussion on crystal form, mention was made of the various shapes in which crystals occur. The crystals are bounded by faces just as a cube, for example, is bounded by 6 square surfaces or faces. The breaking of a crystal parallel to such faces is known as cleavage (fig. 9.3). However, some minerals do not necessarily break along or parallel to crystal faces. For example, quartz, which has a distinct crystal form, exhibits no cleavage. Other minerals may vary as to the quality of cleavage, which is expressed as perfect, good, fair, or poor. Cleavage is a reflection of the internal structure of the mineral. The strength of bonding of atoms varies in different planes in the mineral, and cleavage will occur along those planes where bonding is weakest.

Fracture of a mineral refers to the character of a surface obtained when a mineral is broken in a direction other than a plane of cleavage or along some random direction (fig. 9.4). Minerals such as quartz, which exhibit no cleavage, yield fracture surfaces very readily. Those minerals with good-to-perfect cleavage generally break along cleavage planes and show little or no fracture. Several types of fracture surfaces are recognized. *Conchoidal* fracture exhibits a smooth surface like that found on the fracture surface of glass. An *even* fracture has surfaces that are flat, although not completely smooth, whereas *uneven* fractures show a rougher surface. A *hackly* surface is uneven with many

The Earth's Materials

Figure 9.3 Cleavage (a) biotite, (b) calcite

Figure 9.4 Conchoidal fracture in obsidian.

sharp protrusions, while a *splintery* surface is, as the name implies, splintery or fibrous.

Specific gravity is the ratio between the weight of a mineral and the weight of an equal volume of water. This term is sometimes confused with or erroneously used interchangeably with density. Density refers to the weight of a given object for a given unit of volume, e.g., grams per cubic centimeter or pounds per cubic foot. Specific gravity is actually the ratio between the densities of a substance and water. Water having, by definition, a density of one gram per cubic centimeter, it is generally taken as the standard. Some apparatus such as a balance is required to make a specific gravity determination. In addition, for the determination to be accurate, it is necessary for the mineral to be reasonably pure. However, when such a determination can be made, it is of value in the identification of minerals, especially in working with metals where the specific gravity is unusually high.

Other properties of minerals have been studied, although they are not as useful in identification. These miscellaneous properties include magnetic, optical, and electrical properties in addition to radioactivity, fluorescence, iridescence and opalescence. Some of these properties are of economic significance, others are at present only a curiosity for which a use may be found in the future. A simple key for the identification of minerals is included in the appendix.

9.4 MINERAL CLASSIFICATION

The Earth's Materials

More than 2,000 minerals are presently recognized, and new ones are constantly being added to this number. Some, like the diamond, are extremely rare. The rarity of diamonds is due to the small amounts of free carbon, from which diamonds are formed, available as a raw material at the time and place when conditions are suitable for diamond formation. Other minerals, such as feldspar or quartz, are quite common because they are composed of the more abundant elements (silicon, oxygen, aluminum) found in the earth's crust.

To deal effectively with so many species of minerals, a system of classification is necessary. The system most often used groups minerals into classes that are chemically alike by virtue of the anion which is a negatively-charged nonmetallic ion. Because of its relatively large size, the anion also, to a degree, controls the crystal structure of the minerals. Thus, the system relates minerals that are alike chemically and which also have similar crystal properties as a result of their chemical composition. Table 9.1 lists the major mineral classes in this system and a few of the more common minerals in each class.

Table 9.1
The Mineral Classes and Some Common Examples

Mineral Class	Mineral Name	Chemical Composition
I Native Elements	Gold	Au
	Platinum	Pt
	Sulfur	S
II Sulfides	Pyrite	FeS_2
	Galena	PbS
III Oxides	Hemitite	Fe_2O_3
	Corundum	Al_2O_3
IV Halides	Halite	NaCl
	Fluorite	CaF_2
V Carbonates	Calcite	$CaCO_3$
	Magnesite	$MgCO_3$
VI Sulfates	Gypsum	$CaSO_4 \cdot 2H_2O$
	Epsomite	$MgSO_4 \cdot 7H_2O$
VII Phosphates	Apatite	$Ca_5(PO_4)_3F$
VIII Silicates	Quartz	SiO_2
	Biotite	$K(MgFe)_3(OH)_2AlSi_3O_{10}$
	Orthoclase	$KAlSi_3O_8$

Native elements: None of the approximately 20 elements found in their natural state in the earth's crust occur in large masses. However, their value is of sufficient importance to man that deposits of these materials are actively sought. They are invariably of economic value and include materials which are metallic (gold, silver, platinum), semimetallic (antimony, bismuth, arsenic), and nonmetallic (sulfur, diamonds, graphite).

Sulfides: The sulfides are a combination of a metallic element with sulfur and are important because of the occurrence of several ore-bearing minerals in this group. These minerals include galena (PbS), an important source of lead; sphalerite or zinc sulfide (ZnS), a source of zinc; and chalcocite or copper sulfide (Cu_2S), a valuable ore of copper. Pyrite or iron sulfide (FeS_2) is more commonly known as "fool's gold" because of its similarity in appearance to gold. Pyrite is sometimes mined for the sulfur, gold, or copper associated with it but seldom for iron. Also in this class is cinnabar or mercury sulfide (HgS)—the world's principle source of mercury.

Oxides: The oxides are compounds formed by the combination of one or more elements with oxygen. Among the most common of the oxides are those which contain iron, such as hematite (Fe_2O_3), magnetite (Fe_3O_4) and limonite. Hematite and magnetite are most important to man as ore-forming minerals because they are the major source of iron. Corundum (Al_2O_3), an extremely hard mineral, is in this group and when slight amounts of impurities occur with it, the oxide will under certain conditions form gem-quality minerals such as sapphires and rubies.

Halides: The halides are those minerals which include the negatively charged halogen ions, chloride, bromide, iodide and fluoride. The principal and most common mineral in this class is halite (NaCl) which is familiar to everyone as common table salt. It is found in beds up to several hundred feet thick where it was deposited by evaporation of enclosed seas during the geologic past. Fluorite (CaF_2), not as common as halite, is important in certain industrial processes such as, for example, the smelting of aluminum.

Carbonates: The carbonates are a common group including a number of minerals with similar physical properties. The basic unit is the carbonate ion $(CO_3)^{-2}$. The extensive deposits of limestone and marble are formed principally from calcite ($CaCO_3$) or dolomite ($CaMg(CO_3)_2$) that have been deposited either by organic or inorganic processes.

As a result of structural differences, carbonate minerals fall into two groups: the calcite group and the aragonite group. The calcite (limestone) deposits are important in the manufacture of cement.

Nitrates, compounds in which $(NO_3)^-$ is the basic unit, are very similar to carbonates in structure and are generally included in the carbonate class. Nitrates, particularly soda niter ($NaNO_3$), are softer and more soluble than calcite and are therefore found only in very arid regions. Soda niter is the most common form in this class and is mined in Chile for use in explosives and fertilizer.

Borates: About forty-five borate minerals are recognized, but only four are sufficiently abundant to be of commercial importance, Borax, $Na_2B_4O_7 \cdot 10H_2O$ is probably the most important of these and is found principally in Death Valley, California.

Sulfates: Not only does sulfur occur in the elemental and sulfide form in

The Earth's Materials

Figure 9.5 View of a borate mine near Boron, California. Courtesy of Lawrence J. Herber.

nature but also as the sulfate anion $(SO_4)^{-2}$. Of this group, gypsum $(CaSO_4 \cdot 2H_2O)$ is the most important commercially and the most abundant of the sulfates. It is a material used in the building industry as plaster and for agricultural purposes as a soil conditioner.

Phosphates: The phosphates are a large group of minerals of which only a few are common. The phosphate anion, $(PO_4)^{-3}$ is the basic unit of these minerals. The most common and the most important mineral from an economic standpoint is apatite, a major source of phosphate fertilizer. Turquoise, a copper aluminum phosphate mineral, is valued as an ornamental, semi-precious stone and is found in many places in the world. Arsenic (As) or vanadium (V) may substitute for phosphorus as the central ion in the anion, so arsenates and vanadates are included in the phosphate class.

Silicates: The largest and most common class of minerals are the silicates, of which there are approximately 500 to 600 different types. This diversity of form is not so much the result of the inclusion of different elements as it is the result of the variation in the arrangement of the silicon and oxygen atoms. The silicon-oxygen atoms form tetrahedra (fig. 9.6) which are linked to each other or connected by other atoms producing a number of different crystal structures. For this reason silicates are classified into a number of structural groups according to the geometric arrangement of the silicon tetrahedron and the silicon-to-oxygen ratio.

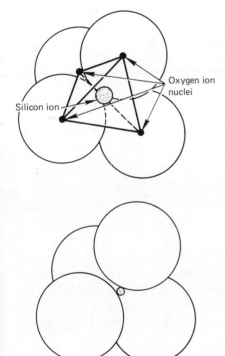

Figure 9.6 The SiO_4 tetrahedron: The small silicon ion fits inside the space made by the close packing of four oxygen ions.

With few exceptions, igneous rocks (formed from the solidification of molten material) are formed from silicate minerals, but only a relative few are really important in igneous rock. These are, however, significant in that they constitute over 90 percent of the crust of the earth and include plagioclase, orthoclase, quartz, muscovite, biotite, amphibole, pyroxene and olivine.

9.5 THE IMPORTANCE OF MINERALS

Of what importance are minerals to us, and why are minerals studied with such diligence and in such detail? Minerals from the earth represent the raw materials for the metals necessary for our social and technological progress. This progress has been dependent upon our ability to make increasingly greater use of the metals we extract from the earth. Through research, we have improved the quality of the metals by developing new alloys, and we have found many new uses for those metals with which we have long been familiar. Within the past 150 years, a wide range of new mineral materials have been discovered and put into use. Many of these materials have substituted for substances commonly used for hundreds of years. For example, aluminum has widely replaced steel as a structural material, just as steel once replaced stone.

The occurrence of minerals in the earth is a reflection of the chemical composition of the earth's crust. Thousands of samples analyzed chemically have revealed that, of the 92 naturally occurring elements, 8 comprise over 98% of the mineral composition of the crust (table 9.2).

Table 9.2
The Chemical Composition of the Earth's Crust by Weight

Element	Percent Weight	Percent Volume	Element	Percent Weight	Percent Volume
Oxygen	46.6	94.27	Sodium	2.8	0.99
Silicon	27.7	0.88	Potassium	2.6	1.66
Aluminum	8.1	0.47	Magnesium	2.1	0.58
Iron	5.1	0.47	All Others	1.4	1.00
Calcium	3.6	0.68			

It is interesting to note that almost 50% of the earth's crust by weight is composed of oxygen, which is also found in the atmosphere as a gas. If the oxygen in the crust is presented on the basis of volume, it occupies almost 94% of the volume occupied by all the elements. The earth's crust (and therefore all minerals) is essentially a packing of oxygen atoms bonded together by silicon, iron, aluminum and other elements.

The relatively large percentages of such metals as iron and aluminum in the earth's crust, and their importance in the development of our society, are readily evident. Iron and aluminum have played a major role in the construction of tall buildings, long bridges, high-pressure devices, and many other similar artifacts of an advancing technology. At the present time, these materials are easily accessible in areas where high concentrations occur in the earth's crust. As greater and greater demands are made upon these resources, we will ultimately be faced with the prospect of severe shortages or we will have to mine low-grade sources at high expense. Minerals are truly a nonreplaceable item and depletion is inevitable. There is little of the land surface that has not at least superficially been explored for ore deposits, except that portion which is under glacial ice and the oceans. Although some rich ore deposits may have been overlooked in these initial explorations, there ultimately will be a limit to our resources, a limit which for some metals, minerals, and fuels may be only a few decades hence. Will we be faced with the prospect of a slowdown in the advance of our civilization or a reduction in our standard of living? Such a prospect need not necessarily come about, since we have several alternatives at our disposal for overcoming this problem.

Before the resources are depleted, we must consider alternate sources of these metals so important to society. One such metal is aluminum, which is commercially produced from the mineral bauxite. Bauxite is a material containing a high percentage of aluminum oxide (Al_2O_3), but a material of which there is a limited supply. A vastly greater quantity of aluminum also occurs in some forms of clay, but in a much lower concentration than in bauxite. Here the problem of extracting the metal from clay is economic rather than technical because of the cost of processing huge quantities of clay necessary to obtain an adequate amount of aluminum. However, when the bauxite is exhausted there is no doubt that aluminum will be extracted from high-alumina clays and other aluminous minerals, but at higher prices.

Currently most mining for mineral wealth occurs close to the earth's surface, but when these deposits are exhausted deep mining will be inevitable. Deep mining engenders many difficulties, not the least of which is the increased heating that is usually experienced at greater depths. Temperatures in the earth's crust generally increase on the order of $1°C$ per 55 meters of depth, presenting serious problems when mining a kilometer or more beneath the surface. However, when the demands of society require it, these problems will be overcome, making deep mining feasible.

New geological techniques capable of finding new sources of metal may make it possible in the future to discover ore deposits now hidden from view by hundreds of feet of surface material. These techniques have been developed mainly for oil exploration but can readily be adapted to the search for new mineral wealth. One method involves the detection of small changes in weight of a known mass due to a variation in density of subsurface rock. An unusual increase in density may indicate the presence of an ore body. Another geophysical technique important enough to be considered is explosion seismology. In this method, an explosive charge is set off in the ground, and the wave patterns passing through the rock are recorded on a seismograph. These wave patterns may be interpreted as indicating various types of rock formations and serve to locate minerals of economic value. Changes in the magnetic patterns of the crustal rock provides still another possible means of finding new ore deposits.

Data acquired from space satellites is another relatively new source of knowledge of the geology of the earth that will be helpful in identifying areas which have a high potential for the existence of mineral resources. These areas can later be investigated from the surface in more detail.

One other method for deriving new sources of metal that has been suggested is to tap the magma bodies (molten rock beneath the surface) and bring the rock to the surface in molten form much as oil is brought to the surface in liquid form. Such an approach is impractical at the present time, since many problems in handling the molten rock must first be overcome. However, there may be a time in the future when this suggestion will be seriously explored.

Another source of mineral wealth that is now in the early stages of exploration is the ocean. Seawater is plentiful, easily obtainable and easy to dispose of. There are none of the problems of land-based mining, such as the removal of huge quantities of overburden to reach the ore, disposal of the tailings in the mining operation, and removal of the slag in the smelting process. However, seawater contains relatively small quantities of the elements, requiring that great amounts of seawater be processed in order to obtain economic quantities of the desired elements.

The inherent value of minerals to society cannot be overstated. All of our industrial and commercial activities, the food we eat, the fiber used in making clothing, and materials used to provide shelter are directly or indirectly dependent upon the existence of mineral wealth. The mineral resources are not inexhaustible; therefore, careful usage and conservation are of paramount importance in maintaining a constant supply. In the future our ingenuity will be taxed to the utmost in an endeavor to sustain our level of existence in the face of dwindling reserves of natural resources.

9.6 ROCKS

Rocks: A rock may be defined as an aggregate of minerals, and it is at this point that mineralogy (the study of minerals) and petrography (the description and classification of rocks) overlap. There are exceptions to this definition of rocks, in that a few sedimentary rock types are formed from the skeletal remains of some organisms and therefore are not necessarily aggregates of minerals.

The geologist is interested in more than just the association of minerals in rock. He is interested in how the rocks were formed and in past events in the earth's geologic history as revealed by the rock record. The process of rock formation is of importance, for this is the basis upon which rocks are divided into three broad categories: igneous, sedimentary, and metamorphic rock. The processes responsible for the formation of rocks are extremely complex, as are the

The Earth's Materials

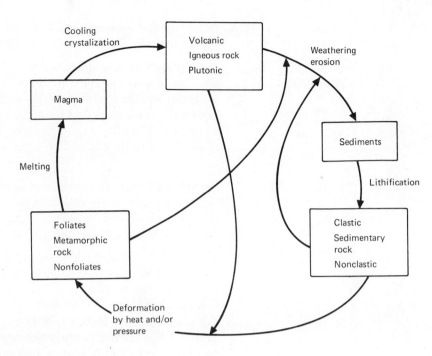

Figure 9.7 The rock cycle.

relationships that exist between the types of rock. Rock formation is thought to be cyclic, and these relationships are illustrated in figure 9.7.

9.6A Igneous Rocks

The generally accepted theory is that igneous (fire-formed) rock is formed as a result of the solidification of molten material. This molten material or magma will form intrusive rock if solidified beneath the earth's surface. Intrusive rock is generally medium to coarse textured, because slow cooling permits the mineral crystals that form the rock to grow to considerable size. Texture is such that the mineral crystals can be easily recognized without benefit of magnification. On the other hand, molten material which reaches the surface, generally as an outpouring of lava during volcanic activity, solidifies rapidly, forming an extrusive igneous rock. This rock will be fine to medium textured, since rapid cooling of the matrix prohibits extensive mineral crystal development. Thus it can be seen that texture (crystal size) is an important property of igneous rock directly related to the manner in which the rock was formed and is important in the classification of igneous rock.

The grading of coarse texture to fine texture includes most igneous rock, but two additional textures should be defined to complete the range covered by this property. These are glassy and porphyritic. The igneous rock obsidian is a typical example of a rock with glassy texture. This is a rock that has solidified so rapidly that mineral crystals do not have an opportunity to form from the molten material.

A porphyritic texture results when the molten material cools at different rates. Slow cooling at first permits the formation of some mineral crystals, particularly those that solidify at fairly high temperatures. If at this time a change in environment occurs that results in rapid cooling, the balance of the molten material will solidify as fine-textured rock. Thus, the larger crystals known as phenocrysts are encased in a fine-textured matrix, forming a porphyritic rock.

Igneous rock may also be differentiated on the basis of mineral composition. Only a few of the several thousand minerals make up a major portion of igneous rock. These include olivine, augite (pyroxene group), hornblende, biotite, plagioclase (anorthite and albite), orthoclase, muscovite, and quartz.

These rock-forming minerals do not crystallize at the same temperature, and this differential rate of crystallization is the basis for the reaction series first described by N. L. Bowen and known as the *Bowen reaction series* (fig. 9.8). Olivine is one of the first to crystallize, along with calcic plagioclase (anorthite). These minerals will form a rock known as peridotite if primarily olivine, or a gabbro when olivine is associated with calcic plagioclase. The gabbro is a coarse-grained rock, indicating that its origin was primarily intrusive. Lava, a molten rock of the same mineral composition as gabbro, would form basalt upon solidification. Quartz is the last of this series of rock-forming minerals to solidify, and in combination with biotite or orthoclase will form a granite from magma, and obsidian or rhyolite from lava.

By a combination of the two properties of texture and mineral composition, it is possible to establish a workable system for classification of igneous rock.

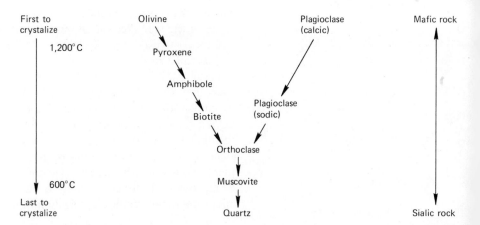

Figure 9.8 This simplified diagram of the Bowen reaction series demonstrates a possible means by which several igneous rock types may originate from the crystallization of a basaltic magma.

One such simple system relates texture to color where color is related to mineral composition. Olivine, augite, hornblende, and biotite are dark. Quartz, muscovite, and orthoclase are light-colored, and plagioclase ranges from light to dark. Various combinations of these minerals will result in light- to dark-colored igneous rock (fig. 9.9).

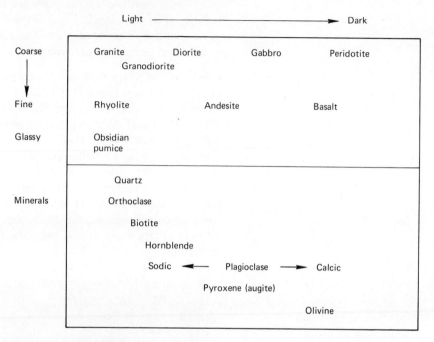

Figure 9.9 Classification of a few common igneous rocks according to color and texture.

9.6B Sedimentary Rocks

Sedimentary rocks are for the most part secondary rocks derived from pre-existing rocks that have been broken down by weathering, or they are salts that have been precipitated from evaporating bodies of water, or they are the accumulation of carbonate shells of living organisms. About 75 percent of the exposed surface rocks are sedimentary, and of this amount about 90 percent is sandstone, shale, and limestone. They are characterized by layering and bedding.

Deposits of clay, silt, sand, and gravel result from the process of weathering and erosion of surface rocks. These particles are transported by the action of water, and to a lesser degree by wind and ice, and are deposited when these forces no longer have the energy to transport the particle (see Chapter 11). The particles may be deposited in beds of sand by action of wind in an arid region, or they may be deposited in lake bottoms, stream beds, or ocean floors by the action of moving water. Sediments of this type, upon becoming consolidated, form clastic (consisting of fragments) sedimentary rocks.

The consolidation of the sedimentary particles results mainly from the cementing together of the individual particles by the precipitation of material from solutions percolating between the grains. Soluble materials such as calcium carbonate are deposited around the individual particles, filling the pore spaces in between. In this manner, the pore space is gradually filled, and the total effect is to combine all the particles into a solid rock. Calcium carbonate, because of its abundance, is one of the most common cementing agents. Silicon dioxide and iron oxide also serve as cementing agents, although they are not as common as calcium carbonate.

Clastic sedimentary rock differ in several ways, but the most obvious means of classification is based on particle size. In 1922, C. K. Wentworth proposed a scheme in which three categories of clastic sedimentary rock were recognized. These were *conglomerates* made up of particles larger than 2 millimeters (mm) and up to several meters in diameter. This category included rounded particles only, and a separate subgroup called *breccia* was set up to include angular particles in the same range of diameters. Sand-sized particles from $\frac{1}{16}$ mm to 2 mm in diameter form the *sandstones*, whereas silt and clay particles less than $\frac{1}{16}$ mm in diameter make up *shales* and *mudstones*. Within each of these broad groupings, the rocks are further classified on the basis of the mineral composition of the particles, color, cementing materials, and other inclusions.

Figure 9.10 Examples of sedimentary rock: conglomerate, sandstone, and shale.

The nonclastic sedimentary rocks are composed of chemically precipitated materials. Of these, limestone is perhaps the most common and the most extensive, consisting of calcium carbonate ($CaCO_3$) and calcium magnesium carbonate ($CaMg(CO_3)_2$). Limestone is deposited as a result of organic precipitation. Organisms living in the sea form protective shells by removing calcium bicarbonate from sea water. When the organisms die, the shells of calcium carbonate ($CaCO_3$) accumulate on the sea bottom, building up great thicknesses of shells that in time become consolidated to form limestone, a rock thus composed primarily of calcium carbonate.

Other chemical precipitates, such as gypsum ($CaSO_4 \cdot 2H_2O$), anhydrite ($CaSO_4$), and halite ($NaCl$) are deposited as the body of water in which these salts are dissolved is evaporated. Interior drainage basins or bodies of water cut off from the sea by geologic processes will evaporate and deposit layers of salts hundreds of feet thick. In the evaporative process, the sulfates are usually deposited first, followed by the chlorides, resulting in distinctive layers of sedimentary rock. On occasion it is possible to see this sequence repeated several times in one deposit, indicating the alternating advance and recession of the sea during geologic times.

Travertine or tufa, both calcareous materials, may be deposited from freshwater springs. These deposits occur in limestone regions where limestone is dissolved by percolating ground water containing carbon dioxide. The carbon dioxide causes the water to be slightly acidic, permitting the water to dissolve the limestone (decomposition). When the water issues from the ground, some of the carbon dioxide is released, thus reducing the acidity of the water and resulting in the precipitation of the calcium carbonate that forms the travertine or tufa.

Coral reefs are limestone deposits resulting from the buildup of tiny calcareous organisms which form large calcareous structures such as the Great Barrier Reef of Australia. Structures of this type have been found in mountains and inland areas, indicating the presence of a sea at some time in the geological past. Chalk is formed in essentially the same manner by foraminifera, free-floating single-celled organisms that form a calcareous shell. Coral reefs, diatomite, and chalk represent only a small fraction of sedimentary rock found on land, but they are of geologic importance in that their presence is evidence of the existence of ancient seas. This evidence contributes to the knowledge of changes taking place on the earth's surface.

Siliceous rock formed by the deposition of chemically precipitated silicon dioxide (SiO_2) is also part of the nonclastic sedimentary group. This sedimentary rock takes several forms. Among the more common ones are chert, flint, and jasper. Diatoms, which are free-floating, single-celled organisms, are responsible for deposits of siliceous diatomite. Diatomite represents the accumulation of the tiny siliceous shells of diatoms, which sink to the sea floor when the organism dies.

9.6C Metamorphic Rock

Rock, once it has formed, will be reasonably stable unless there is a considerable change in its environment. A change in conditions may cause a rock to be altered or metamorphosed into a different type, depending on its original mineral composition and the environment responsible for the metamorphism. The principal factors causing metamorphism of rock are heat, pressure, and some chemically active fluids operating over an extended period of time. Metamorphism does not necessarily imply that the rock returns to a molten state, since recrystallization may come about without the rock actually becoming a fluid. There may, however, be plastic flow, as evidenced by the textural pattern of some metamorphic rocks.

Several types of metamorphism which bring about the formation of metamorphic rocks are recognized:

1. *Contact metamorphism* is a form in which heat is primarily responsible for the alteration. A mass of molten rock or magma intrudes the surrounding consolidated rock and the attendant high temperatures cause changes in the rock in contact with the molten mass. This contact zone is called an *aureole*, which may be a few inches to several hundred feet thick. The aureole is of value to man, since it is in such formations that valuable metals like lead, zinc, and gold are found. The great prospectors of the past, seeking their fortune, would search for aureole formations that had been exposed by uplift and erosion. Copper ore deposits of Butte, Montana, were formed this way.

2. *Regional metamorphism* refers to rock alteration that has taken place over a large area as a result of the accumulation on the surface of great quantities of sediments or of igneous rock from volcanic activity. In this instance, great pressure as well as heat is applied to the rock, causing alteration in its structure. Sedimentary deposits as much as 9,000 to 18,000 meters thick have been measured. The weight in such deposits, increasing about 8 kilograms per square centimeter for every 30 meters of depth, will at 9,000 meters result in considerable pressure. At these depths the temperature also becomes considerable, with a rise of about $1°C$ for every 55 meters of depth. Under these circumstances, rock could be altered, and subsequent uplift of the rock mass and extensive erosion would expose this metamorphic rock at the surface.

3. *Dynamic metamorphism* is the alteration of rock resulting mainly from unequal pressure. Under conditions of severe local stress on rock brought about by movement along fault zones, rock will be altered mainly by pressure. The original rock may be pulverized under such stress and recrystallized by the pressure to form metamorphic rock.

4. *Hydrothermal metamorphism* is caused by high temperature gases and liquids which are released from magma into the surrounding rock. These fluids will influence the rock by reacting with the minerals beyond the aureole of contact metamorphism. In this manner, new substances may be introduced to form new minerals, or some material may be withdrawn from the existing mineral which, upon recrystallization, results in a new form of rock.

Each of these conditions under which metamorphism occurs results in characteristic textures of metamorphic rock. Regional metamorphism, being the most extensive, results in the development of the most common forms of metamorphic rock. These rocks can generally be divided into two groups: the foliates, which have a layered appearance, and the nonfoliates, which are more homogeneous. Foliation is a layered structure resulting from segregation of different minerals into layers as a result of metamorphic processes. Examples of a few metamorphic rocks and the rocks from which they may be derived are given in table 9.3.

Table 9.3
Examples of Common Metamorphic Rock

Name	Original Rock	Type of Metamorphism
Foliates		
Gneiss	Shale (Sedimentary)	Regional
	Granite (Igneous)	
Schist	Shale (Sedimentary)	Regional
	Rhyolite (Igneous)	
Slate	Shale (Sedimentary)	Regional
Nonfoliates		
Quartzite	Sandstone (Sedimentary)	Regional
Marble	Limestone (Sedimentary)	Regional
Verde Antique	Serpentine and Dolomite	Hydrothermal

9.7 WEATHERING

The loosening of rock from the earth's crust and its gradual breakdown into smaller and smaller particles is the result of the destructive process of *weathering*. The obliteration of markings on statues and the discoloration of freshly exposed rock surfaces are all evidence of this process. Weathering must not be confused with erosion, for weathering acts in place while erosion implies motion and the transport of surface material from one place to another.

The weathering processes include two distinctly different types of activity, and while it is convenient to discuss them separately, it should be understood that they work simultaneously and in conjunction with each other in the gradual destruction of the materials on the earth's surface.

9.7A Disintegration

Disintegration is a result of mechanical weathering processes, mainly frost action and temperature change above and below the freezing point. One type of frost action is *frost wedging*, which results from the freezing and thawing of water in cracks and crevices in the rock structure. Water will increase slightly in density as it cools until it reaches a temperature of 4° C (39° F). At this point, water has reached its maximum density, and further cooling results in a decrease in density that continues as the water forms ice at 0° C.

To see how this results in the breaking of rock it may be well to follow the sequence of events as water freezes in a rock crevice. The temperature of the water cools by a loss of heat to the atmosphere. The surface of the water cools first since it is in contact with the atmosphere and becomes slightly more dense than the water beneath it. This causes the surface water to sink to the bottom to be replaced by slightly less dense (and slightly warmer) water which in its turn is cooled. This process continues until the temperature of all the water in the crevice is 4° C. The surface water continues to cool but now becomes less dense and cools until it freezes, forming a plug that seals off the water in the crevice. As the remaining water in the crevice continues to cool, it expands, exerting a pressure (with a theoretical maximum of 2,100 kilograms per square centimeter) upon the rock, eventually causing it to crack further and finally to break. Action of this type is common in mountainous areas and northern latitudes where temperatures alternate above and below freezing a good portion of the year.

Figure 9.11 Example of frost wedging. Large rock has been split by the action of ice.

Another form of mechanical weathering by frost action is *frost heaving*. This activity is most common in unconsolidated material, particularly the surface soil. Farmers in certain areas find it necessary each spring to "harvest" a crop of rocks resulting from frost heaving. Alternate freezing and thawing gradually forces rocks up through the soil to the surface by the same mechanism which produces frost wedging. Water accumulates beneath a rock, and expansion caused by the water freezing will produce a pressure. Since there is least resistance toward the surface, the movement is in that direction. With a thaw, more water trickles beneath the rock, carrying with it small soil particles which fill in the space. The next freeze produces expansion and further movement, so that eventually the rock will be forced to the surface. Frost heaving is often responsible for breaking up a road constructed on a poorly drained roadbed, or the cracking of a building where provision has not been made for drainage beneath the foundation.

Weathering has been erroneously attributed to temperature changes alone, particularly in arid regions. Alternate heating and cooling, with a consequent expansion and contraction of rock material, was once thought to be responsible for peeling off layers of rock in a process known as *exfoliation* (fig. 9.12). An experiment in 1936, in which a piece of polished granite was subjected to alternate heating and cooling to simulate several centuries of this activity, indicated that temperature alone was not, in fact, responsible for exfoliation, since no visible changes occurred on the granite from the heating and cooling process.

Figure 9.12   Exfoliation on surface of large rock that has been subjected to weathering processes.

But a fine spray of water was introduced periodically in another portion of the experiment to simulate rain, and this resulted in a marked change in the rock. The polished granite became dull, feldspar minerals became altered, and small cracks appeared that could eventually lead to exfoliation. Expansion of the rock resulted from an increase in the volume of the minerals, particularly feldspar, brought about by chemical reaction with water and carbon dioxide. Thus exfoliation, a "mechanical" process of weathering, was accomplished by chemical means.

Certain biological activities also contribute to mechanical weathering. This includes burrowing by small animals and insects—not a significant contributor to weathering—and also growth of roots of trees, shrubs, and grasses in the rock mass. Almost everyone is familiar with the effect of tree roots on sidewalks and roads and with the way certain grasses force their way through crevices in concrete. Similar events occur in nature, with tree roots and grasses forcing rock apart as the roots penetrate deeper into crevices.

9.7B Decomposition

Decomposition, chemical weathering, is dependent on the presence of large amounts of water—the universal solvent—and is therefore the dominant form of weathering in humid regions. Several types of chemical reactions such as solution, oxidation, carbonation, and hydration are involved in decomposition.

Solution, or the dissolving of rock, is the principal means by which limestone and similar rock are attacked. These materials do not dissolve readily in pure water but are quite reactive in acids. Water, when combined with carbon dioxide from the atmosphere, forms dilute carbonic acid, which, in turn, reacts with limestone (calcium carbonate) to form calcium bicarbonate as shown in the following reactions:

$$CO_2 + H_2O \longrightarrow H_2CO_3$$
$$\text{Carbon Dioxide} + \text{Water} \longrightarrow \text{Carbonic Acid}$$

$$H_2CO_3 + CaCO_3 \longrightarrow Ca(HCO_3)_2$$
$$\text{Carbonic Acid} + \text{Calcium Carbonate} \longrightarrow \text{Calcium Bicarbonate}$$

These reactions are common in humid regions. In areas of widespread limestone deposits, the solution process will result in the formation of underground channels and caves which in some instances are quite extensive.

Oxidation, typified by the rusting of iron, is a form of decomposition familiar to everyone. Almost any metal left outdoors will quickly acquire a dull coat of oxide resulting from the combination of the metal with oxygen from the atmosphere in the presence of moisture. Iron is a common element and in combination with oxygen illustrates the process of oxidation very nicely. Iron may originally be precipitated as pyrite (FeS_2) deep within the earth. When erosion has exposed the pyrite, oxidation will occur, because iron has a greater affinity for oxygen than sulfur.

Hydration, the chemical combination of water with another substance, is a form of decomposition that results in the alteration of the feldspar minerals. These minerals are quite common in many rocks, thus making hydration an important reaction in weathering. Again, hydration is dependent on the presence of water and is a form of decomposition found mainly in humid regions. The reactions that take place are quite complex, and the resulting products are secondary minerals (mostly clays) found in soil. Most of the secondary minerals are readily washed away, and eventually only the sand-sized quartz particles remain. Exfoliation, discussed under mechanical weathering, is thought to occur as a result of decomposition of minerals (feldspars) by hydration.

The action of carbon dioxide on minerals is also recognized, but usually it is discussed in conjunction with other forms of decomposition. Carbonation is the changing of oxides of sodium, calcium, or potassium into carbonates by the action of carbonic acid in water. When feldspars are broken down by the hydration process, calcium, sodium, and potassium are released. These elements readily combine with carbon dioxide to form carbonates, which eventually are precipitated from solution to form sedimentary rocks such as limestone.

The rate at which weathering of rock takes place varies tremendously, depending on climatic conditions and the nature of the rock itself. This fact is well illustrated by Cleopatra's Needle, an obelisk of granite now standing in New York City's Central Park. The obelisk, erected in Egypt over 35 centuries ago, was inscribed with hieroglyphics which withstood the dry heat of the desert with almost no sign of deterioration. Less than 100 years ago (1879), the obelisk was brought to New York, and during the ensuing period the granite has weathered so badly in the humid climate that the inscriptions are now almost illegible. Frost action as well as moisture is thought to be necessary in the weathering of granite, because a similar obelisk in frost-free but humid London has not deteriorated to nearly the same extent. Granites will weather very slowly in a dry desert climate, somewhat more rapidly in a humid climate, and at a much accelerated rate, relatively speaking, in humid climate where frost action occurs. Limestone is also affected by moisture; it is best preserved in arid climates and deteriorates rapidly (by solution) in humid climates.

Chemical weathering occurs at a somewhat more rapid rate in warm, humid climates than in cold regions. At high altitudes or in the polar or subpolar regions, frost action occurs and disintegration is the major form of weathering. In temperate regions both physical and chemical weathering forces act upon the rock, depending on the season and rainfall pattern. Rock generally decays very slowly in desert regions, indicating the importance of the role played by water in the weathering process.

9.8 ECONOMIC IMPORTANCE OF WEATHERING

In the discussion on the importance of minerals, mention was made of rich ore deposits, or ore bodies, without a suggestion as to the manner in which these deposits came about. It appears that weathering, which is essentially a destructive process of tearing down and altering the rock structure, is in some instances responsible for the concentration of materials or ores of value to man. The removal of certain minerals by solution leaves behind highly concentrated ores which are much more economical to mine than the original material. In this manner, rock in the Lake Superior area was leached (that is, removed by the dissolving action of water), converting ore containing 25 percent iron into ore containing 50 percent iron. Iron compounds are quite insoluble and tend to remain in place. This is not true of copper, and one method whereby copper is concentrated is by weathering. During weathering, copper sulfide (chalcocite) is subject to removal as copper sulfate in solution and is redeposited at a lower depth. This process results in the concentration of copper in deposits which may at some later geologic time be exposed by erosion, and the ore can be mined for its copper content. The process, of course, is of little value to man at the present time, but it indicates the manner in which the currently utilized copper deposits were formed in the past.

The soil is one of the most important benefits we derive from the weathering process. It is one of our most valuable resources, for it is from the soil alone that we obtain the food and fiber essential for life on this planet. Weathering of rocks and minerals directly results in the formation of the soil, and even while that soil is being used for agricultural purposes, the weathering continues unabated, releasing more nutrient elements required by plants. It can readily be seen that soils vary tremendously as a result of weathering conditions and the type of rock from which the soil was formed.

The weathering of rock is to a great extent controlled by climate, and this is also true of soil formation. Thus, as with rock weathering, the rate of soil formation is rapid in warm humid regions and slow in desert regions. Climate also controls the growth of vegetation, which in turn has an impact on soil formation. Two types of vegetation are recognized as important—namely, trees and grasses—each of which has a different effect on the soil. Basically, plants contribute to soil formation by the extension of root growth in the soil, and they ultimately add the organic matter that becomes the principal source of nitrogen for subsequent plant growth.

The type of parent rock will determine to some degree the type of soil that results, but again this is modified by the effect of climate. Minerals also contribute to soil fertility and to some extent influence the texture of the soil.

Soils formed on land with steep topography differ from those formed on level surfaces, even though climate and parent rock are similar. Erosion, or the removal of surface soil, is much more active on slopes, causing the formation of shallow soils, while deep soils are formed on level land surfaces under similar climatic condition. However, water does not penetrate into the soil on slopes as readily as on level land; therefore the leaching effect of water on sloping land is reduced.

All these factors contributing to soil formation are operable over extended periods of time, thereby making time itself a soil forming factor. Soils form rapidly from certain parent rock such as shale or sandstone, while those formed from igneous or metamorphic rock take longer. At the same time, soils in desert regions require greater lengths of time to develop. Whatever the circumstances, many hundreds to many thousands of years are required for the development of a good, productive soil. Our carelessness can result in the loss of this soil in just a few years.

9.9 SUMMARY

The Earth's Materials

The study of minerals goes back to antiquity, when people accredited minerals with magical properties. However, the growth of the mining industry in Europe gradually required a more scientific approach to mineralogy. Minerals began to be more clearly described, and this gave rise to the science of mineralogy, a forerunner of geology. Minerals were more precisely defined in terms of crystal structure and chemical composition as well as other related physical characteristics.

Of the approximately 2,000 recognized minerals, less than a dozen are important in the formation of rocks. Rocks are essentially divided into three categories dependent on their mode of formation. Thus, igneous rock is generally considered to form as a result of the solidification of molten material. The mineral composition and the rate of solidification determine the type of igneous rock that results. Sedimentary rock is formed in two ways: one by the cementing together of rock fragments such as sand into sandstone, and the other by the precipitation of salt to form nonclastic sedimentary rock. Alteration of existing rock by heat and pressure results in metamorphic rock. Metamorphism may come about by the intrusion of a molten mass into the overlying structure or from the accumulation of thousands of feet of rock material on previously exposed surface rock.

Minerals and rock are acted on by forces that result in their ultimate destruction. This destruction is brought about by the physical and chemical processes of weathering that result from exposure of the rock to conditions dissimilar to those under which the rock was formed. The process of chemical decomposition and physical disintegration concentrates some materials into ore deposits of economic value; it is also responsible for soil formation, and it breaks down mountain masses to small particles which may then be acted on by the forces of erosion.

QUESTIONS

1. Define the characteristics of a mineral.
2. In one sentence each, list and define seven physical properties of minerals used for identification of minerals other than defined in question one.
3. Which mineral classes are the source of the following elements: iron, gold, mercury, phosphorus (fertilizer), nitrogen (fertilizer)?
4. Briefly describe three potential methods whereby metal necessary to man's development may be obtained.
5. Which minerals are important in the formation of igneous rock?
6. Name a sedimentary rock and its composition formed in fresh water; salt water; on dry land.
7. What conditions are necessary to form metamorphic rock from other forms of rock?
8. Of what value is the process of weathering to people?

10 Mountain Building

How mountains are formed.
Volcanism—volcanic activity, products of volcanic activity.
Types of volcanoes.
Intrusive volcanic activity.
Mountains formed by folding and faulting.
Epeirogeny.
Earthquakes—what are they?
Earthquake waves and the intensity of earthquakes.
Tsunamis—seismic sea waves.
Influence of mountain building on society.

Mountains have, from the very beginnings of history, inspired people to wonder about their origin and to engage in considerable speculation as to how they were formed. Ancient tribes and primitive people today living in mountainous regions have legends which attempt to explain the manner by which mountains came into existence. Most of the writings in classical times make mention of mountains, but little is said about their origin, with the exception of the origin of volcanoes. Even Pliny the Elder, prolific writer that he was, says little concerning the origin of mountains, but he does tell *why* he thought they were made. He felt that nature formed mountains for her own use to control the violence of rivers, break the force of waves, and generally to strengthen the structure of the earth.

Not until the nineteenth century was a more realistic approach taken, when field observation replaced speculation. Faulting, folding, and erosion were seen as factors in mountain and valley building. The realization that the processes were gradual (an idea of uniformitarianism), and not the result of a single catastrophic event, was also recognized.

Mountain Building

Figure 10.1 View of the east flank of the Sierra Nevada mountains of California.

The existence of plains, mountains, and plateaus indicates the presence of forces which mold the surface of the earth's crust into a landscape of almost unlimited diversity of form. An understanding of these forces, responsible for the formation of landscapes, should bring about a greater appreciation of the scenic environment. Such an understanding permits the observer to visualize what events brought about the present landforms and to foretell the ultimate result of the continuing interaction between the forces involved.

Mountains provide an infinite variety of scenery, from low, rolling hills, which give a feeling of serenity, to the great mountain ranges, which provide some expression of the awesome power of nature. Mountains are usually considered to be a portion of the earth's crust which rises above the surrounding earth's surface as a result of some internal force. Some distinction is made between hills and mountains by applying the term "mountain" to any elevation 600 meters or more above the surrounding countryside, whereas relief of less than 600 meters are considered "hills." This is not a rigid rule, since the term "hills" is often applied to landforms of greater than 600 meters elevation; at the same time, some geologists may feel that a hill should be defined as somewhat less than 600 meters.

From a more-than-casual examination of the landscape, it becomes evident that there are two opposing processes at work, each contributing to the development of a variety of landforms. These processes are, first, *mountain building*, and second, *erosion*. Erosion will be discussed in the following chapter.

In the discussion on mountain building, it must be realized that the earth's crust constantly undergoes changes resulting from the interaction of volcanism, faulting, folding, and erosion. Most portions of the earth's surface have undergone several periods of change by this type activity, resulting in regions where the crustal structure is quite complex. It is, of course, more convenient to discuss the interacting processes of mountain building and erosion separately, but it must be recognized that the interaction does exist.

10.1 THE MOUNTAIN-BUILDING PROCESS

What are the processes responsible for mountain building? To answer this question it will be necessary to recall the earlier section on continental drift. In this discussion, mention was made of the *plates* of which the earth's crust is mainly composed. These crustal plates are constantly shifting about due to sea-floor spreading, and they are constantly changing size and shape. Movement causes the plates to collide with or sideswipe each other, and it is this activity that is thought to be responsible for mountain building, especially in those areas near continental margins.

Earlier geologic studies indicated that mountain-building activity has been associated with the formation of *geosynclines*, which are large linear troughs resulting from the subsidence or sinking of the earth's crust. Over long periods of time, thick deposits of sediments accumulated in these troughs, which later were folded into mountain systems. However, geologists have become aware that geosynclines are not all alike, and a unifying theory for mountain building could not be based solely on the geosynclinal theory. However, it does appear that where continental plates come in contact and mountain building occurs, some geosynclinal activity may also take place and in this instance be related to mountain building.

Several types of crustal plate contact are recognized and are considered to be related to mountain-building activity. These include: (1) continent-to-continent collision, (2) trench-continent contact, and (3) rifting. The first two are compressional features, the third is a tensional feature.

Continent-to-continent collision is thought to produce a mountain range as a result of the leading edge of two continental plates approaching each other at a velocity of less than six centimeters per year. The oceanic trough developed between the two plates as they move will be filled with sediment, and the sediment and oceanic crust will be squeezed and thrust upward toward the oncoming continent. The approaching continents will also be folded and thrust-faulted (see section 10.3), and the edges buoyed to great heights to form a wide zone of crustal deformation near the edges of the continents. The collision of India and Asia by such a process is thought to be responsible for the formation of the Himalaya Mountains (fig. 10.2).

It is theorized that when the combined speed of the two colliding plates is greater than six centimeters per year, neither can absorb the impact by buckling. In this case the oceanic plate is forced beneath the continental plate and is

Mountain Building

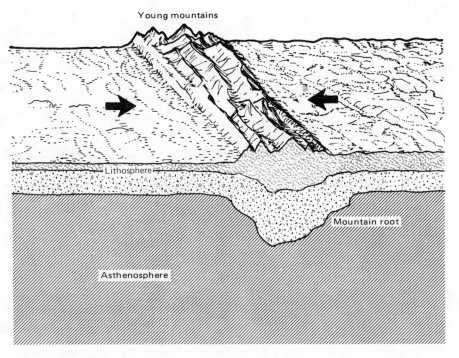

Figure 10.2　Collision of two leading edges of two plates at less than 6 centimeters per year results in both plates buckling and raising a mountain range between them.

destroyed by melting in the asthenosphere. This process creates an oceanic trench such as the ones formed in the western Pacific and along the west coast of South America. The plunging oceanic crustal plate is thought to initiate the generation and intrusion of magma and volcanic activity in the adjoining continental crust. The high temperatures in the crust may produce a doming effect, which may take the form of a volcanic mountain chain such as the Andes in South America, or an island arc such as the Aleutian Islands. Thrusting and crustal thickening may occur, and the result would be extensive metamorphism of the crustal rock (fig. 10.3).

Rifting, responsible for sea-floor spreading (described in section 8.5), occurs mainly along the mid-ocean ridges, which are, in effect, mountain ranges along the ocean floor. The hot, low-density material being extruded is buoyed up by denser, cooler material to form a high ridge. As the sea floor spreads and the rock cools, it becomes denser and sinks into the crust to form the lower elevation flanks of the mid-ocean ridge.

Considerable volcanic and earthquake activity is associated with the kinds of movement described above. The border of the Pacific Ocean is particularly active and is called the "ring of fire" due to the numerous volcanoes located there. Earthquake and volcanic activity is also prevalent along the mid-Atlantic Ridge and in the Mediterranean area. The earthquake activity is generally

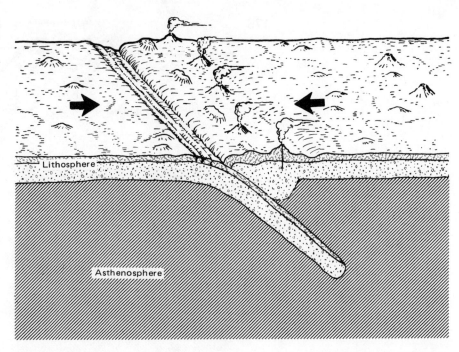

Figure 10.3 Collision of two plates when the combined speeds of the plates is greater than 6 centimeters per year causes one plate to plunge under the other resulting in the formation of an oceanic trench and volcanic activity.

related to crustal shifting, whereas volcanic occurrences result from the generation of magma where appropriate temperatures and pressure conditions exist in association with crustal movement.

The above discussion suggests a mechanism for the formation of mountain systems formed in the last 200 million years, but what of mountain ranges that predate this period? Needless to say, it may be assumed that the same driving force was responsible for older mountain ranges such as the Appalachian Mountains or the Ural Mountains. The Ural Mountains were located in the interior of Laurasia and may represent evidence of the formation of predrift continents from previously existing continental blocks. Thus, it may be speculated that the continental masses existed in several combinations of form and number during the earth's 4.5 billion-year history, and the existence of ancient mountain systems affords the geologist clues as to the nature of these combinations.

It can be seen that mountain building results from a number of complex processes related to continental drift but for simplicity of discussion we can list and describe three types of mountains:

1. *Volcanic mountains:* formed as a result of the accretion of lava, cinders, and other volcanic products.

Mountain Building

2. *Folded and faulted mountains:* formed as a result of crustal movement causing intensive deformation. In this category may be included complex mountains in which igneous activity, along with folding and faulting, is a part of the process.

3. *Fault block mountains:* formed by blocks of the crust that are bounded by steep faults and are being elevated or depressed with respect to the surrounding region.

10.2 VOLCANISM

The nature of volcanic activity has been a source of wonder to man for thousands of years. According to Greek mythology, Aeolus, controller of the winds, imprisoned winds in a cave beneath volcanic islands where eruptions have been taking place since before the memory of man. Later, Aristotle suggested underground fire was formed by the friction of the air blowing through caverns until it burst into flame and erupted as volcanoes, while others during this period thought the earth's interior was a raging inferno. In the seventeenth century, Nicholas Steno described volcanoes as resulting from fire burning in the earth's interior, causing pressures which forced out ashes and lava from some underground source. Abbé Moro in the early eighteenth century studied the nature of sixteen volcanic islands and concluded that all mountainous islands were volcanic, and since mountains on the mainland were no different than mountains found on islands, they too must be of volcanic origin. Moro also saw sedimentary deposits as the accumulation of volcanic cinders and ash. In the middle of the eighteenth century, the action of volcanoes was ascribed to the burning of coal and petroleum, which turned the mountain into a furnace, melting the rock. A great deal has been learned about volcanic action since that time, and the true mechanism of volcanic activity, while not yet completely clear, is beginning to be understood.

10.2A Volcanoes in Action

The term *volcanism* is used to embrace all aspects of volcanic activity. This includes the generation and movement of magma beneath the earth's crust, as well as the forceful ejection of lava and volcanic debris at the surface (fig. 10.4).

Volcanic activity takes on a variety of forms, the most familiar of these being the eruption of gas, lava, and solid material from the *crater* of a cone-shaped hill or mountain. The cone shape is the result of this activity and is not a prerequisite. The initial stage in the development of a volcanic cone is a *fissure* (crack) or a *vent* (opening) in the ground through which the products of volcanic activity erupt. From the accumulation of these products the cone-shaped mountain is formed.

The formation of Paricutin, a volcanic cone located about 200 miles west of

Figure 10.4 A cinder cone.

Mexico City, is typical of this type of activity. The initial formation was preceded for several weeks by earth tremors, which increased in frequency until on February 19, 1943, several hundred separate earthquakes were recorded. On February 20, Dionisio Pulido heard loud explosions while working on his farm and noted cracks in the ground through which sulfurous vapor and dust were being ejected. This activity increased in violence so that by the next day, a cone-shaped mount approximately 10 meters high had been formed of fine dust and ash particles. Day by day, the activity increased. Incandescent rock was thrown high in the air with greater violence—some of the particles as large as 12 to 15 meters in diameter.

Lava began to flow, covering a large area, one flow becoming more than a mile in width and more than six miles long. This flow overran a town, completely destroying all structures except for the church tower, which now extends above the surrounding lava. In the meantime, great clouds of vapor were given off, including water vapor, carbon dioxide, and sulfur dioxide. It was estimated that Paricutin gave off 16,000 tons of water per day as vapor and steam at the height of its activity. Within three months, the cone was 180 meters, and by the time the volcano had been active for one year, the cone was about 300 meters high.

Activity continued for nine years after the formation of the crevice in Pulido's field, and during this period his farm was destroyed along with many square miles of surrounding territory. At the end of this time, Paricutin stood approximately 425 meters in height. Many small cones similar to Paricutin exist in the same general locale, but only one other has been active during historic times. This is Jorullo, which erupted in the midst of a plantation forty-five miles southwest of Paricutin in 1759. Paricutin is not impressive as volcanoes go, but it is nevertheless of great importance because it is the first volcano in modern times to be observed from its birth.

Mountain Building

Figure 10.5 A lava flow.

The Hawaiian volcanoes, particularly Kilauea, have also been the source of much information on the mechanism of eruptions. These volcanoes, although much larger than Paricutin and well established, are quite active, and their eruptions are relatively nonviolent, making them ideal for study.

10.2B Volcanic Products

Studies on volcanism of the type described permits a detailed examination of surface igneous or extrusive activity. Igneous rock below the surface has resulted from intrusive activity which involved the movement and solidification of magma beneath the surface. It is convenient to discuss extrusive and intrusive volcanic activity as two separate entities, but in fact one form of activity (extrusive) is an extension of the other (intrusive).

Products of extrusive volcanic activity may be simply categorized as solid, liquid, and gas. The more or less solid fragmental rock, sometimes called *pyroclastic rock*, are particles blown out by volcanic explosions and subsequently deposited on the ground. These particles range in size from large *blocks* or *bombs* weighing many tons to very fine dust particles. Blocks are angular masses of hardened lava, while bombs are variously shaped igneous rock formed by partially congealed lava spinning through the air. Particles that are larger than 32 millimeters in diameter are classified as blocks or bombs and are generally found close to the base of the volcano. *Cinders* are smaller volcanic particles ranging from 32 millimeters down to 4 millimeters in diameter. The volcanic *ash* includes all particles less than 4 millimeters in diameter; ash is of worldwide significance, since during major eruptions small ash particles reaching the upper atmosphere completely encircle the earth.

The liquid portion of volcanic products include the molten lava, which varies greatly in composition. The composition will determine the type of igneous rock that will form when the lava solidifies.

Gases released by volcanic activity are an important part of the eruptive products, with water, in the form of steam, as perhaps the dominant fraction. From 60 to 95 percent of the gases released from typical volcanoes has been identified as water. Carbon dioxide is usually the second most abundant gas, with sulfur dioxide, ammonium chloride, hydrogen sulfide, nitrogen, hydrogen, chlorine, carbon monoxide, argon, fluorine, and others making up the balance.

Gases involved in igneous activity are called the volatile components, and are released as the magma approaches the surface and begins to crystallize. These gases are considered to be, in part, a source of the atmosphere. Water emitted by volcanic activity may be from several sources, one being from the magma itself. Estimates vary as to the actual water content of the magma, but a consensus appears to center around 2 percent. This water is designated *juvenile water* and is water that has never been a part of the hydrologic cycle. Another source of water from volcanic activity is *connate water*, or water that was trapped in the pore spaces of a sedimentary rock at the time the rock was formed. Hot volcanic material passing through these sedimentary layers will cause the release of this water, permitting it to reach the surface again. A third source of water is *ground water* (see Chapter 11), or water found in the zone of saturation close to the surface. This water may also be vaporized as hot magma passes this zone to reach the surface.

10.2C Types of Volcanoes

Volcanoes are classified in a variety of ways or groupings. One such system categorizes volcanoes as explosive eruptions, intermediate eruptions, and quiet eruptions. There is no clear-cut distinction between the groups, the differences being arbitrarily drawn. However, the classification system is useful, and a few examples in each group will serve to illustrate the nature of the different types.

Some magmas contain more gas than others, and the amount of this gas has a direct bearing on the type of eruption that occurs. As magma approaches the earth's surface, pressure within the mass is reduced and gas is released from the magma. If no outlet to the surface exists, the gases accumulate and build up pressure until an opening or vent is formed, in some instances with considerable energy.

Explosive eruptions occur in somewhat this manner, although the flow of seawater into the vent of erupting island volcanoes is sometimes an external contributing factor to the degree of violence of the eruption. Perhaps the most spectacular volcanic explosion ever to have occurred within recorded history took place in 1883 when the island of Krakatoa blew up. Krakatoa was a volcanic island in the Straits of Sunda between Java and Sumatra. On August 26, the mountain began to erupt most violently, an event witnessed by the crew of the *Charles Bal*, a British ship which passed within 15 kilometers of the island. On the following day, four tremendous explosions occurred, resulting in the disappearance of the 800-meter mountain; only three tiny islands, fragments of the original island, remained to mark the spot. Although almost

20 cubic kilometers of material was erupted into the atmosphere, very little rock comprising the original island was found. However, great quantities of pumice was observed on the surrounding islands and floating on the ocean. Pumice is a frothy igneous rock of cellular structure which therefore has a sufficiently low density to permit it to float until it becomes water saturated. The pumice was ejected by the explosion, and it is believed that the cone of the volcano collapsed within itself. The noise of the explosions was heard as far as 4,800 kilometers away. The shock wave on the atmosphere was recorded on barometers around the world, and in this way it was shown to have traveled around the world seven times before it was too weak to record. A catastrophic sea wave caused by the explosion approached the coasts of Sumatra and Java within a short time after the eruption. The wave, surging inland and cresting at a height of about 35 meters, was responsible for drowning 36,000 people. Ash from this volcano circled the earth in the atmosphere for more than two years before it completely settled. Activity in the area indicates that Krakatoa is now actively rebuilding an underwater cone.

The collapse of a volcanic cone such as Krakatoa results in a structure called a *caldera*, of which Crater Lake in Oregon is an almost perfect example. The caldera has a circular and generally larger-than-normal crater with a diameter four to five times as great as the depth. The crater itself is about 1,200 meters deep and about 9 kilometers in diameter. The lake, which has no outlet, is approximately 600 meters deep, resulting from the accumulation of rain and melting show. Within the lake, Wizard Island is a small cone built up in the caldera subsequent to its formation.

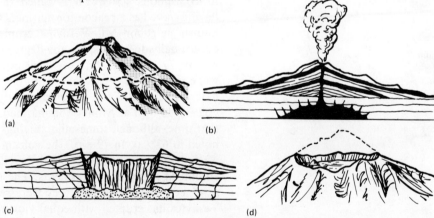

Figure 10.6 Possible means of development of Crater Lake caldera. a) Mountain before collapse, b) cross section of mountain during active period, c) collapse of mountain to form a caldera, d) present appearance.

The Crater Lake caldera was formed upon the destruction 5,000 to 10,000 years ago of a 3,500-meter volcanic mountain called Mt. Mazama. The portion of the mountain that has disappeared represented about 70 cubic kilometers of material, although only about five cubic kilometers of debris is found in the surrounding area. The balance possibly collapsed into the cavity left by the flowing lava to form the caldera as it appears today. There is not complete agreement on this fact, as some geologists maintain that Mt. Mazama blew up, scattering debris over many square miles of the surrounding landscape.

The *nuée ardente*, or *fiery cloud*, is a special type of explosive eruption and is typified by the event which took place on Mt. Pelée on the island of Martinique in 1902. On May 4, the usually smoldering volcano began to erupt more violently, terminating in a blinding explosion on the morning of May 8. Hot gases had built up a tremendous pressure against the central plug in the volcano, until finally the gases succeeded in breaking through the side of the cone. A great explosion produced a huge cloud of highly incandescent dust and gas at temperatures up to 850° C. The cloud, being heavier than air, rolled down the slope at high speed, engulfing the city and harbor of St. Pierre about five miles away. The entire population of more than 25,000 people, plus a great many refugees from the surrounding area who had been frightened by the increased volcanic activity of the previous few days, were annihilated. Little or no lava flowed during the early violent phase of this eruption, although following the explosion a *volcanic dome* formed a cap over the vent by the flow of a viscous lava.

Of a less violent nature are such volcanoes as Vesuvius and Stromboli, which are classed as volcanoes of intermediate violence. Although classed together, the two differ in activity as much as the activity of Krakatoa differed from that of Mt. Pelée. Stromboli sputters and erupts mildly in an almost continuous fashion, having performed in this manner for hundreds of years. Because of its continuous activity, it is called the "lighthouse of the Mediterranean" having served as a beacon for mariners for many centuries. Stromboli's better-known neighbor Mt. Vesuvius erupts at intermittent intervals, and only occasionally does it display any degree of violence. The first recorded eruption was in 79 A.D., and this was well documented by Pliny the Younger who, while relating the events that led to the death of his uncle Pliny the Elder, also described much of the volcanic activity. Mt. Vesuvius, formed in an extinct volcano known as Monte Somma, had not been noted for eruptions prior to that time, although some mild stirrings, along with earthquake activity, were noted in 63 A.D. In 79 A.D., the volcano erupted, sending out huge quantities of cinders and ash which buried the cities of Herculaneum and Pompeii and entombed many of the residents. The two cities remained hidden for nearly seventeen centuries before they were rediscovered. Since the initial eruption, Vesuvius has erupted with equal violence about ten times up until the seventeenth century, when it began its almost continuous mild activity with an occasional violent eruption.

As previously stated, the explosive activity of volcanoes is dependent largely on the amount of gas in the magma, and if this amount is small and an easy outlet is provided, the eruptions are of a nonviolent type. These quiet eruptions result in the buildup of gently sloping volcanic cones (shield volcanoes) of the type which formed the Hawaiian Islands. The Hawaiian chain of islands extend northwestward for 2,500 kilometers from the island of Hawaii, the

Mountain Building

largest one in the group. These islands are of volcanic origin, rising almost 9,000 meters from the floor of the Pacific. The lava that forms the islands is exceptionally fluid, flowing readily and spreading thinly over the surface. In most cases, the slope is quite gentle, ranging from less than 10° to almost level.

On continental areas, fluid lava flowing from fissures can cover hundreds of square miles with a flat layer of basalt called *plateau basalts*. No cones are formed, although the flows may be several thousand feet thick, filling valleys and covering ridges, and generally forming a flat plateau. In the northwestern United States, such a series of flows covers approximately 500,000 square kilometers and extends into five states. Exposures in canyons cut by the Columbia River show thicknesses up to 1,200 meters, built up gradually by individual flows of 4 to 6 meters in thickness. Extensive study of the area indicates that there are in fact two separate periods during which flows occurred. The Columbia plateau is the result of a number of flows occurring 25 to 30 million years ago, while the flow covering northern California and Nevada and extending into Idaho is of more recent origin, having been formed ap-

Figure 10.7 Small volcanic intrusions south of Monument Valley. Dike in foreground seems to parallel joints in sandstone. By John S. Shelton.

proximately 1 million years ago. Other fissure flows of this type may also be found in India and Iceland.

At the present time, there are about 500 active volcanoes on the earth, and many others recognized as dormant. Most of the activity is concentrated around the border of the Pacific Ocean, earning it the title "ring of fire." Other volcanic areas include the Indonesia-Malaysia region and the Mediterranean coast. In addition to these, many extinct volcanoes and evidence of volcanic activity are found in regions now free from volcanism, indicating that volcanic activity occurred on parts of the earth where there is no activity today.

10.2D Igneous Intrusive Activity

Much of the evidence of ancient volcanism is of intrusive activity which has been exposed by crustal uplift and erosion. One such intrusive mass that occurs on the surface is the volcanic *plug* or *neck*. These necks are the remnants of volcanoes which remain after the less-resistant rock of the cone has been eroded. Generally, these necks are 150 to 200 meters in diameter but may exceed 1 kilometer. The rock in a neck is frequently of a coarser texture than that which forms from lava, since it has cooled more gradually. Occasionally the rock may be of a glassy texture, particularly in the upper portions of the neck; or the rock may be a breccia resulting from the shattering of the surrounding rock mixing with lava when the volcano was active.

Radiating from the neck may be a series of wall-like structures called *dikes*, although dikes may form wherever magma intrudes into cracks resulting from the stress produced by the intruding magma. Dikes are tabular masses of rock which are *discordant*, meaning that these intrusive masses cut across the rock strata as the magma intrudes into fissures or cracks in the earth's crust. Dikes vary in width from less than an inch to several miles in thickness and may extend for miles. A noteworthy example is the Great Dike of Rhodesia in Africa which has an average width of 9 kilometers and has been traced for over 500 kilometers.

Sills are also tabular masses of rock, but they are *concordant*—that is, they are sheetlike bodies of rock formed from magma which intruded between and parallel to the strata of surrounding rock. Most sills are connected to and supplied with magma by dikes. Some sills cover only a few acres while others cover many square miles and range in thickness from a few centimeters up to a thousand meters though most are less than 30 meters thick.

Other igneous intrusive masses not tabular in form are referred to as *massive plutons*. If a large mass of magma intrudes between sedimentary beds (as when a sill is formed) and lifts the overlying beds into a domelike structure, then the solidified magma forms a type of pluton known as a *laccolith*. The mass is supplied from beneath or from the side, and it is implied that the solidified magmatic mass has a flat base resting on sedimentary rock or other pre-existing rock.

185

Mountain Building

The largest igneous intrusive rock or pluton is the *batholith*, which is believed to be the source of supply for dikes, sills, laccoliths, and volcanoes at higher levels. Batholiths are aligned with mountain systems and may be hundreds of miles long and many miles deep. For example, the Sierra Nevada batholith forming the Sierra Nevada Mountains in California is 650 kilometers long and from 65 to 110 kilometers wide. Batholiths differ from laccoliths in that the batholiths have no known bottom and are thought to extend to great depths. Similar to batholiths but smaller in size is the *stock*. Usually a stock is an intrusive body of unknown depth but less than 100 square kilometers in area.

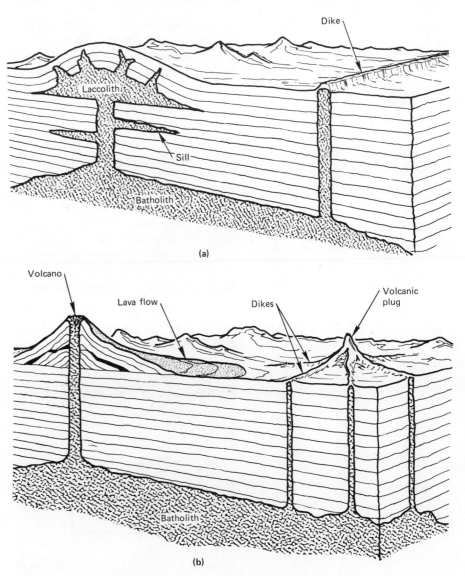

Figure 10.8 Relationship of intrusive (a) and extrusive (b) igneous masses.

The presence of batholiths close to the surface indicates the existence of magma bodies within the crust of the earth. The conclusion should not be drawn that the earth is molten at a shallow depth, for this is not the case. The earth is essentially solid down to a depth of approximately 3,000 kilometers or to the outer surface of the core. However, localized masses of magma do occur close to the surface (within 100 kilometers) and are the source of all volcanic activity.

10.3 FOLDED AND FAULTED MOUNTAINS

Before discussing the formation of folded and faulted mountains it is necessary to describe the deformation structures that form mountains as a result of crustal movement. These crustal movements are caused by continental drift or by the ever-present force of gravity acting upon the crust and subjecting the rock to stresses which tend to deform it. The resulting structures from these activities can be classed as (1) folds, and (2) faults.

10.3A Folding

Folds are simply wrinkles or bends that are produced in rock under great pressure due to compression. Folds may be a few centimeters across, so that they may readily be seen in some rock specimens, or they may range for a number of meters, as can be seen in roadcuts made through sedimentary rock. Or folds may range for many kilometers, as in some mountain systems. In the latter case it would be necessary to trace the pattern of the rock for an extended distance in order to determine the exact form of the folds. *Anticlines* are folds in which the earth's crust is up-arched and a *syncline* is the downfolded segment of the crust where a trough is formed. This would lead one to suspect that the anticline forms the top of a mountain ridge and the syncline the bottom of a

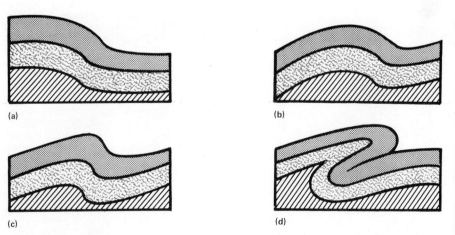

Figure 10.9 Types of folds: a) monocline, b) symmetrical syncline and anticline, c) asymmetrical syncline and anticline, d) recumbent fold.

valley, but this is not always the case. In the Appalachian Mountains, some of the ridges occur in the synclinal portion of the fold and the valleys on the anticline as a result of the erosion pattern. A third type of fold is the *monocline*, which is the offsetting of rock strata by a steplike bend. The inclined beds of a monocline, or one limb of an extensive fold whose relationship to other beds is not determined, are sometimes identified as a *homocline*. In other cases, the folds are more complex, being tilted or even overturned, but the terms anticline and syncline will still apply.

Folds may also occur as a roughly symmetrical upfold, forming extensive *domes* in which the beds dip generally outward in all directions from a central high point. Circular domes are far less common than elongated domes, which grade in form to anticlinal structures (fig. 10.10). Generally, domes are found in association with other folded structures, but occasionally domes may be isolated in regions of normally level topography. A single such oval dome formed the Black Hills of South Dakota, which owes its rugged appearance to erosion. The erosional process has been useful in exposing the internal structure of the dome for examination, thus permitting geologists to determine the nature of the deformation.

Opposite to a dome is a *basin*, which may form as a structural depression by the downwarping of the earth's crust. In this instance the strata dip from the rim downward toward the low center of the basin. Care must be exercised in

Figure 10.10 The Black Hills of South Dakota were formed by the erosion of a dome formation.

the use of the term basin, as it also applies to erosional basins. The two types of basins may have a similar surface appearance, but the structural features will differ. An erosional basin will not show strata generally dipping toward the center of the depression.

Another structural feature related to folding are *joints* (fig. 10.11). Joints are features in the rock resulting from tensional stresses set up by the gradual crustal movements. The joints are generally vertical or inclined to the bedding and display little or no movement parallel to the plane of the joint. Joints may also be formed by the contraction of cooling lava. The columns, as exemplified by basalt flows that formed the Devil's Postpile, are typical of this type of *columnar jointing*.

Figure 10.11 Devil's Postpile shows columnar jointing of basalt flows.

10.3B Faulting

Mountain Building

The movements of the earth's crust that create folds are usually slow and gradual, and rock masses can adjust to the compression without significant rupturing. In other cases, movement is such that the rock mass ruptures, causing the displacement of the separate segments of the crust with respect to each other. Fractures in the crust, along which this type of displacement has occurred, are known as *faults* (fig. 10.12). In classifying faults, it has been

Figure 10.12 View of fault zone in Dixie Valley, Nevada.

found convenient to use mining terms to describe the blocks on either side of the fracture or fault. The miner saw the overhead portion of the tunnel in which he worked as the *hanging wall*, and thus the block above the fracture of an inclined fault became known as the hanging wall. The portion upon which the miner walked was called the *foot wall*, and this same term was applied to the block beneath the fracture (fig. 10.13).

Geology

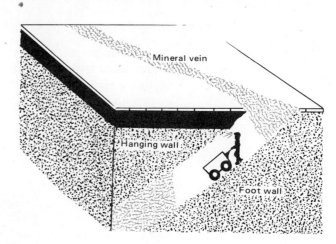

Figure 10.13 Relationship of foot wall and hanging wall are displayed in this diagram of a miner's tunnel.

Inclined faults are then classified according to the manner in which the hanging wall moves with respect to the foot wall. A *normal fault* or *gravity fault* is one in which the hanging wall has moved downward relative to the foot wall (fig. 10.14). This does not imply that only the hanging wall has moved, since the same relationship would exist if the foot wall were moved upward. A *thrust fault* or *reverse fault* is one in which the hanging wall has moved upward with respect to the foot wall, and again either block may move so long as the appropriate relationship is maintained. Displacement along an inclined fault results in the exposure of a surface called a *fault scarp*. Scarps may show a displacement ranging from just a few feet to hundreds of feet, resulting from the accumulated movement of one side or the other along the fault (see fig. 10.14B).

Horizontal or lateral movement also occurs along a fault, as in the case of the San Andreas fault. In this instance, the block west of the fault is moving in a northwesterly direction relative to the area to the east of the fault. Such a fault is known as a *strike-slip* fault.

10.3C Folded Mountains

Most of the great mountain systems of the world are primarily the result of folding and faulting activity, although some igneous intrusive activity and volcanic activity may contribute to a small degree. Usually the folded and faulted mountains are composed of thick layers of sedimentary and metasedimentary rock and have deep roots. There is a rather complex geological history in the development of such mountains which follows a fairly well-defined sequence of events. This sequence may be described by relating the events that resulted in the formation of the Appalachian Mountains.

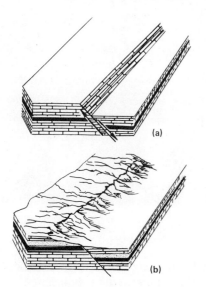

Figure 10.14 Diagram shows: a) The downward movement of the hanging wall of a normal fault, b) scarp as it appears after erosion.

Mountain Building

Initially, an elongated trough or geosyncline, approximately 3,000 kilometers long and 500 kilometers wide, began to form in what is now the eastern part of the United States. The exact mechanism for the development of the geosyncline is not known for certain, but one explanation suggests that the Atlantic Ocean has opened, closed, and reopened. In the Precambrian era (prior to 600 million years ago) the proto-Atlantic existed between what is now Africa and North America. This was followed by a period during which the ancestral Atlantic began to close and in the process formed the trough or geosyncline 400 to 500 million years ago. While downwarping occurred, sediments from higher elevations adjacent to the trough were being deposited in it and eventually built up thick sedimentary deposits 10,000 to 12,000 meters thick in the trough. The accumulation was gradual as the downwarping continued and usually took place in shallow marine waters which alternately invaded and retreated from the trough during its quarter-billion-year existence. At times a portion of the geosyncline was a swamplike area, and it was under these conditions that the coal beds of Pennsylvania and West Virginia were laid down. On these occasions, the surface of the geosyncline was above sea level and the marshland as well as the broad deltalike areas existed in a nonmarine environment. The slow subsidence was occasionally interrupted by some crustal deformation and volcanic activity.

Subsidence and deposition finally ceased near the end of the Paleozoic era (see table 12.1). Folding and faulting began due to great compressional forces resulting from Africa and North America colliding. This caused the deformation of the thick sedimentary rock deposits into large folds. Further compression resulted in rock breaking and thrust faulting, in turn resulting in anticlines becoming overturned and large thrust masses overriding the underlying beds. By this activity, great folds and thrust masses, one on top of another, were moved many miles. Such movement resulted in older rock lying on top of younger rock. Careful examination of the resulting layering enabled geologists to interpret the sequence of events.

The folding and faulting was followed by several hundred million years of quiescence, during which widespread erosion reduced the mountain system to a broad, featureless plain. During the Cenozoic era, upwarping reoccurred, elevating the Appalachians to their present levels of approximately 1,200 meters. Very little mountain building is now noted, and the Appalachians are again subject to the erosional processes which will again ultimately reduce the mountains to a featureless plain.

This cycle of mountain building and erosion is typical of the events occurring in the history of folded and faulted mountain systems. From this description, it is possible to see that the sequence, in simple terms, involves downwarping in the formation of the geosyncline and the deposition of sediments therein, a period of crustal deformation in which there is widespread folding and faulting, and an extended period of quiescence during which extensive erosion occurs, reducing the mountains to a plain.

10.3D Epeirogeny

In addition to mountain-building activity, there are also *epeirogenic* movements in which land masses of continental magnitude are raised or lowered with little or no evidence of folding or faulting. A classic example of vertical movement of this type may be seen in the marine sedimentary rocks that border the Grand Canyon at an elevation of 2,100 meters. These deposits have been raised to this elevation in an almost horizontal attitude just as they were laid down beneath the sea. Marine terraces sometimes observed along coastlines are evidence of the type of uplift seen, for example, in Palos Verdes, California (fig. 10.15). Thirteen terraces have been traced in this area that show the combined effect of beach erosion followed by uplift. On the eastern coast of the United States, particularly in the New England area, drowned valleys and numerous small islands indicate that slow subsidence is taking place without folding or faulting. It must also be recognized that the slow rise in sea level contributes somewhat to the submergence of these areas. Epeirogenic movement may be slow, taking millions of years, or it may be rapid enough to have taken place during recorded history. Such an example is provided by a Roman ruin near Naples, Italy. The structure was presumably built by early Romans on dry land but holes bored by *Lithophagus*, a sea mollusk, in columns 5 meters above the floor of the structure indicate submergence and subsequent uplift have all taken place during recorded history.

10.4 EARTHQUAKES

Movement of the crust in the process of mountain building results in tremors which are called *earthquakes*. They have been feared by man since the beginning of his time on earth, and for good reason: earthquakes are considered to be the most destructive of the natural phenomena and are often more terrifying than volcanoes.

The study of earthquakes in the modern sense, *seismology*, originated only a little over a hundred years ago when geologists first realized that earthquakes were the result of abrupt movements of large blocks of the earth's crust. Prior to this, many divergent views were held as the cause of earthquakes. Classical philosophers considered earthquakes to be the result of "elemental forces," with the elements themselves (fire, air, water, and earth) responsible for the action. According to Thales (ca. 600 B.C.) the movement and wave action of the sea was the cause of earthquakes. He came to this conclusion when he viewed the devastation brought about by waves during great storms. Anaxagorus (ca. 450 B.C.) believed that fire within the earth, rising rapidly to the surface and bursting through with great violence, caused the shocks. Anaximenes (ca. 550 B.C.) attributed the cause to great masses of rock collapsing within the earth, while Aristotle thought they were due to air trapped and compressed beneath the surface and forcing an exit.

Mountain Building

Figure 10.15 Marine terraces reveal uplift of coastal area of Palos Verdes, California.

Modifications of these ideas were still held in the middle of the eighteenth century. At this time, John Michell, an English clergyman, wrote about earthquakes, particularly about the great earthquakes of Lisbon in 1755. He held to the old idea that fires within the earth caused earthquakes but put forth the new concept that the motion of earthquakes was due to the movement of elastic waves in the crust and that these waves moved outward from their source and gradually faded away. This was the first suggestion that the earthquake tremor was caused by movement of large blocks of the crust either horizontally or vertically along a fault.

The San Andreas fault in California was recognized as a zone of such movement some time before the San Francisco earthquake of 1906. Surveys of the fault starting in 1874 showed that the eastern block was moving southeastward and the western block was moving northwestward. Tremendous stresses were set up which overcame the friction between the two blocks on April 18, 1906, resulting in the San Francisco earthquake. Maximum horizontal displacement in this event was 7 meters, while vertical displacement was less than 1 meter.

The origin of earthquakes has been established as being caused by movement of crustal rock, a movement that can be related to continental drift. The vast majority of earthquakes occur along the mid-ocean ridges where sea-floor spreading occurs or along the trenches where the sea floor turns down. Shallow earthquakes (within 70 kilometers of the surface) appear to be associated with sea-floor spreading activity along the mid-ocean ridges. The axis of the mid-ocean ridge is offset at many points by transcurrent fracture zones. Thus, between two axes of offset crustal material the crust moves in opposite directions. Where such movement takes place a shallow earthquake results.

Intermediate-depth (70 to 300 km) and deep-focus earthquakes (deeper than 300 km) coincide with the approximate location of ocean trenches where the crustal plates turn down beneath the land masses. These earthquakes appear to occur on a plane which dips at a 45° angle from the floor of the trench downward beneath the continental mass. Movement of the ocean crust as it slips beneath the continent along this plane is thought to be the cause of earthquake tremors. The plates may also move horizontally past one another along such great faults as the San Andreas in California.

10.4A Earthquake Tremors

The abrupt movement along a fault sets up strong vibrations in the bedrock. These vibrations are heard as deep rumbling in the early stages of an earthquake and take the form of waves passing through the rock. There are three types of earthquake waves recognized. The first of these has a motion similar to the motion caused by a vibrating tuning fork on the surrounding air or the motion of water if the tuning fork is immersed in water. These waves move most rapidly through rock, and because of their high velocity (5.5 to 13.5 kilometers per second), are the first manifestations of the shock to reach a distant point. They are called longitudinal or primary (P) waves. The second type of wave motion caused by the shock is an oscillating type of motion in which the rock particles oscillate at right angles to the direction of wave propagation, much like that of a rope shaken up and down when tied at one end. This transverse wave or secondary (S) wave travels slower (4.0 to 7.0 kilometers per second) than the P waves, but it has a greater amplitude of motion. These waves, also called shake waves, travel only through solids, and because they do not travel through the earth's core they have provided the principal evidence for concluding that the earth's core is in a fluid state.

The P and S waves differ from the third type of wave motion in that P and S waves originate at the center of maximum earthquake activity many miles

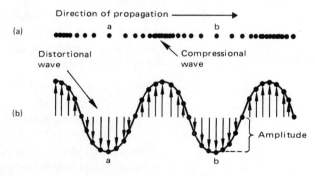

Figure 10.16 The primary wave (a) generated by earthquake activity is a longitudinal wave whereas the secondary earthquake wave (b) is a transverse wave.

Mountain Building

below the surface. This point of origin is not precisely a point but rather a limited area or zone on a fault surface called the *focus*. The long waves, or L waves, are generated by the arrival of the P and S waves at the earth's surface or *epicenter*, a point directly above the focus. The L waves travel along the surface, have the slowest velocity (4.0 kilometers per second), but are responsible for the major portion of the structural damage wrought by earthquake activity.

10.4B Seismographs

The occurrence of earthquake waves can be recorded on instruments called seismographs. The recording will appear as a series of wavy lines showing first the arrival of the P waves, followed by the more intense vibration of the S waves, and finally by the L waves which have the greatest amplitude. A critical part of the seismogram, as the recording is called, is the interval between arrival of the P waves and the S waves. This interval is used to determine the distance from the epicenter to the recording station. Once the distance is found, a circle drawn on a globe with the distance as radius yields a circle, the circumference of which will fall on the epicenter of the earthquake shock.

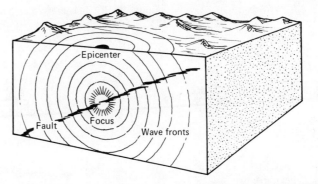

Figure 10.17 The focus is the zone within the crust where rock slippage and shock waves originate. The epicenter is a point on the earth's surface directly over the focus.

Unfortunately, the seismogram does not yield direction, so it is necessary for at least three recording stations to contact each other for distances. The resulting circles will all intersect at the epicenter. Modern communication methods provide rapid pinpointing of earthquake disaster areas and the location of earthquake activity in uninhabited regions or beneath the sea.

10.4C Earthquake Intensity

Measuring the intensity of an earthquake is difficult. There is a tendency to ascribe a higher degree of activity to an earthquake in a populated area where damage is easily assessable than in an uninhabited area. One of the earlier organized attempts to develop a scale resulted in the Rossi-Forel scale of

Geology

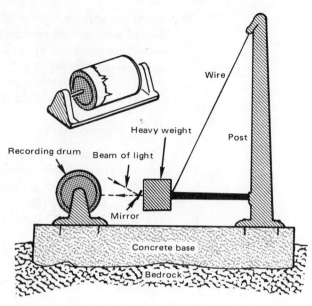

Figure 10.18 Schematic diagram of the horizontal unit of a seismograph.

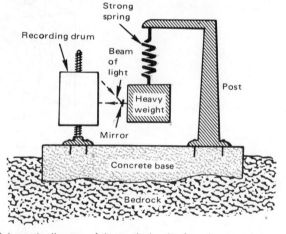

Figure 10.19 Schematic diagram of the vertical unit of a seismograph.

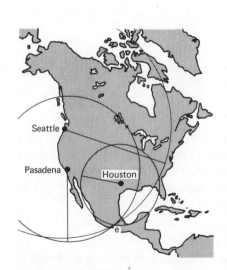

Figure 10.20 Pinpointing the location of an earthquake requires information from at least three seismic stations.

earthquake intensity, in which the activity was categorized into 10 classes, with Roman numeral I the least damaging and X the most intense. This scale, developed about 1875, was used for over 50 years, and intensities were based on surveys made after earthquakes occurred. Thus, on the basis of dishes rattling, chimneys toppling, and plaster falling, isoseismic lines (lines connecting points of equal intensities) were drawn on a map to indicate the zone in which the earthquake activity occurred.

A similar scale, the Mercalli scale, modified for use in the United States in 1931, was subdivided into 12 intensity classes similar to those of the Rossi-

Forel scale. Both of these scales were based on subjective information from correspondence and personal interviews, no doubt colored by the individual's tendency to exaggerate, and were therefore not entirely satisfactory.

In 1935, C. F. Richter developed a standardized scale based on seismogram recordings which has proved to be superior to the Mercalli or similar scales. The magnitude is based on a recording of earthquake activity at 100 kilometers (about 62 miles) from the epicenter. Each unit on the scale represents about 60 times the energy of the unit preceding it. Thus an earthquake of magnitude 7 will release 60 times the energy of an earthquake of magnitude 6, or more than 210,000 times the energy of one of magnitude 4.

The study of earthquakes in the last 65 years has revealed that approximately 80% occur in a zone bordering the Pacific Ocean and 15% in the region extending from the Alps to the Himalyan Mountains. The balance are distributed through the mid-Atlantic and mid-Indian ocean ridges and in the Pacific islands. C. F. Richter has suggested that there may be as many as a million earthquakes each year but that most of these are too weak to be felt. In a study made by Gutenberg and Richter, they found that between 1904 and 1946, an average of two strong earthquakes with a magnitude of greater than 7.7 occurred each year, and that 120 potentially destructive earthquakes (M. = 6.0 to 7.7) occurred annually. In addition, slightly over 150,000 earthquakes were estimated to have occurred having a magnitude from 2.5, which is just strong enough to be felt, to a magnitude of 6.0.

The prediction of earthquakes has been the concern of seismologists for some years, but as yet no satisfactory solution to this problem has been reached. Stresses along active faults can be measured, making it possible to foretell that an earthquake will occur in the future, but to predict the exact moment is as yet impossible. Many people believe that weather is a contributing factor and that earthquakes occur under certain meteorological conditions commonly known as "earthquake weather." Studies of this phenomenon have revealed that earthquakes have occurred under all types of weather conditions, from hot and humid to cold and crisp, and no correlation between weather and earthquakes exists. It has been suggested that changes in barometric pressure may influence crustal movement, but this also has not been substantiated.

10.4D Tsunamis

Periodically, giant "tidal waves" strike the shores around the world, creating havoc as a wall of water rushes over the land. These tidal waves, or tsunamis, are not related to tidal action caused by the sun and moon, but rather they are the result of submarine earthquakes where the displacement of the crust generates energy for the formation of the tsunamis. The dimensions of a tsunami are tremendous compared to an ordinary sea wave. Waves created by wind, even under stormy conditions, are less than 300 meters from crest to crest. A tsunami may be 800 to 1,000 kilometers from crest to crest and travel across the water at hundreds of miles per hour. This represents tremendous power; when it reaches the shore, it may be a wave 3 to 30 meters high, though one 70-meter wave was recorded in Kamchatka in 1737.

The first indication on the shore of an approaching tsunami is a sharp swell, followed by a tremendous withdrawal of water from the shore, exposing a large expanse of beach normally not seen at even the lowest tide. This is followed by the onrush of a series of giant waves separated by intervals of fifteen minutes to an hour. The third to the eighth waves are usually the largest.

A disastrous tsunami in the Hawaiian Islands in 1946 prompted the development of a warning system throughout the Pacific with its center in Honolulu. The center receives information about earthquakes and possible tsunamis, and it is able to estimate the rate of approach and time of arrival of the wave at the Hawaiian Islands. Although a number of tsunamis have occurred in the Pacific since 1946, the warning system has prevented the loss of life in Hawaii.

10.5 MOUNTAIN BUILDING AND SOCIETY

Does volcanic activity have an influence upon society? Certainly it is easy to see the adverse effect on human activity when we consider Dionisio Pulido's plight upon the formation of Paracutin, or the consequence of the eruption of Mt. Vesuvius on Pompeii. Loss of life and the destruction of property can be the fate of people who live near active volcanoes. But there is also a positive aspect to volcanic activity. While lava flows are initially barren for long periods after their formation, the weathering process eventually breaks down the volcanic rock into highly fertile soil. Volcanic ash, on the other hand, may be utilized for agricultural purposes shortly after an eruption. Land covered by volcanic ash in Hawaii quickly came into production again, usually within a few years after the eruption had deposited the ash. Volcanic islands also benefit from volcanic activity, in that it is a means by which the islands are enlarged—in some cases significantly so. Volcanoes as such add little or nothing to the mineral wealth of the earth, except for diamonds formed in kimberlite rock found in the pipes of ancient volcanoes in South Africa. Intrusive activity (see Chapter 9), on the other hand, does create conditions whereby some ores are concentrated to a degree where it becomes economically feasible to mine them.

The existence of mountain ranges also has its influence on society. Planning the right-of-ways for railroads and roadways is much more difficult and the building more expensive in mountainous areas than on the plains. Man seeks level areas where it is easy to build. Never has he developed areas of large population densities in mountainous terrain. Mountain systems influence the weather by blocking weather fronts, causing one side of a mountain range to be humid while the other side is arid. This also influences man's decision in deciding where to settle.

Mountain systems do, however, provide conditions for making available many of the materials useful to society. Folded regions indicate great stress, and the pressures thus generated have changed limestone to marble, shale to slate, and bituminous (soft) coal into anthracite (hard) coal, all useful resources to man. Fault planes are zones of considerable disturbance that furnish an environment where ore-forming solutions can accumulate in the shattered rock to give rise to ore deposits. Water from underground sources may be blocked by the shifting of the earth at fault zones and rise to the surface to become available to man. Petroleum becomes trapped in the porous rock along fault zones and may be more easily accessible than in the rock in which it was originally formed.

10.6 SUMMARY

For millions of years, the forces of erosion have been operating to reduce the land surface to sea level. That this has not come about is the result of forces within the earth's interior which continue to elevate the land. These forces appear to be responsible for movement of crustal plates which collide or sideswipe each other, resulting in the crustal deformation that produces mountain systems. This activity is responsible for volcanic rock formation and intrusive igneous activity. Extrusive rock results from the eruption of volcanoes and the flow of lava upon the land surface. Intrusive igneous rock masses are due to the solidification of magma beneath the surface, which results in igneous masses that are revealed when the surface material above it is eroded. In some instances, intrusive igneous masses, forcing their way to the surface, elevate the land overhead.

Crustal movement without volcanic activity is evident from the folding and faulting revealed at the earth's surface. Folding is a manifestation of slow crustal movement resulting in the deformation of rock. In some instances, folding is so extensive that it forms mountain systems over a period of millions of years. In this way, the crust is elevated, countering the effects of the erosional processes. Faulting results in uplift of segments of the crust along cracks in the earth's surface. This movement, revealed by earthquake shocks, may result in mountain systems if the movement is vertical. The causes of this type of movement, along with folding and volcanic activity, are thought to be related to sea-floor spreading and continental drift.

QUESTIONS

1. Describe briefly the several different forms of crustal contact related to the formation of mountains.
2. List and in a phrase describe the products of extrusive volcanic activity.
3. How may a caldera be formed?

4. How do a batholith, a laccolith, and a stock differ?
5. How do folding and faulting activities differ?
6. Describe the crustal movement along a normal fault; a thrust fault.
7. Briefly, in outline form, describe what is thought to be the development of the Appalachian Mountains.
8. Earthquake activity is, for convenience, divided into three categories according to depth of focus. What are these and where do they generally occur?
9. Describe the activity of a tsunami.

11 Erosion

Erosion as a counterbalance to mountain building.

Erosion by gravity—types of activity—influence on society.

Erosion by wind—types of activity—influence on society.

Erosion by ice (glaciers)—how glaciers may form—glacial activity—influence on society.

Erosion by water—stream flow—types of activity—stream formation.

Ground water erosion—water in the subsurface rock—underground erosion by water—influence on society.

Erosion by sea wave action.

Countering the building up of the continental masses by volcanic activity and mountain building are the processes of erosion, which slowly and inexorably work to level the face of the earth. The slow dissolving action of water, or the downslope movement of rock particles over long periods of time, are examples of the leveling process which is directed toward moving material from higher elevations toward sea level. Erosional activity, it has been established, occurs at a rate that would cause the earth to be leveled in 25 to 30 million years. That the earth is more than 4.5 billion years old and not yet leveled indicates some degree of balance between mountain-building activity and the erosional processes.

Erosion is accomplished by mass wasting, which is the downslope movement of surface material in response to the force of gravity; by wind, generated by the rotation of the earth and by thermal activity within the atmosphere; by water; and by glacial activity. Erosion by any one of these agents is preceded by weathering of the rock surface, which forms a residual layer of unconsolidated rock and soil. In the erosional process, this weathered material is set in motion, transported from one location to another, and finally deposited.

Each erosional agent accomplishes these actions in its own unique manner and forms erosional landforms unique to the particular agent or agents that shaped them.

Although a small degree of erosion occurs beneath the sea, erosion is primarily thought of as a land-altering process. The erosional process ceases to be effective on land when sea level is reached, and therefore sea level may be considered the principal *base level*. The concept of base level applies mainly to stream water erosion but may be appropriately used where mass wasting, wind, and glacial activity occur as well. Major streams act as temporary base levels for their tributaries, and lakes may act as temporary base levels for streams flowing therein. These features are temporary in the sense that ultimately they will no longer exist in their present form as erosion proceeds to reduce the landscape toward the principal base level.

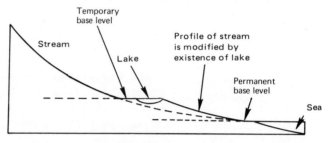

Figure 11.1 Lakes may act as temporary base levels. The sea is the ultimate base level.

Erosion goes on at an extremely slow pace, and while this pace is measurable, the entire cycle of landforms that develop at one site as a result of erosion cannot be observed in a short time because the changes occur too slowly. However, a careful study of the landscape has enabled geologists to piece together the progression of changes that take place and arrange these in a continuous series which differ only slightly one from the next. This progression of change is known as the *cycle of erosion*. This concept permits geologists to determine at any point what form a landscape had in the past and predict what it will look like in the future. It must be understood that this can only be done in a generalized way, since many variables enter into such predictions. These variables include such factors as climatic differences, slope, and variations in the properties of rock and their resistance to erosion. More than one erosional agent may be active in any single region and their activities are interrelated. However, for convenience of discussion each will be dealt with independently.

11.1 THE EROSIONAL WORK OF GRAVITY

Movement of the surface layer of weathered rock, induced by the action of gravity, is called *mass movement* or *mass wasting*. This type of movement is prevalent on slopes and is influenced by the steepness of the slope, the amount

Erosion

of loose surface material available, and the degree to which the material is saturated by water. Water is an important adjunct to the action of gravity on surface material, since water adds weight to the material and acts as a lubricant, lending an impetus to downslope movement. Countering this action is the influence of the roots of plants, which tend to bind the surface material and impede its downslope progress.

The degree of activity prompted by the force of gravity varies from a very slow movement of a small amount of material to very rapid movement of a whole mountainside, including the bedrock. Because of variations in this type of activity, different systems of classifying mass movement have been devised for the purpose of better describing this phenomenon. One system that has proved most adequate divides mass movement into three broad categories: slow flowage, rapid flowage, and sliding.

11.1A Slow Flowage

Slow flowage, or *creep*, as it is sometimes called, refers to the slow and continuous downslope movement of surface material, particularly the soil. This type of movement is generally not evident to the casual observer, but its influence is noticeable over extended periods of time. Tilted fence posts, curved tree trunks and, in some instances, displaced roadbeds and houses give evidence of this downslope movement commonplace in hilly country.

What prompts this downslope movement? Steepness of slope, amount of loose material available, and the presence of water, as previously mentioned, are

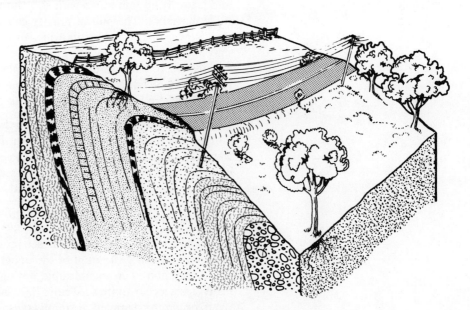

Figure 11.2 The effects of creep on bedrock and surface features are shown.

involved; the movement caused may amount to one to two feet in ten years. The movement is generally greater at the surface than in the subsurface. This can be seen in the curved structure of steeply inclined strata as it appears at the surface on a steep slope. This phenomenon is probably due to the fact that there is less restriction on movement at the surface than in the subsurface, permitting surface material to move downslope at a faster pace.

Several mechanisms have been suggested to account for creep. One is frost heaving (see sect. 9.7A), where particles are forced upward by the expansion of water when freezing, and then when the ice melts, are caused to settle back but slightly downslope due to the force of gravity (fig. 11.3). Similar action occurs in clay-rich soils where the soil expands when wet and contracts when drying, with a resulting gradual downhill movement of the surface material.

Rock creep, included in this category, is the slow downslope movement of individual rocks which have become separated from rock outcrops by frost action and the weathering process. The accumulated pile of rock particles of various sizes at the base of the slope is known as a *talus* (fig. 11.4). The material in the talus itself is subject to creep due to the steepness of the talus slope and the heaving action of ice.

11.1B Rapid Flowage

Rapid flowage takes the form of more rapid movement of surface material, usually occurring on steep slopes and activated by the work of water and ice. Typical of this form of activity is the mudflow often brought about by a sudden thaw or heavy rainfall at high elevations. Water from melting snow and ice or from rain mixes with soil and rock, adding greatly to the weight of these materials. The pull of gravity causes the mixture of mud and water to flow downslope, usually following an existing stream channel. The mixture may be quite fluid, consisting of 25 to 30 percent water, and may move as rapidly as 3 to 5 meters per second, depending on the slope and the amount of material involved. Mudflows are common in the western United States, particularly where little vegetation exists to hold the surface material in place.

Flows similar to this may also occur on the steep slopes of volcanic mountains, moving the loose (unconsolidated) ash and cinders by the action of flowing water. In these instances, it is difficult to separate the action of water from the influence of gravity, indicating that in many instances there is an overlapping activity by these forces in the erosional process.

11.1C Sliding

The most spectacular form of surface movement is the rapid *sliding* of surface material and a large amount of the underlying bedrock. Landslides, although not as common as creep and rapid flowage, are much more destructive and in many cases occur quite suddenly. Side effects of this type of activity include rivers becoming dammed, forming lakes; communities in the path of the slide

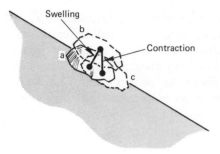

Figure 11.3 Diagram shows how the downslope movement of a particle is accomplished by the creep mechanism.

Erosion

Figure 11.4 A talus slope forms from the rock accumulated at the base of a slope.

being destroyed; and highways and railroads becoming blocked. Slides leave distinct scars on mountain slopes and, in some cases, a rather hummocky landscape where slide material has come to rest.

Landslides occur where the rock strata lies parallel to the steep slope surface. The rock structure may be weakened by earth tremors either from natural causes or by explosions set off in mining and quarrying activities. Water seeping through the strata weakens the rock structure further by weathering, by adding weight to the total mass, and by acting as a lubricant. When the internal

Figure 11.5 A landslide scar on slope above Lake Tahoe, California.

friction of the rock is overcome by the pull of gravity, movement begins, in some cases resulting in large masses of material sliding downslope.

Slides may also occur beneath the sea surface in the form of *turbidity currents* as a result of earthquake activity or other disturbances. Turbidity currents are flows of muddy water which move downward relative to the surrounding water due to the greater density of the mud. Sediments deposited by large rivers flowing into the sea will accumulate near the continental slope. These sediments mixed with water may become dislodged as a turbidity current into one of the submarine canyons. The movement is considered to be quite similar to a landslide or mudflow on land and may attain a very high velocity. Turbidity currents have not been of any significance in human activities in the past. However, the future exploration and exploitation of the sea may reveal turbidity currents to be one of the principal hazards in undersea work.

11.1D   Mass Movement and Society

The action of gravity on the earth's surface is an important erosional factor, but it is not always as obvious as the work of other erosional agents. However, the action of gravity does influence significantly the works of man. Man's

Erosion

activities, in turn, sometimes contribute to conditions which lead to movement of surface material by gravity.

When a landslide occurs in a remote area, its effect is of little note unless the material dams a stream or blocks a highway. Landslides in inhabited regions become events of great importance because of the destruction generally associated with them. Occasionally one reads of slides that bury a village, killing many persons, such as the one that occurred in Salvador, Brazil, in 1967. In this coastal city, 1,200 kilometers northeast of Rio de Janeiro, a row of laborers' houses, plus a crowded bus and several other vehicles were buried by the avalanche.

The 1903 Turtle Mountain slide in Alberta, Canada, is a classic example of a landslide; it is frequently described because of the high velocity involved in the movement of the material. An estimated 30 million cubic meters of material moved a distance of approximately 4 kilometers in slightly under 2 minutes. The avalanche killed 70 people as it roared like a giant wave through the town of Frank in its passage across the valley floor and 100 meters up the opposite slope.

Events such as these two examples may be brought about by natural causes, although the Turtle Mountain incident may have been aggravated by the coal mining operation near the base of the mountain. Major slides sometimes are the direct result of man's own efforts to rearrange the landscape to suit his needs, thereby disrupting the natural balance. Immense slides into the Culebra Cut of the Panama Canal caused the canal to be intermittently closed from 1914, when it was officially opened, until 1920. Almost 50 percent of the material excavated at the Gaillard Cut of the Panama Canal was the result of slides, adding greatly to the expense of construction. Highways are frequently closed by slides where roads are cut through hilly terrain. Elaborate and expensive (to the taxpayer) measures must be taken to reinforce the cut areas and to provide drainage for the removal of excess water, the principal culprit in many slides.

Removing water or preventing it from entering a potential slide site is one of the chief tasks in the effort to prevent landslides in areas where they could be harmful to man's activities. One such effort was made in the Ventura oil fields in Southern California, where a large block of land moved approximately 30 meters and sheared off 23 oil wells in the process. An extensive drainage system was built to remove water from the slide area and thus reduced, at least in part, the danger of further landslides occurring.

The expense of such activities is tremendous and affects the general populace only indirectly, because government agencies and large corporations are involved in the above examples. It must be pointed out, however, that the individual can be seriously affected by rapid flowage or landslide activity. Examples abound but a few will suffice to illustrate.

Prior to World War II, building in the Los Angeles area was for the most part restricted to the flatlands, where construction was easy and landslide problems were minimal. There were a few regions where slides occurred, such as along the Pacific Palisades and at Portuguese Bend on the Palos Verdes Peninsula. These areas periodically performed, enough to remind builders of the potential dangers. However, memories were short, and even an awareness of the possible loss of life and financial ruin did not deter some from building on or near these slide areas. The result was that in 1958, 150 homes were damaged or destroyed in the Portuguese Bend area owing to large-scale earth displacement. This activity was repeated in 1969 and 1970.

During the 1950s most of the flatlands were built up and population pressure plus the aesthetics of hillside homes caused the hills surrounding the Los Angeles basin to be developed at an accelerated pace. Huge earth-moving machines permitted the building of roads and the cutting of pads for houses on a large scale without due regard to the nature of the rock beneath the surface or the effect on the surrounding area, particularly downslope. While some provision for drainage was made, they were in many instances inadequate. The results for some individuals was financial disaster. Homes and swimming pools slid downslope or were filled with mud during the periodic heavy rainstorms that occurred in the area.

To combat this problem, Dr. Awtar Singh, an expert in soil mechanics, is attempting to develop a method for determining the "creep potential" for a

Figure 11.6 Views (above and at right) of mudslides in Glendora, California (1970) showing damage to homes. Courtesy of Glendora Police Dept.

Erosion

Figure 11.6 (Continued)

slope. He foresees the time when it will be possible for a hillside developer to accurately gauge the possible slide danger and take remedial steps to reduce the hazard to a minimum.

Even on relatively stable slopes, creep will have an adverse effect on structures for which a proper foundation has not been fashioned. The effect may not become apparent for a number of years, and then only so gradually that it may go unnoticed for a considerable period. Doors and windows may become difficult to open and close due to slight misalignment of the structure, and after many years greater structural damage becomes evident.

One example is that of Durham Castle, built of stone several hundred years ago on a steep slope in northern England. Over the years, cracks began to appear in the structure, and they eventually became so wide that remedial steps were required to prevent the collapse of portions of the castle. It was found that the foundations rested on loose surface material which was slowly moving downslope. Because this downslope creep was not uniform, the structure was twisted, and great cracks began to appear. Only at considerable expense and difficulty was it possible to save the castle by extending the foundation to greater depths and reinforcing it for greater strength.

Does society learn from these experiences? This is a difficult question to answer. As a result of a rash of sliding incidents, commissions are set up, committees are formed, and recommendations are made but frequently not followed. Geologic reports resulting from these studies are generally on file with zoning boards or building departments but are seldom referred to. In the final analysis, it is up to the individual to be knowledgeable or to seek advice before committing himself to a course of action where earth movement could result in ruin or tragedy.

11.2 WIND EROSION

Aeolian erosion, or more commonly, wind erosion, is more effective as an erosional agent in arid and semiarid regions where there is a comparative lack of vegetation than in humid areas. The dust storms generated in the southwestern United States during the mid 1930s and in Russia in the late 1950s is evidence of this fact. Vegetation had been removed by plowing the land, and when subsequent droughts dried the surface soil, the loose particles could be readily moved by the wind.

The wind-erosion process has three major phases: initiation of movement of surface material and its transportation; deflation and abrasion; and deposition. Each phase is influenced by the force of the wind, the nature of the ground surface, and the physical properties of the surface soil.

11.2A Movement of Surface Material by Wind

The movement of soil begins with the most erodible grains on the most exposed positions of the surface becoming dislodged by the direct force of the wind against the soil particles. The larger particles are moved mainly by a sliding or rolling action along the surface. In a moderate wind (approximately 40 kilometers per hour), this type of motion would apply to particles about one millimeter in diameter. Smaller particles will be impelled forward in a jumping

Erosion

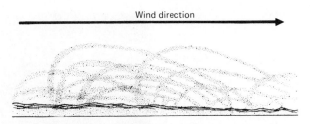

Figure 11.7 Paths of loose sand particles in saltation over surface.

motion known as *saltation*. The height of this type of motion will depend on the size and density of the soil particles, the roughness of the ground surface, and the velocity of the wind. Particles may rise up to several feet off the ground, depending on the initial energy imparted to the particles by the force setting them in motion. As a particle rises and falls through the air, it gains momentum from the pressure of the wind against it and continues to gain velocity until it strikes the ground. Here the particle may rebound and continue its saltating movement, or by striking other particles, set them in motion by imparting energy to them.

Fine dust initially is extremely resistant to movement by the direct force of the wind against the ground. Fine particles resist movement in part because they cohere to the ground surface due to the presence of moisture in the ground. Incipient cementation also contributes to this lack of movement, as well as the fact that the particle is below the zone of turbulent flow of air. Air movement, in a paper-thin zone above the ground, is practically zero, with the thickness of the zone dependent on the height of irregularities of the surface. Dust clouds form as a result of saltating sand grains striking clay and silt particles and bringing them into the airstream. Dust may be raised in a like manner by vehicles, animals, or other moving objects.

11.2B Deflation and Abrasion

Once the particles are moving, the erosional work of the wind begins to take effect. Erosion by wind takes two forms: deflation, which is the removal of sand, silt, and clay, resulting in a lowering of the land surface, and abrasion, which is the sandblast effect that windblown particles have on stationary objects.

Deflation occurs where little or no vegetation exists in arid regions during extended periods of drought. Sand, silt, and clay are moved, leaving behind the larger fragments of rock which eventually form a protective covering on the ground known as *desert pavement*. The desert pavement prevents further deflation unless the rock fragments are disturbed.

If such a rock covering is not formed, deflation can continue for an indefinitely long time, resulting in an extensive basin. Such a depression is the Qattara depression in the Sahara Desert. This depression, fashioned largely by wind,

Figure 11.8 Desert pavement.

is about 18,000 square kilometers in area and is coated by salt. Salt has accumulated by evaporation of water, creating an extremely rough surface and making it an impossible area to traverse. Estimates of deflation in the Great Salt Lake Desert would indicate that the eastern end of the desert has been lowered about 3 meters by wind action during the 10,000 years or more since the lake disappeared from this area. Deflation may be limited by the presence of water beneath the surface, which binds the particles sufficiently to prevent further blowing. However, even this protection may be gradually eliminated by evaporation, permitting the wind to continue deflationary action but at a slower pace.

Abrasion, the sandblast effect of windblown particles, results in the wearing away of surface features as well as the further breaking up of individual particles by continuous wear. Abrasion is not as important as deflation in its effect on the total land surface, but it is more spectacular in that the abrasive action of sand on surface features is more evident. The essential conditions for wind abrasion are a supply of particles capable of being moved by wind and a strong wind. Abrasion results in the rapid wearing away of soft rock, causing in some instances the formation of odd-shaped structures. Imbedded hard materials begin to stand out in strong relief as the surrounding softer material is worn away.

Erosion

Abrasion is most effective close to the ground where most of the coarse sand moves. Because of this, cliffs tend to be slightly undercut and telephone poles and fence posts are worn away at the base. The abrasive action of wind on rock is in part furthered by weathering of the rock. Weathering weakens the surface rock making it more susceptible to abrasion. Were it not for weathering, many of the features now attributed to abrasion could not occur.

11.2C Deposition

Windborne material may be transported for many miles before a slackening in the wind results in the deposition of this material. Sand-sized particles, being heavier, are deposited first, and the smaller silt and clay particles gradually filter down to form a thin blanket on the surface of the land or fall into the sea to settle on the bottom as sediments. In this manner wind has, in the past, covered ancient abandoned cities with sand until they were completely hidden from view.

Sand dunes are by far the most noticeable landform created by wind action, although not the most important. Their occurrence is not restricted to deserts, as is commonly thought, but are also found along seacoasts where a supply of sand is available for dune formation. Dunes form as a result of some interference in wind movement, by a rock or vegetation, causing the wind to slacken and the sand to be deposited, forming a mound around the obstruction. The velocity of the wind, the size of the particles, and the amount of obstruction present are factors which control the type and size of the dunes. In the initial stage, a dune is just an oval mound of sand forming around an obstruction, but this shape gradually changes as the dune grows. The windward side develops as a gentle slope usually not more than 15° from the horizontal, while the lee side is quite steep, up to a maximum of 34°, which is the angle of repose for dry sand.

Sand dunes average from 3 to 15 meters in height, although in the Great Dunes National Monument, Colorado, some attain a height of 150 meters, and in the Sahara Desert some reach 300 meters. The dunes tend to migrate in the direction of the prevailing wind unless stabilized by some obstruction such as extensive vegetative growth developing on the surface of the dune. Vegetation is sometimes established artificially on dunes to halt their migration over valuable property. Without interference, dunes will migrate from 7 to 15 meters per year, depending on the size of the dune and the force of the wind.

Evidence of ancient windblown deposits may be found in some sandstones. These sandstones exhibit variation in inclination and direction of strata resulting from changing environmental conditions while the deposits were being formed. These sandstone formations are useful in tracing events in the earth's history, since climatic conditions may be inferred from the presence of dunes as well as the direction the wind was blowing when the dunes were being formed.

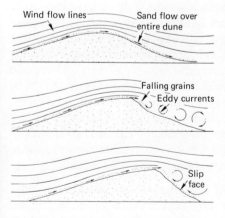

Figure 11.9 Formation of a sand dune.

Figure 11.10 Barchan near Tule Wash, west of Salton Sea, California, looking north. By John S. Shelton.

11.2D Wind Erosion and Society

The effect of wind erosion as a geologic phenomenon is not of as great a consequence as is, for example, water erosion. However, the effect of wind erosion on society is serious and in some instances extensive. Agriculture, of all of human activities, is most significantly affected by wind erosion. Agriculture is a basic industry, important in determining the manner in which a society develops. Any factor that causes, even to a limited degree, deterioration of agricultural activity will have a profound influence on the social order.

The effect of the wind is such that in many instances farmers have lost entire crops, but more serious than this is the fact that the fertility of the soil has sometimes been greatly reduced by wind action. This has been brought about by the loss of the fine soil fractions (silt, clay, and organic matter) gradually being sorted out from the sands and carried great distances. Silt, clay, and organic matter represent the portion of the soil which contains needed plant nutrients, and their removal will result in a marked reduction in soil fertility. The sands, which represent the coarser fractions of the soil, are infertile and are no more than a skeleton of the soil. In areas where wind erosion has been serious, crop yields have been lower for many years after erosion occurred. In some instances, the land appears to have been permanently damaged by wind erosion.

In addition to the reduction in fertility of soil, wind erosion also creates other unpleasant conditions. Railroads and highways have been buried under drifted soil. Traffic accidents have been more frequent during severe soil blowing. Vegetation, including shrubs and trees, have been smothered; homes,

Erosion

especially in rural areas, have been ruined by soil drifts. Nor are the cities immune to this activity. The "black dusters," as the dust storms of the 1930s were called, moved vast quantities of surface soil from the Plains region eastward over the densely populated areas of the Atlantic coast. During one storm in 1934, the sun was almost blotted out by dust from the central United States. This dust was swept out over the Atlantic ocean, engulfing ships more than 800 kilometers at sea. Measurements indicated that as much as 300 million tons of soil was moved during this one storm, which is the equivalent to an amount sufficient to cover over 100,000 acres one foot deep. The housecleaning chores brought about by such an event must have been phenomenal.

Erosion by wind does not usually have such widespread consequences, except when human activity has created adverse conditions in agricultural regions, such as in those instances described above. In this case, wind erosion control can be practiced by modifying those factors which influence wind erosion. The strength of the wind itself can be somewhat reduced close to the ground surface by providing extensive shelter belts of trees or shrubs planted in rows at right angles to the direction of the prevailing wind. Protecting the soil surface with some form of vegetative covering or roughening the soil surface to slow down the wind velocity and to trap drifting soil has also been found effective. These methods are practical and useful in reducing the damage by wind erosion in many rural areas that formerly were effected during the "dust bowl" era.

11.3 GLACIERS

Ice, as such, does not contribute appreciably to the erosional process, but rather it is one of the agents responsible for physical weathering. Ice becomes an erosional agent only when it occurs in large masses such as glaciers.

Figure 11.11   Influence of ice and glacial activity on landscape.

A glacier is defined as a large mass of ice formed by the compaction of accumulated snow. Fallen snow is usually light in weight and highly porous, with the pore spaces filled with air. The snow partially melts or evaporates in the presence of air. The resulting water or water vapor percolating through the remaining snow refreezes, resulting in the pore spaces becoming filled with ice. The whole mass takes on a granular texture and markedly increases in density. When the density reaches approximately 0.8 grams per cubic centimeter (snow has a density of 0.05 to 0.1 gram per cubic centimeter), the mass may be defined as ice, at which stage the ice is more compacted than the original snow mass. For the ice mass to be a glacier, it must be at least 50 meters thick. At this point sufficient pressure is exerted at the bottom of the mass by the weight of the ice to cause the ice to achieve a degree of plasticity, permitting it to move away from the point of snow accumulation. Glacier ice occurring on land should not be confused with sea ice or pack ice which results from the freezing over of an open polar sea.

Glaciers presently occur in the remote regions of the Antarctic and on Greenland, and at high elevations where the accumulation of snow in the winter season exceeds the loss of ice and snow by melting and evaporation during the summer. These regions are generally far removed from man's normal habitat so that the opportunity for seeing a glacier is, for most people, not very great. However, the evidence of past glacial activity in populated areas is quite common and testifies to the power of ice as an erosional agent. A study of the manner in which glaciers sculptured the land leads to an understanding of the extent to which glaciers once covered great areas of Canada and the northern United States, with extensions of glacier activity further south at high elevations. Simultaneously in Europe, glaciers covered Scandinavia and Great Britain and reached into northern France, Germany, Poland, and Russia.

11.3A Types of Glaciers

The glaciers that covered great parts of North America and northern Europe were *continental glaciers*, also referred to as *continental ice sheets* or *ice caps*. These glaciers covered great areas by moving in all directions from points of maximum snow accumulation. At the present time, the Greenland ice cap, covering 1.1 million square kilometers and up to 3,000 meters thick, and the Antarctic glacier, 11 million square kilometers and as much as 4,200 meters thick, are the only continental glaciers in existence. All other glaciers, found at high elevations in various parts of the earth, are *valley glaciers* (or *mountain glaciers*). These glaciers are more numerous than continental glaciers and originate from snow fields at high elevations. The ice from valley glaciers moves downslope, generally following preexisting valleys as the easiest path to lower elevations. Valley glaciers vary in length from 1 km to over 150 km. The Hubbard Glacier in Alaska, for example, is reported to be about 120 kilometers long. Some valley glaciers flow out onto an open plain and spread out to join other valley glaciers in a continuous sheet of ice called a *piedmont glacier*.

Erosion

Figure 11.12 Areas of the Northern Hemisphere covered by glaciers during the past million years (Pleistocene epoch).

11.3B Glacial Origin

During the past several million years (known geologically as the Pleistocene epoch or Ice Age), glaciers advanced at least four times over the North American continent, covering vast areas with glacial ice thousands of feet thick. Each advance or stage lasted up to 100,000 years from start to finish and alternated with interglacial intervals of varying length. The most recent glacial stage (Wisconsin stage) receded from what is now densely populated areas only 10,000 to 15,000 years ago. Other stages include the Nebraskan stage, which ended 900,000 years ago; the Kansan stage, which ended 600,000 years ago; and the Illinoian stage, which ended 250,000 years ago. The names for the glacial stages were taken from the states where deposits of these stages were

Geology

Figure 11.13   A valley glacier entering the sea.

first noted, although deposits of all four stages may be found in one area, one on top of another.

Many theories for the origin of glaciers of this type have been suggested, but all have been subject to objections of one kind or another. Any theory must not only be able to explain the presence of glaciers that have occurred during the past million years, but also glacial activity that had occurred 250 million years ago, 600 million years ago and 2 billion years ago.

From the foregoing it can be seen that glaciers do not occur at regular intervals; in fact the infrequent and irregular occurrence suggests an exceptional combination of several minor factors, anyone of which by itself would be inadequate to cause glaciers to form. However, common to all theories of glacial occurrence and essential for glacial development are increased precipitation. Conditions must be such that more snow falls in the winter than can melt away in the summer. Although none of the theories for glacial origin are completely satisfactory, they are interesting and for this reason several are discussed.

1. A reduction in the amount of solar energy received by the earth has been given as a reason for a decrease in temperature leading to glacial development. One theory, to account for such a reduction in solar energy, suggests that during an era of extensive volcanic activity great quantities of ash was thrown into the atmosphere, thereby partially blotting out the sun. This activity reduced temperatures and provided particles around which moisture could supposedly condense to form snow in sufficient quantities to produce a glacier.

However, investigation of the earth's history has shown that while there have been occasions of great volcanic action, there have been no attendant glacial occurrences.

Another theory suggested fluctuations in solar radiation as a means by which temperatures on earth could be lowered. Fluctuations of the sun's radiation of as much as 3 percent on a short-term basis have been observed, and it is thought by a few astronomers that larger fluctuations may be possible. Such variations in solar radiation, if of ample magnitude and of long enough duration, could reduce temperatures to a level permitting glaciers to form. However, no evidence exists to substantiate such large-scale changes in solar radiation, nor is there evidence of any correlation between reduction in solar radiation and the development of glaciers.

2. Another factor suggested as having an influence on earth's temperatures is the amount of carbon dioxide in the atmosphere. Certain of the sun's radiations penetrate the atmosphere readily. These radiations are absorbed by the earth, only to be reflected as infrared radiation and absorbed by water vapor and carbon dioxide in the atmosphere. This absorbed energy from the infrared radiation is retained in the atmosphere as heat. A relatively high carbon dioxide content (average content 0.03%) in the atmosphere would increase surface temperatures while a lower-than-average carbon dioxide level would result in reduced temperatures suitable for glacial development.

3. Maurice Ewing and W. L. Donn suggested a theory for the formation of glacial ice that depended on the alternate freezing over and complete thawing of the Arctic Ocean. When no glaciers existed and the ocean level was at its highest point, warm water from the Atlantic Ocean flowed into the Arctic over a shallow sill extending from Greenland to Norway. This caused the ice cap on the Arctic Ocean to melt. The then open sea was a rich source of moisture for weather fronts moving south over North America and Europe. The moisture provided sufficient snow to cause the glaciers to form. Formation of the glaciers resulted in the lowering of the sea level, thus reducing the flow of warm water into the Arctic Ocean. The resulting cooler sea temperatures in the Arctic Ocean permitted the sea to freeze, thereby reducing the flow of moisture which supplied the glaciers. Once this moisture supply was reduced, glaciers melted faster than they were supplied, causing the glaciers to gradually disappear. The rise and fall of the sea level and the advance and retreat of the glacier has been repeated several times, and there is reason to believe that the present interglacial stage is a prelude to another ice age.

Ewing and Donn's theory depends on the migration of the earth's continents. It is their contention that the high-latitude location of the Arctic Ocean resulted in the formation of the polar ice cap and the Pleistocene glaciers. If the north pole had been at one time located in an open sea, the free movement of water would tend to moderate the temperature of the ocean and prevent glacier formation. This would account for the lack of glacial activity for 200 million years prior to the Pleistocene ice age. Two hundred million years ago, according to this theory, the poles were located in a position which permitted the formation of glaciers in South America, South Africa, India, and Australia. Under the present circumstances, the advance of another ice cap on the North American continent may be expected, although the time of its occurrence cannot be predicted.

11.3C Glaciation

Glaciation is the process wherein the earth's surface is altered through erosion and deposition of rock material by the glacial ice. Erosion is accomplished by the plucking or quarrying action of ice and by the abrasion of the land surface by materials carried by the ice.

Plucking or *quarrying* is the process in which blocks of rock are removed from the bedrock by ice. This process is generally initiated by frost wedging, which is considered to be a form of physical weathering rather than erosion. Frost wedging separates rock particles from the bedrock. The loose rock then becomes encased in ice and is pulled from its position and carried along by the glacier (fig. 11.14).

The rock material entrapped in the ice may vary in size from huge boulders to microscopic particles, and it is this material that acts as an abrasive as the glacier moves. The larger particles gouge the bedrock, leaving grooves which mark the passage of the glacier, while at the same time the rock particles are

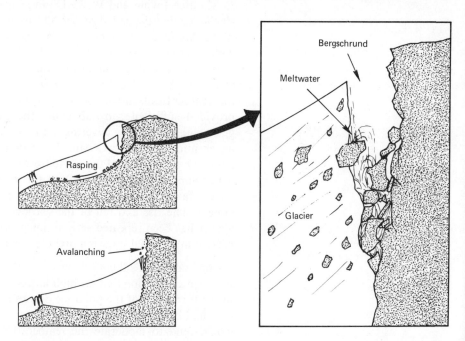

Figure 11.14 Progressive erosion at the head end and bottom of the glacier is accomplished by plucking and quarrying activity.

Figure 11.15 Glacial polish on porphyritic granite west of Tuolumne Meadows. "Lumps" are phenocrysts of orthoclase. By John S. Shelton.

worn until flattened on one surface. In this way the glacier grinds and smooths the land over which it travels, deepening the valleys and rounding off irregularities on the surface. The plucking and quarrying action is more effective in eroding the upper reaches of the glacial valley, while abrasion is the dominant form of erosion further downstream.

Eventually the rock material accumulated by the glacier as it travels downstream will be deposited. Deposition occurs when the glacial ice melts and can no longer support the load.

The erosion by both valley and continental glaciers results in a number of land forms unique to glaciers. Erosion caused by valley glaciers usually results in the development of a U-shaped valley—a valley that is deep, flat bottomed, and with relatively straight steep sides. This contrasts with young river valleys which are generally V-shaped and narrow, or with mature river valleys which are wide and shallow.

Frequently, small tributary glaciers feed into a larger valley glacier. The small glacier does not erode its valley to the depth that the larger glacier does because of the lesser weight of the ice. Thus, after the glaciers disappear, the valley floor of the tributary glacier is at a higher elevation than the floor of the main valley at the point of juncture. The tributary valley is known as a *hanging valley*. The small valley from which the Bridal Veil Falls in Yosemite National Park flows is a typical hanging valley.

Figure 11.16 U-shaped valley formed by glacial activity.

The plucking action of the glacial ice forms a steep-walled amphitheater-shaped enclosure called a *cirque* at the upper or head end of the glaciated valley. This is considered the point of origin of the valley glacier. The glacier, once formed, generally works upstream as well as downstream. In doing so, the cirque wall becomes higher and steeper as the excavating action proceeds toward the mountain peak. Heights of 650 meters or more for the steep walls of cirques are common in many mountain ranges. Depressions in the base of the cirque would ultimately fill with water, forming a lake or *tarn*, as it is called. If cirque development takes place from several sides of a mountain peak, a jagged pinnacle or *horn* will result, of which the Matterhorn in Switzerland is a typical example (fig. 11.18). A jagged ridge or *arête* may be formed in the same manner by cirque walls intersecting from opposite sides of a ridge. If the ridge is intersected excessively at one point, a pass or *col* results.

11.3D Deposition By Glaciers

Much of the topography formed by valley glaciers and continental glaciers is the result of deposition of the rock material carried by the ice. Rock material is carried mostly on the bottom of the glacier and consists of any material picked up by the ice as the glacier moves. The material carried by valley glaciers also includes debris added from the valley walls which accumulates on top of the glacier. As the glacier progresses down the valley it erodes the valley wall, causing rock to fall from above and collect on the ice.

223

Erosion

Figure 11.17 Bridal Veil Falls in Yosemite is typical of a hanging valley formed by glacial action.

Drift is a term that is applied to all such rock debris carried directly by glacial ice or by glacial meltwater. This term dates from the time, more than a hundred years ago, when it was thought that all such material had been deposited by water. Now more specific terms are used, one being *till*, which applies to all

Geology

Figure 11.18 Cirque formed at the head end of a mountain glacier.

Figure 11.19 Horn formed by several mountain glaciers.

unstratified material of all size laid down directly by ice. The other term is *glacio-fluvial deposits*, which include stratified and sorted material deposited by glacial meltwater.

A general term applied to land forms composed of glacial till is *moraine*. *Ground moraine* is probably the most common form of till deposits, occurring as a veneer of rock debris over large areas once occupied by a glacier. In a few areas, glaciers have deposited ground moraine as oval-shaped hills, called *drumlins*. These are streamlined hills ranging up to 1.5 kilometers in length, 1 kilometer in width and up to 65 meters in height. The drumlins in one locality are generally more or less parallel and are oriented in the direction in which the glacier moved. Bunker Hill in Boston is perhaps the most famous example of a drumlin.

Terminal moraines are ridges or belts of till marking the farthest advance of a glacier. These are also called *end moraines* and may range from a few meters to 30 or more meters in height. Generally, they extend across the entire front of the glacier but the action of streams formed from the melting ice may alter or destroy the terminal moraines completely. Terminal moraines are formed as the glacial ice continues to move from the source toward the terminal end of the glacier. However, due to the balance between the rate of melting and the rate of ice movement, the glacier as a whole is stationary. During this period, more and more till is brought to the terminal end of the glacier by the moving ice, building up the moraine.

Figure 11.20 Moraines below Convict Lake, Sierra Nevada Mountains. By John S. Shelton.

Recessional moraines are similar to terminal moraines in form. However, the recessional moraines are deposits located toward the source of the glacier with respect to the terminal moraine. These deposits were laid down while the glacier was generally retreating. Temporary halts in the retreat of the glacier resulted in the deposition of till and the forming of recessional moraines.

Lateral moraines result from the accumulation of till along the side of a valley glacier. The moraine is composed mainly of material that fell on the glacier from the valley walls, and with melting of the glacier this material was deposited along the sides of the valley as lateral moraines. *Medial moraines* form when two valley glaciers join into a single ice stream, or piedmont glacier. At the point of juncture of the two glaciers, the lateral moraines will combine to form a stream of till located toward the center of the valley glacier which can be traced down the valley for some distance.

Glacio-fluvial deposits generally occur beyond the boundaries of the glacier and are important in reshaping the landscape as the glacier retreats. These deposits are formed from material carried and deposited by glacial meltwater. *Outwash plains* are probably the most extensive of these deposits, forming as a result of the meltwater carrying much sand and gravel and depositing this material over an extensive plain beyond the front of the glacier. The water is continuously diverted, resulting in a fairly even-surfaced plain sloping gently away from the end of the glacier. The waterborne material will cover previous land forms which may at some future time be reexposed by erosion.

As the glacier retreated, large blocks of ice were sometimes separated from the main glacier and became isolated from it only to be covered by outwash material. When the block of ice melted, a depression called a *kettle* was left which could range in size from 5 meters to 15 kilometers in diameter. If appreciable meltwater flow continued in the area where the kettle was formed, there was a possibility that it would become filled with waterborne deposits. In some areas, a large number of these kettles formed as the glacier retreated, resulting in a pitted outwash plain.

Meltwater seeping out from beneath a stagnant glacier formed tunnels in the ice. These tunnels served as channels through which a considerable amount of water later flowed, carrying great quantities of gravel. The gravel accumulated in the tunnels, deposited there by the meltwater. When the ice vanished, the gravel remained as a stratified ridge or *esker* on top of the ground moraine. Eskers are often many miles in length and may be branched, indicating the pattern of water flow beneath the glacier. The esker is usually steep-sided and from 3 to 30 meters in height.

One form of waterborne deposit, which has proved of value to the geologist in tracing the movement of glaciers and the length of time involved, is the *varve* (deposit of the season). A varve is a sedimentary bed or lamination laid down within one year's time in a lake that is usually frozen in the winter. Two distinct layers accumulate each year, a thick, light-colored layer deposited

Erosion

Figure 11.21 Ponds in glacial deposits south of Plymouth Harbor, Massachusetts. By John S. Shelton.

in the summer during active water flow, and a darker layer of organic matter and clay which deposits from the still water beneath the ice in the winter. Varves range from a fraction of a centimeter to $\frac{1}{2}$ meter in thickness, although a thickness of less than 5 centimeters is most common. Much of the information on the recession of glaciers has been obtained by studying varves.

11.3E   Effect of Glacial Activity on Man

Glaciers are of little direct significance to man at the present time, except to the geologist who investigates glacial phenomena. However, the landforms left by previous glacial activity, especially in areas now densely populated, are of some significance to society. Many lakes and waterways in the north central United States owe their existence to the last glacial ice cap which covered this area as recently as 15,000 years ago. These lakes are of value both for commercial traffic and recreation and have contributed greatly to the development of that region. Vast quantities of ground water are stored in the glacio-fluvial deposits, especially where this material has filled deep preglacial valleys. The water, when pumped, serves as domestic and industrial water supplies for many large cities. These deposits also provide the raw material for construction, for it is from glacial deposits that sand and gravel are obtained in many areas of the Midwest.

Agriculture has benefited in some areas from glacial activity. Outwash plains and old glacial lakebeds formed the prairie lands, some of the most productive farmland in the United States. On the other hand, New England was not so blessed, for here the glaciers left a stony cover on the land, making it difficult to cultivate.

Geology

What changes could be expected if another ice age occurred? This can only be answered on the basis of events that have occurred in the past. Much of the land that was blanketed by ice was depressed by the weight of the ice. Ice has a density of about a third that of rock, but thicknesses of 1,000 to 1,200 meters were not uncommon, adding tremendous weight to the land surface. The present-day Greenland glacier is up to 3,000 meters thick in places and depresses the land upon which it rests below sea level.

The sea level is also affected by the advance and retreat of glaciers. At the present time, about 10 percent of the earth's surface is covered by glaciers (Greenland and Antarctic), and it has been predicted that if these completely melted, the sea level would rise as much as 75 meters. Since 1890, the rate of glacier retreat has accelerated somewhat, and the sea level is rising at the rate of about 12 centimeters per century. A report issued in 1965 by the President's Science Advisory Committee states that the increasing amount of carbon dioxide accumulating in the atmosphere by heat-producing fuels may have a profound effect on glaciers in the near future. By the year 2000, the temperature may have been raised sufficiently by the greater amount of atmospheric carbon dioxide to cause the Antarctic ice cap to melt in a period of between 400 and 4,000 years, raising the sea level approximately 75 meters. If the melting of the world's ice caps was accomplished in 1,000 years this would mean a rise in sea level of 7.5 meters per century, enough to inundate every major seaport in the world. Such an eventuality would result in major economic dislocation and necessitate the shifting of large segments of the population. But what if the reverse would happen and the North American continental glaciers were to form again? Millions of cubic miles of water would be removed from the seas, thereby lowering sea level a predicted 75 to 90 meters below its present level. The advance of the glacier would destroy everything in its path—all vestiges of human effort would be obliterated, and much of the existing landscape changed, first by the plucking and gouging action of the ice and then by the deposition of iceborne material as the glacier retreats again.

11.4 WATER EROSION

Water or *fluvial* erosion is the most important form of erosion, because it is responsible for the movement of more surface material than the combined effect of gravity, wind, and ice. The landscape undergoes constant change through water erosion, although in many instances this change is not particularly noticeable. Movement of the water over the land and the erosional effect of water have been extensively investigated and for convenience can be classed into three broad categories: stream flow, ground water, and wave action. Stream flow or the movement of water over land surfaces has the most obvious and the most extensive erosional effect.

11.4A Stream Flow

Erosion

Early Greek scholars (including Aristotle) considered the problem of stream flow and were of the opinion that streams and rivers originated from many great reservoirs that existed within the earth. Aristotle found it impossible to believe that the source of the stream was rain alone. He felt that a portion of the water resulted from the condensation of air beneath the crust, and part from the condensation of vapor from an unspecified source. The controversy on the actual source of water continued into the Middle Ages, during which time the Bible was extensively quoted as supplying the proper answer. A passage from Ecclesiastes 1:7, as an example, reads as follows:

> All rivers run into the sea, yet the sea is not full: unto the place from which the rivers come, thither they return again.

Many other explanations for the cause of stream flow were given during the Middle Ages. In 1674, Pierre Perrault published a book in which he attempted to show the relationship of rainfall to stream flow. Perrault measured the drainage area of the river Seine from its source to Aignay-de-Duc and determined the average rainfall in this region for the years 1668, 1669, and 1670. He then measured the amount of water that flowed through the Seine canal at Aignay-de-Duc each year and found the amount equal to one-sixth of the total rainfall. He concluded from this that the rainfall alone was responsible for the water flowing in the stream. The balance of the water was evaporated into the atmosphere again or sank into the ground to become a part of the ground water supply.

In the development of stream flow, running water begins its action as rainwash on slopes, which quickly concentrates into rills and then flows into gullies. The gullies may become small streams which drain into larger streams, and these, as tributaries, in turn drain into a river. In this way the stream performs the very important function of removing surface water from the land and in a somewhat organized manner returns the water to the sea. The main river, with its many branches and tributaries, is a drainage system. The region drained by this system is a drainage basin, separated from adjacent drainage basins by a drainage divide. In some cases the drainage divides are ridges easily recognizable as divides, while in other instances the uplands are so broad and flat that the actual dividing line is obscure. The continental divide in the Rocky Mountains is a line separating the drainage eastward toward the Mississippi basin and westward drainage toward the Pacific Ocean. The continental divide does not delineate the drainage basin of a single river but rather the eastward and westward flow of all streams originating in the Rocky Mountains.

River basins form as a result of erosion by water within the basin. In the process, channels are eroded in which the rivers flow. The fact that most rivers erode their beds to form the valleys in which they flow was not recognized until late in the eighteenth century. Previous to that time it was thought that

Geology

valleys were large cracks or crevices in the earth's crust along which water found a convenient channel for exit to the sea.

The patterns of erosion resulting in the formation of the river basin is influenced by the nature of the underlying rock. If the rock is uniform in its resistance to erosion, the headward or upstream erosion of the river tributaries will form a dendritic pattern reminiscent of the branching of a tree. If the rock exhibits unequal resistance to erosion, as may occur where sedimentary rock layers have been folded, the less resistant rock will be worn down more readily, forming a rectangular or trellis pattern. The rock in these instances are said to exercise structural control in the formation of the stream pattern. Other drainage patterns of lesser importance exist such as, for example, the radial pattern formed by streams radiating from a central point such as a volcanic peak. These patterns are not directly the result of the rock's resistance to erosion but rather are determined more by the slope of the land form.

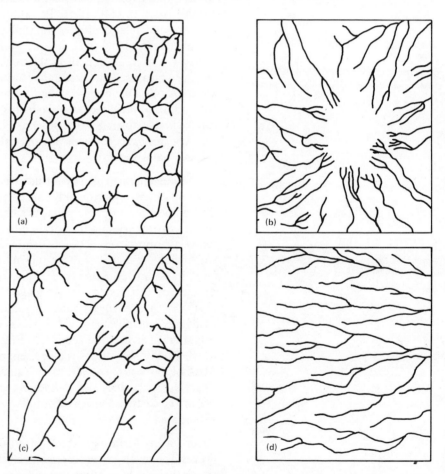

Figure 11.22 Water erosion will develop drainage patterns dependent upon the rock structure. Diagrams show the following patterns: a) dendritic, b) radial, c) trellis, d) parallel.

11.4B Work of Streams

Erosion

The stream performs three important functions in the process of draining water from the surface of the land: erosion, transportation, and deposition of rock material.

Erosion by a stream is accomplished by the *hydraulic action* of the water, the *abrading action* on the bottom and sides of the channel by the sediments carried by the stream, and by *solution*.

The hydraulic action of the water will remove particles from the rock as a result of the actual force of the moving water. This is not an important form of erosion but it does occur where fast-moving streams move over poorly consolidated rock or unconsolidated deposits. The abrading action of erosion is accomplished by sediments carried by the stream striking against the bottom and sides of the channel with sufficient force to break off new particles. This is the manner in which rock more resistant to erosion than the poorly consolidated rock mentioned above is eroded by stream action and it is perhaps the most important aspect of stream erosion. Erosion by solution is chemical erosion—the water dissolves the rock as the stream flows by. This type of erosion will occur where the channel bottom is composed of limestone or other soluble form of rock.

At what rate does the eroding process take place? This is dependent on the materials available for abrading the channel bottom and sides and the velocity at which the water moves. The rate of water flow or velocity is measured as the rate in meters per second at which water moves past a given point. Stream velocity is governed by a number of factors, principal of which are the gradient or downstream slope of the stream channel and the volume of water flowing down the channel. It is easy to recognize that the steeper the gradient of the stream bed, the faster will be the velocity of flow. However, it is more difficult to recognize how an increase in the volume of water flowing in a stream can result in an increase in velocity of flow. In times of flood, gradient does not change, but the velocity of water movement is greatly increased. This is due to the increased volume of water flowing in the channel without a proportionate increase in friction of the water against the bottom and sides of the channel. Velocity of stream flow is also governed to a certain extent by such factors as roughness of the channel bottom, channel shape, and the load of material that the stream is carrying.

Rock particles being transported by stream action are moved in several different ways. Larger rock fragments are generally rolled along the channel floor, whereas somewhat smaller particles are moved in a hopping fashion called *saltation*. Small particles are carried along by the current in *suspension* without settling to the bottom. In addition, rock material is transported in *solution* from which it may ultimately be precipitated to form new rock.

The size of the particle and the amount of material transported by the stream is a function of the velocity of stream flow. The greater the velocity of flow,

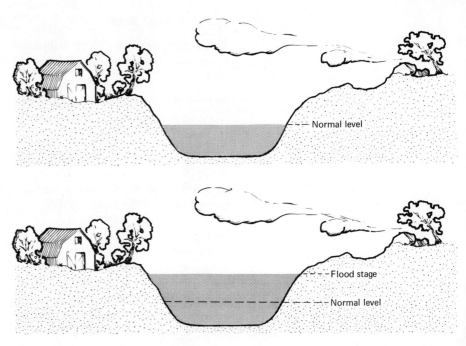

Figure 11.23 Raising the water level in a stream (increasing the volume) does not result in an equivalent increase in frictional surfaces in contact with the water. The water velocity is therefore increased with an increase in volume of water moving in the channel.

the greater is the *competence* of the stream—which is a measure of the largest-sized particle that the stream can move.

While competence indicates the maximum size of the particle moved by a stream, *capacity* refers to the maximum amount of sediment a stream is capable of carrying. A stream will not be carrying its capacity if sediments are not available in sufficient quantities. Under these conditions, erosion is taking place as the stream picks up additional material from the channel floor. The actual amount of material the stream carries is its *load*, measured as a unit of weight per unit of time, e.g., tons per day.

Rock particles are not carried along by the stream indefinitely but are eventually deposited when the velocity of flow slackens. The larger particles, moved during periods of flood, will settle to the channel floor when the flood subsides. Smaller particles are deposited later as the velocity of flow decreases further. The velocity may decrease as the river flows into a relatively quiet sea, or because the river channel broadens or a flood subsides. A loss of water from the stream due to the percolation of water into the ground, loss by evaporation, loss by removal for irrigation, domestic or industrial purposes all contribute to a small degree to reduction in the velocity of streams.

11.4C Stream Formation

Erosion

Through the erosional process, a process which follows a predictable pattern of change, rivers will gradually shape the valleys in which the rivers flow. In the initial or *youthful stage*, the velocity of stream flow is relatively high and erosion of the stream bed is rapidly taking place with little or no lateral erosion occurring. Rapids are common, especially toward the source of the stream. Streams in this stage occur in mountainous regions where the gradient is steep.

The river valleys formed under these circumstances are characterized by steep sloping sides forming a V-shaped valley. Gravity contributes to the erosional process, shaping the valley by the constant movement of particles down the sides of the slope into the stream. Avalanches occasionally contribute large amounts of material to the stream load and sometimes block the stream to form a temporary lake.

In some areas, society is dependent upon rapid-flowing streams as a source of power, and man has harnessed the energy to drive dynamos to generate electricity. The stream flow may be modified somewhat by this usage, but with careful planning little change in the natural developmental processes of the stream will occur. Young river valleys sometime are important because they provide the only access through a mountain system. Highways and railroads through such canyons may be expensive to build and maintain, since the river occupies the entire valley floor, but the price may still be less than the cost of going around the mountains.

As the erosional process continues, the downcutting may decrease as a result of a reduction in velocity of water flow or because a particularly hard stratum of rock is encountered. Lateral erosion then becomes the dominant form of erosion, causing the valley to widen. The river now slackens its pace and begins to wander back and forth across the widened valley in a series of looplike bends called *meanders*. This meandering results in both erosion and deposition of material in the stream bed. Initially an obstruction or rock outcrop deflects the current from one bank of the stream to the opposite bank. Erosion occurs where the stream turbulence is greatest, supplying material for deposition where turbulence is low. As the stream current swings from one bank to the opposite, the meander increases in size. At this stage in stream development, the river is in a state of equilibrium with its tributaries in that the stream is capable of transporting the rock material being supplied by the tributaries.

The river is now in the early *mature stage*. Lateral erosion continues so that the stream no longer flows in a relatively narrow valley but curves and meanders over a broad area. Erosion against the outside of each meander (the *cutbank*) is compensated for by filling in with sediments on the inside (*slip-off slope*) of the curve. At the same time, erosion on the downstream portion of the curve is compensated for by deposition on the upstream side. This combination of activity tends to broaden the valley and causes the meander gradually to

234

Geology

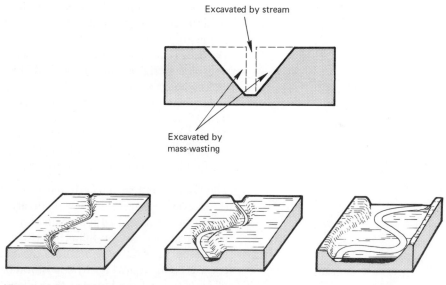

Figure 11.24 View and diagram of a young stream valley.

Erosion

Figure 11.25 View of a mature stream valley.

shift downstream. Eventually the lateral cutting of the stream will produce a valley floor several times the width necessary to accommodate the asymmetrical meanderings of the stream. The side slopes, originally high and precipitous, are now worn low and inconspicuous, and the valley is in the late maturity stage of river-valley development. The valley floor now represents a *flood plain* over which the river periodically flows during the flood stage. The flood plain is the recipient of much sediment deposited by the river as each flood stage ebbs. During a flood, the channel of the stream is subject to change and the stream can abandon a meander loop by cutting across the intervening neck of land, forming a crescent-shaped lake called an *oxbow*. These lakes may become filled with sediment from subsequent flooding, or if no independent source of water is supplied the oxbow lake, it will become marshy and eventually dry up.

Erosion continues, but at a slower pace, and the land relief becomes subdued. Some geologists consider the stream valley to be in the *old-age stage* when the flood plain is 5 to 8 times the width necessary to accommodate the meandering river.

The ultimate depth, or base level, to which a stream will erode its channel is determined by the level of the body of water into which the stream flows. A river may serve as a temporary base level for its tributaries. Lakes also serve as temporary base levels, but sea level or slightly below is the ultimate or

236

Geology

Figure 11.26   Meander curve and oxbow lake development in an old stream valley.

principal base level—the lowest point to which a stream can erode the land. Downcutting of a valley will not proceed below the base level until the base level itself is lowered.

11.4D   Features of Stream Deposition

The rate at which a stream erodes is governed by the velocity and the volume of water flowing in the stream. A reduction in the velocity of flow causes the waterborne sediments to be deposited into a variety of characteristic landforms. This deposition process will build up the level of a periodically inundated flood plain by depositing a thin layer of mud and silt each time the river flood recedes. The greater portion of the sediments are deposited near the river bank where the initial reduction in velocity takes place. This results in the buildup of a *natural levee* which is higher than the flood plain beyond the river bank.

Erosion

As rivers enter the sea or flow into a lake, the loss of velocity of stream flow results in the deposition of almost all the sediments. The final distribution of these deposits may be influenced by wave action and longshore currents. The coarser materials accumulate at the mouth of the stream, and the finer textured particles are carried further out to sea to form a triangular structure called a *delta*. The Nile delta is a classic example, being roughly triangular in shape. Similarity to the symbol for the letter delta in their alphabet caused the ancient Greeks to apply the name to this geologic feature. Frequently, a river may divide into many smaller streams, called *distributaries*, as the river passes through the delta into open water. Where there is no interference by longshore currents, development of fingerlike projects from the delta is possible. These can be readily seen in an aerial view of the Mississippi delta.

Figure 11.27 Humboldt River west of Elko, Nevada. Direction of flow is toward lower left. By John S. Shelton.

The land counterpart of the delta is the *alluvial fan*. When an intermittently flowing stream emerges from a canyon unto a relatively broad plain, the stream loses velocity and deposits the material it has been carrying. The coarser material is deposited first at the apex of the fan located at the mouth of the canyon, and the finer textured material is carried out onto the plain. Constant shifting of the stream as it emerges from the canyon results in the characteristic fan shape.

238

Geology

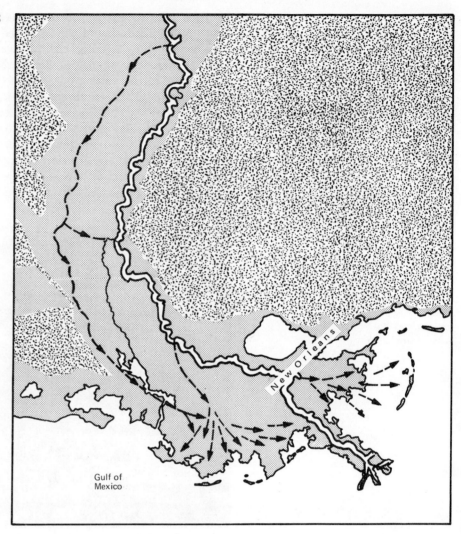

Figure 11.28 Diagram of delta formed by the Mississippi River showing various courses taken by the stream.'

11.4E Water Erosion and Society

River valleys have played an important part in man's past, for it was on the flood plains of such rivers as the Tigris and the Euphrates that civilization probably had its beginning. The flat lands bordering the rivers were fertile and easily tillable, and the river provided irrigation either by periodic flooding or artificially by means of systems built by early engineers. The rivers also provided an easy means of transportation and contributed to the development of commerce, a basic ingredient in the structure of a progressing society.

Erosion

Figure 11.29 Alluvial fans in Death Valley, California.

The early settling of the United States also developed along the many waterways for much the same reason that caused ancient people to seek the rivers. Towns sprang up along the navigable streams, and the boundaries of states and private property were often fixed at midstream. The shifting of channels caused many legal problems until the boundaries were fixed at a specific point in time and maintained despite change in the river's course.

Geologic erosion of the type that develops river valleys is a long-term process. Man is more concerned with the immediate effect of the erosion of his cultivated land. Since ancient times, man has been aware that rain was responsible for removal of the thin layer of soil necessary for the growth of his food. Throughout the world, some farmers have always sought, by one means or another, to prevent this loss. In the hills above the ancient city of Antioch in Syria are the remains of erosion control structures that probably predate the Christian era. In Peru, 400 years ago, the conquistadores found the Incas farming steep Andean slopes on terraces walled with stone, the cost of which, computed on the basis of present labor costs in the United States, would amount to $30,000 to $40,000 per acre.

In the United States, alert farmers have tried to protect their land against erosion since the earliest colonial times. By 1769, George Washington was experimenting with conservation farming practices at Mount Vernon. Patrick Henry is said to have declared soon after the revolution that "since the achievement of independence, he is the greatest patriot who stops most gullies."

Erosion not only scars and alters the physical landscape, but it also has serious economic and social implications, especially when aggravated by human activities. Many communities grew and prospered on an agricultural economy, and the prosperity in turn prompted the intensification of cultivation. More land was put into production, oftentimes land not suitable for such use. Land on slopes that was previously forested or in grassland (watershed), when plowed became subject to more serious erosion than occurred under natural conditions. Deterioration of this land by erosion and the deposition of silt from this land on the more productive land in the valley led to reduced yields and lower farm incomes. In extreme cases, a submarginal agricultural economy eventually resulted. Farms were abandoned, migration from the rural community occurred, and there was a disruption of the tax base and general community disintegration, creating serious economic and social problems in the area. It would appear that high prices associated with prosperity, even though prevailing for only a short period of years, fostered land use which over a longer term proved to be inadvisable and uneconomical. The accelerated erosion that occurred as a result of the injudicious cultivation practices was much more damaging than natural erosion would ever be.

Natural erosion as well as the accelerated erosion brought about by man's folly has other effects in addition to the ruination of agricultural land. The destruction of watershed areas by erosion has been a direct cause of flooding, often with disastrous consequences. Flooding and the attendant increased erosion has resulted in silting up streams and reservoirs, thus destroying recreational sites and water storage facilities. For example, a dam built in 1904 on the Dan River in Virginia, for purposes of providing water and electric power had its capacity reduced by 80 percent by 1930 as a result of the deposition of erosional debris. Even the mighty Boulder Dam, it is estimated, would become completely filled with sediment in 200 to 300 years if preventive measures were not taken.

From the above discussion, it can be seen that the physical effects of erosion are not confined to land impoverishment alone. In the case of water erosion, the effects extend to adjacent lower areas and to alluvial plains far and near, as well as to stream channels, reservoirs, and harbors, where a large proportion of the material removed from eroding lands eventually comes to rest. We can only begin to visualize the magnitude of the erosion process when we realize that more than one-half billion tons of solid matter is carried annually into the Gulf of Mexico by the Mississippi River alone.

Solid material carried by such rivers as the Mississippi is deposited at the mouth of the stream to form a delta. The main channels or distributaries build the accumulation of coarser material farther and farther out to sea, building the delta up to and above sea level. Periodic floods raise the land level further and occasionally change the course of the stream, a condition which creates hazards for those who live on the delta. In the process, deltas are continually being enlarged. The Mississippi delta is building seaward at the rate of approximately 6 miles per century. In some instances, this activity has caused cities, originally built as seaports, to become inland river ports. Such a city is Adria on the Po River, originally a seaport, now 14 miles inland. Most of the Netherlands occupies the combined deltas of the Maas, the Rhine, and the Scheldt Rivers, indicating how large some deltas may become.

The large river deltas have been of economic importance from the earliest historical times. Usually flat, they support an extensive agricultural economy as important now as it was in the past. Much of Egypt's agriculture is centered around the Nile delta, while the Mekong delta of Vietnam is an important rice growing area in southeast Asia.

11.5 GROUND WATER

Water not lost through evaporation or by runoff soaks into the ground to become a part of the water beneath the earth's surface known as the ground water. The upper surface of this underground water body is called the *water table*. Above the water table is the *zone of aeration* where air and some water occupy the pore spaces between the rock and soil particles. Below this point is the saturated or *ground-water zone*, where all space between the rock particles is filled with water. The water table may occur at the ground surface or be hundreds of feet below the surface. The water table follows the contour of the topography to some extent but is modified in that the water table level is further beneath the surface under the hilltops than under the valleys. The surface of marshes, some lakes, and most permanent streams are at the approximate water-table level, with the level rising and falling depending on the rainfall. The fluctuation of water level is also apparent in wells and in the appearance and disappearance of intermittent streams.

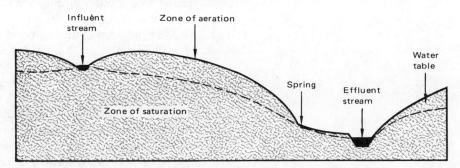

Figure 11.30 Diagram shows ground water relationships and water table.

Water moves through some rock despite its apparent solid appearance. Sandstone is *permeable*, that is, the pore spaces are interconnected and permit water to move very slowly through the rock. Shale is fine-grained and is generally impervious. Some water is absorbed, but because of the diminutive size of the pores, little movement results. In some regions, a permeable rock, or *aquifer* may be sandwiched between two impervious layers of rock. If these rock

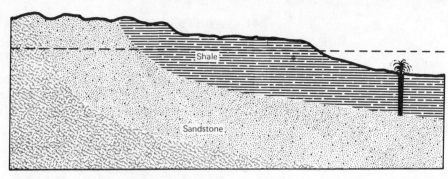

Figure 11.31 An artesian system showing source of water at high elevations being tapped by well in the valley. Sketch is not to scale.

strata are sloping, water may enter the permeable layer at high elevation and percolate downward toward a lower elevation. This will result in a hydrostatic pressure being built up in the aquifer at the lower depths, permitting water to be forced to the surface through some opening such as an *artesian well* or *artesian spring*. Under ideal conditions, pressure may be sufficient to cause water to spout many feet in height. The largest artesian system in the United States underlies the Great Plains region and occurs in the Dakota sandstone. This aquifer absorbs water on the east slope of the Rocky Mountains and is the chief source of water for many parts of the Great Plains region. So many wells have been drilled into this system that legislation was necessary to curb further drilling and thus permit existing wells to maintain adequate water levels. A well in Calais, France, dug in 1126 A.D. still maintains a good flow, as the rate of withdrawal does not exceed the supply.

Underground flow of water through the rock results in the dissolving of certain minerals. This is particularly true of limestone, and the dissolving action leads to several interesting underground features. Dissolving the rock along an underground joint or crack in the limestone may result in the formation of a cavity in the rock beneath the surface. Eventually the surface material covering the cavity may collapse forming a depression in the ground called a *sink*. Sinks range in size from less than an acre to several square kilometers and are common in areas where large-scale limestone deposits occur. Where sinks are numerous, the landscape is called a *Karst topography* because of its similarity to the Karst plateau in Yugoslavia. Parts of Kentucky and adjacent states have this type of topography formed in limestone deposited 200 to 500 million years ago (Paleozoic era) when that area was beneath the sea.

The process of dissolving minerals by the underground flow of water may result in the formation of caves. These caves will usually form at or below the water-table level. At some later point in geologic time when the land is uplifted, the cavern is drained and becomes accessible for exploration if an opening to the surface is available. Any underground solution channel large enough to be explored is called a cave, although only the great enlargements, or "rooms," are given any widespread attention.

Figure 11.32 Karst topography, near Bedford, Indiana. By John S. Shelton.

Several structures in caves may be mentioned. These are formed when the water-table level in the cavern is lowered so that mineral deposition is substituted for the dissolving action. Drops of water emerging from the roof of the cavern hang for a time before dropping. During this interval a small amount of calcium carbonate is precipitated around the margin of the drop. Drops falling incessantly will continue the process until an icicle-shaped deposit is formed. This deposit, hanging from the roof of a cave, is called a *stalactite*. Water dripping down onto the floor of the cave will build up a similar deposit beneath the stalactite called a *stalagmite*. Continued growth of these structures may eventually unite them into a *pillar* or *column*.

11.5A Ground Water and Society

Ground water is a principal source of water for human use in many areas of the world. This is particularly true in the rural areas where, as anyone who has lived on a farm or ranch knows, a well for domestic and irrigation purposes is vital. To be assured of a good supply of water throughout the year, it is necessary to drill a well deeper than the dry-season level of the water table. In humid regions, the water-table level may be only a few feet below the surface

Figure 11.33 View in cave shows stalactites (ceiling) and stalagmites (floor).

and a shallow well suffices. However, in arid regions it may be necessary to drill hundreds of feet, thus making the drilling of a well an expensive item.

When water is being pumped, the water table level in the vicinity of the well is drawn down in what is called a *cone of depression*. A deep well may draw water down to such a level that it causes shallow wells close by to run dry. The degree of drawdown is dependent upon the nature of the rock from which the water is being pumped. If the material has a low permeability, the cone of depression will be steep, but if the well is in highly permeable sandy material, the drawdown will be shallow.

The quality of water is a prime consideration, especially when the water is to be used for domestic purposes in a town or city. Water movement through underground rock causes some of the rock to be dissolved. The amount of mineral matter that goes into solution is dependent upon the solubility of the rock through which the water passes. If a small amount of mineral matter is dissolved (less than 60 p.p.m.), the water is rated as soft water; large amounts of soluble material, particularly calcium and magnesium compounds, will cause the water to be rated hard.

Hard water serves admirably when used for irrigation purposes but creates some problems when used domestically. Every housewife is familiar with the fact that soap will not lather in hard water and that hard water causes lime

Erosion

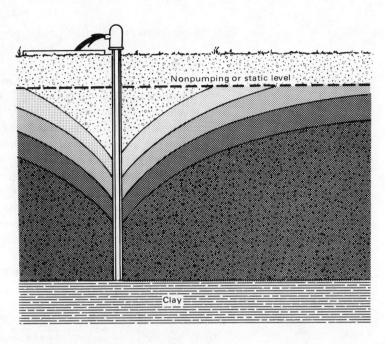

Figure 11.34 Cone of depression (well drawndown) formed by pumping water from the well.

deposits to be formed in plumbing fixtures. Many water companies have installed water softeners to reduce this problem and many homes have domestic water softeners to serve the same purpose.

One of the problems with the use of wells for domestic water supply, especially in rural areas, is contamination. Drainage from stables, pigsties, and cesspools into the ground has sometimes found its way into the domestic well, creating a danger of typhoid fever and other diseases. Care in placing the well in relation to sources of pollution will help to reduce the problem. Slow percolation through the ground purifies contaminated water, but the process requires at least several hundred meters, or more in some cases, to make certain that the purification process is complete.

Municipal water supplies are also subject to contamination from the dumping of industrial wastes into rivers and streams. Drainage from these sources into wells have resulted in pollution of municipal water supplies, particularly from certain detergents (see Chapter 20).

Ground water is the only source for a large percentage of domestic and industrial users. Because of this, the study of ground-water movement and the problems of contamination are vital to society. The growth in the use of herbicides, fertilizers, and other chemicals, as well as the dumping of liquid refuse on the ground and into streams, poses an increasing threat to the quality of our drinking water.

11.6 WAVE ACTION

Surface erosion and ground-water activity are the result of snow or rainfall on the land surface. The action of wind-driven waves is another form of erosion affecting shorelines of oceans and large lakes. Characteristic land forms along a seacoast result from the steady pounding of the waves, eroding the zone between the highest level reached by storm waves to depths of 35 to 65 meters below the sea surface depending on the degree of wave action. This action undercuts the headlands projecting into the sea, causing landslides and forming precipitous *wave-cut cliffs*. The debris from these landslides is carried by wave action to be deposited, forming beaches, sand bars, and spits. *Beaches* are considered those areas between the low-tide line and the storm-wave level covered with unconsolidated material ranging from boulders to sand. The beach is a part of a *wave-cut terrace* planed off by the eroding action of the waves as the wave-cut cliffs are steadily eroded landward. The material thus eroded is a source of sand for beach or bar formation.

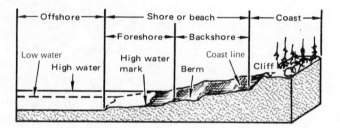

Figure 11.35 Features of the shore zone.

Bars are formed by deposition of sand in shallow water. In some instances the bar is submerged, but in any event it acts to protect the mainland from the sea. Bars that act as a barrier are not built up as a result of erosion from the land as was once thought. Rather, the bars are formed from the seaward-side wave action. If the gradient or slope of the bar is gentler than the slope formed by wave action at that depth, then the waves remove sand from the bottom, making the slope steeper, and deposit the sand closer to the shore, forming the bar closer to the beach. Sand eroded from the land by wave action is also deposited on the bar, but it is not thought to be a factor in the original building of the bar. Longshore currents eroding a headland carry sand which forms projections from the land called *spits*. Spits are usually hook-shaped with the end curved toward the land by wave action.

The general tendency of wave action is to erode the shoreline into a straight, featureless coastline. The fact that this has not occurred is due in part to the variable nature of the materials of which the land is composed, but mainly it

Erosion

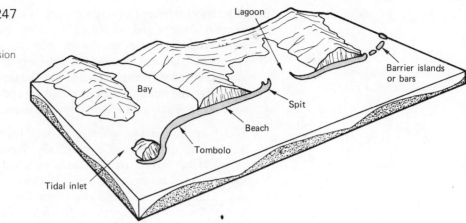

Figure 11.36 Features formed by deposition along a coastal area.

is due to the rise and fall of the shoreline. The emergence and submergence of the land may occur as a result of variation in sea level, or it may be the result of raising and lowering of the continental mass itself. Both a change in sea level and altitude of the land may be taking place simultaneously, as when the great mass of the glaciers depressed the land while at the same time a lowered sea level resulted from the water accumulated in the ice mass. This tends to complicate shoreline erosion and provides for a continually changing set of circumstances under which this type of erosion takes place. Even at present, the sea level is gradually rising, while the west coast of the United States is slowly emerging and the east coast is submerging.

11.7 SUMMARY

Erosion, the process whereby land is leveled by the natural forces of gravity, wind, ice, and water, has been functioning for millions of years. Gravity causes the slow downslope movement of surface material, although periodically segments of the steep land will slide down rapidly, including in the movement some portions of the bedrock. Action of this type has occasionally resulted in changing the course of rivers or in the formation of lakes by damming streams.

Wind causes the erosion, movement, and deposition of sand, silt, and clay particles where this type of material is unprotected by vegetation. This windborne material abrades structures close to the ground surface in a manner similar to a sand blast. Sand dunes are characteristic deposits resulting from wind action; they take a number of forms, depending on the nature of the prevailing wind and the existing landscape.

Glaciers are defined as large masses of moving ice and are classified as continental glaciers and mountain or valley glaciers. These large bodies of ice move from a point of maximum snow accumulation, radially in the case of continental glaciers, downslope in response to gravity in the case of valley glaciers. The landscape is modified by the grinding action of iceborne material, the quarrying action of the ice, and by the deposition of large quantities of rock and fine-textured material into moraines and stratified deposits.

Water erosion is the major erosional force, removing surface material in tremendous quantities and eventually depositing it on some low-lying area or in the sea. Rivers are the means by which the land surface is drained, and the process of transporting water to the sea results in the modification of the land. Underground erosion also takes place as a consequence of the dissolving action of water as it percolates through soluble rock. This activity creates the numerous caves and sinks found in the limestone regions. Wave action operating on the coastline tends to change the shore by erosion and deposition and is responsible for the formation of beaches, bars, and in some instances, the clifflike nature of the coastal areas.

QUESTIONS

1. Describe the three categories of erosion due to gravity.
2. Cite an example near your home of damage caused by mass movement of some kind.
3. By what means is solid material moved by the wind?
4. What human activity suffers the most damage from wind and what is the nature of the damage?
5. What are the landform characteristics unique to glacial erosion? Describe each one in a sentence.
6. What effects do the aftermath of glacial activity have on man?
7. Briefly describe the sequence of stream valley development.
8. Describe and diagram the nature of an artesian system.
9. What are some of the problems related to using ground water as a source of domestic water?
10. What activities prevent a shoreline from becoming a straight featureless coastline due to erosion?

12 Historical Geology

Understanding geologic activity with respect to time.
Deciphering the rock record.
The fossil record.
The Paleozoic Era.
The Mesozoic Era.
The Cenozoic Era.

In the foregoing chapters on geologic processes, the discussion dealt with the processes which shaped the earth's crust and the landforms that resulted from this activity. If we accept the concept of uniformitarianism as one of the fundamental assumptions of geology, the conclusion can be reached that these processes have been operating since the earth was formed and, as a result, the surface of the earth has been continually modified.

Man is interested in these changes, not only those relating to events during recent time, but changes and events that have occurred since the very beginning of earth's history. He is also interested in the evolution of the many organisms that have existed on the earth and in the factors that have modified them or caused them to disappear entirely from the face of the earth.

One of the problems in discussing geologic history is that an appreciation of the lengths of time involved is difficult to grasp. The Civil War was an event which, from our point of view, occurred a long time ago, yet this length of time is but an instant when compared with geologic events that took place a million or a billion years ago. Even one million years is relatively short in the geologic sense, but small continuous changes occurring over such a period of time become significantly larger. For example, a rise in sea level of 1 millimeter per 100 years would be of no importance during the few thousand years of recorded history, but if continued over a million-year period, it would be quite noticeable.

12.1 THE MEASUREMENT OF GEOLOGIC TIME

Time is an important factor in the geologic process because the topography is dependent upon the length of time the processes have acted on the earth's surface. For this reason, some thought should be given to time and its measurement as applied to geology.

Time may be thought of in two ways. The first of these, relative time, places a series of events in proper sequence without reference to the length of time between each event. For example, in our history, the Civil War is an event that occurred prior to World War I. Relative time places these events in proper sequence, but nothing is said about whether they occurred five years or fifty years apart. Some estimates may be made as to the time between them by a close examination of weaponry used, social conditions of the two periods, and such other historical data as can be accumulated. Examination of this information would reveal that these two events are closer to being fifty years than five years apart. At the same time, the study of the same kind of data relating to the westward movement in the United States would permit placing that event between the Civil War and World War I.

The same technique may be used to place geologic events in the proper time sequence. The law of superposition (see Chapter 8.1) states that those layers on top are younger than those beneath. Careful analysis of data such as this permits the geologist to arrange geologic events in their proper sequence. Some estimate may be made as to how long ago such events took place, but these estimates would not be very precise.

The second way of considering geologic time is in terms of absolute time. Absolute time gives a more precise date to an event, be it human history or geologic history. Absolute time measures whether a geologic event occurred 1 million or 1 billion years ago, and therefore it not only gives a sequence of events, but also lends some precision to the length of time between these events.

Techniques devised for determining absolute geologic time include measuring the increase in salinity of the ocean, measuring the accumulation of sediments, measuring the rate or erosion of the earth's surface, and determining the age of rock by radioactive decay of some elements. The first three techniques now have rather limited application; however good results have been obtained by these techniques in the past. The discovery in the twentieth century of the uniform radioactive decay of certain elements has proved to be of real value in determining the age of rocks.

Measuring the increase in the salinity of the oceans is useful for determining the age of the earth's oceans, but even here its use is of doubtful value. A wide variety of salt concentrations are found in different parts of the sea and the method does not take into account the precipitation of salt in the oceans. Nor does it measure the time the earth was in existence before the oceans were formed.

The rate of accumulation of sediments has also been suggested as a technique for dating the age of rock. Some early geologists thought the thickness of rock was directly proportional to the length of time necessary for it to accumulate, and that by multiplying the thickness of the sedimentary rock unit in centimeters by the length of time it took to accumulate one centimeter, some definite age in years could be calculated. However, sedimentation is not a uniform process. Thick deposits may be laid down in a short period, followed by extensive periods of slow deposition, thus making it an unreliable method for determining the age of rock over an extended length of time.

Some use has been made of the rate of erosion in predicting changes in landforms, e.g., the movement of Niagara Falls, which advances upstream at the rate of 1 to 2 meters per year. By using such a method, it has been estimated that erosion continuing over the entire surface of the earth at its present rate would level the land surfaces to sea level in about 25 million years if mountain building and volcanism ceased (see Chapter 11). However, the use of erosion as a tool for measuring the age of rock strata is unreliable, since climatic changes, variation in the hardness of the erosion surface, and the topography of the land (slope) upon which erosion is taking place will influence the rate at which erosion occurs.

Only radioactive age determination has proved to be a reliable "geologic clock." The rate of decay of radioactive elements is constant regardless of their environment, and it is possible to obtain absolute values that may be interpreted in terms of years.

12.2 THE GEOLOGIC ROCK RECORD

Only by careful observation of the rock and an understanding of how the rock units were formed is it possible to obtain a picture of the appearance of an ancient region and the conditions during that period. A great deal of this information is found in sedimentary deposits which are most sensitive to changing conditions. Much sedimentary rock such as limestone, now exposed on the land surface, is similar to deposits now being formed in shallow seas, indicating that portions of the land were submerged at one time or another.

Sediments forming sedimentary rock reflect the environment in which they are deposited. If the environment changes or the nature of the sediments being supplied varies, then the nature of the layers of sedimentary deposits will change. This causes recognizable changes in stratification or layering and results in sequential strata or layers that can be identified and traced. Information of this nature, gathered in the field, is necessary for mapping, and the maps are useful in penetrating the secrets of the rock records. This is not a simple task, for the geologic map, unlike a roadmap, must include information on subsurface features. Information on the subsurface is difficult to obtain, but a number of techniques have been developed to obtain such data, as well as techniques for describing the rocks and for correlating the rocks by age from one area to another. Direct examination of cores from water wells and oil wells, and study of the strata exposed by erosion, are techniques used to decipher the rock record. Well logging, a method for recording variations in conditions found in wells, is also employed. These recordings are obtained

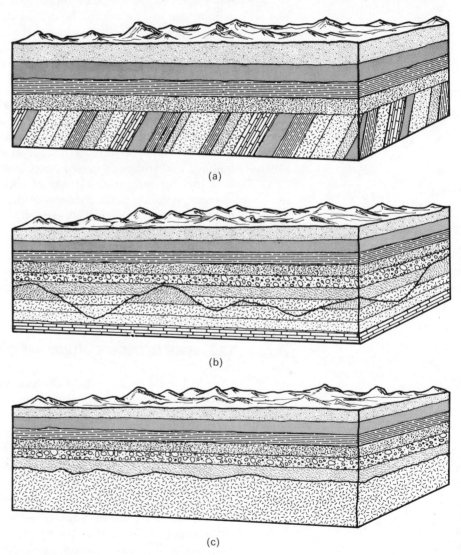

Figure 12.1 Examples of unconformities: a) angular unconformity, b) disconformity, c) nonconformity.

while drilling a well and may be in the form of radioactive logs, electrical logs, drill-time logs or temperature logs. This information gives some indication of the underground rock conditions. Well shooting has also been used: This is a technique in which velocities of seismic waves, created by an explosive charge at or near the surface, are recorded. Information from this form of logging also gives clues to the nature of subsurface rock.

The record revealed by deposits of sedimentary rock is not always continuous but may be broken by an erosion surface or a surface of nondeposition. The break indicates a time interval for which there is no record of geologic activity in that area, and the break is called an *unconformity*. An unconformity conveniently separates older rock units from the younger overlying strata, and it was used as a means of separating groups of rock strata into identifiable units when mapping of geologic structures was begun in the nineteenth century.

Unconformities take several different forms. For example, older sedimentary rock may have been folded and the surface eroded. This surface was subsequently overlaid by younger sedimentary rock, resulting in an *angular unconformity* (fig. 12.1a). A break may also have occurred between two parallel strata of sedimentary rock. In this instance an irregular erosion surface appears between the older sedimentary rock and the younger strata of sedimentary rock which is deposited on the erosion surface. Such a break in the rock record is called a *disconformity* (fig. 12.1b). In instances where older rocks of plutonic (igneous) origin underlie sedimentary deposits, the unconformity is called a *nonconformity* (12.1c).

Interpretation of the data and the development of a geologic map is the next step in the process of unraveling records of the past. A simple example (figure 12.2) may serve to illustrate how this can be accomplished. Rock material below the unconformity was first deposited horizontally, then folded. In order for erosion to occur, the folding process must have been accompanied by a certain amount of uplift. The period of erosion was followed by a period of deposition and the erosion surface was buried by layers of sediments which formed the rock above the unconformity. If the new strata are of marine

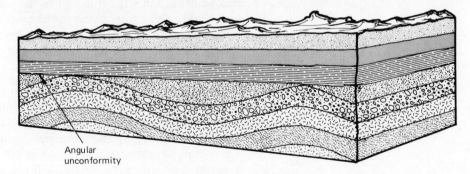

Figure 12.2 A block diagram showing subsurface rock units which are used to unravel the occurrence of past geologic events.

origin then the land must have been submerged beneath a shallow sea before deposition occurred. If these strata are of nonmarine origin, it is possible that the elevation was sufficiently low to permit streams to deposit sediments rather than to erode.

12.3 THE GEOLOGIC FOSSIL RECORD

Fossils are the remains or evidence of remains of animals or plants preserved by natural causes in the rock. The relative paucity of fossils could lead to the conclusion that animal and plant life were somewhat scarcer in the geologic past, but this is not the case. Most plant and animal tissue is rapidly destroyed by bacterial action and by weathering, thus indicating that rather special conditions are necessary for the preservation of organic material. The most frequently preserved parts are the hard parts, including teeth, bones, shells, and other forms of exoskeletons, and the woody tissue of plants. Rapid burial of these hard parts reduces or eliminates bacterial action and weathering and favors fossilization. The most likely place where fossilization may occur is in the sea. Water, which contains less oxygen than the atmosphere and thereby reduces the possibility of decay, along with the abundance of sea life and continuous sedimentation, make the sea the greatest source of fossil material. Swamps, quicksand at river banks, and asphalt deposits have trapped and preserved many land animals. The La Brea tarpits in Los Angeles are a good source of fossil material, having yielded the bones of many hundreds of specimens.

The impressions left by plants or animals in rock are also recognized as fossils. Percolating ground water may dissolve the organic structure of a mollusk, for example, leaving a void in the rock which is known as a *mold*. A *cast* will be formed if subsequent water percolation deposits some mineral such as quartz in the void, thus causing the mineral deposit to take the form of the original organic structure. The black imprints of thin objects such as leaves or fish scales are often found in mudstone or shale. These markings result from a residue of carbon left behind when the soft parts are distilled away, or they may simply be an impression in the mud which preserves the structural features of the leaf. Footprints and the impression of reptile scales have also been preserved in this same manner and are valuable in reconstructing the nature of animals now extinct.

In many instances, the fossil appears to be "turned to stone." This results from a process of replacement of the organic matter by mineral matter. The organic matter is gradually replaced by quartz or calcium carbonate deposited from water containing these inorganic materials. For example, woody stems are commonly preserved by replacement, forming petrified wood. Porous bones are similarly preserved by a process of *permineralization* in which the pore spaces are filled with mineral matter deposited by the ground water without altering the bone structure.

Historical Geology

Fossils rarely appear in igneous or metamorphic rock. However, segments of trees have been preserved in lava which cooled before the tree completely burned, and distorted traces of fossils have been found in marble. These fossils have not contributed a great deal to the knowledge of past life. The preponderance of fossil finds that have yielded valuable information for correlating rock units occur in sedimentary rock.

The study of fossils has been valuable in determining the conditions under which rock units were formed. Marine fossils contained in rock now at high elevations indicate that these sedimentary deposits were formed in the past when the land was submerged beneath a shallow sea. This explains the presence of marine fossils in deposits at 7,000 meters in the Himalaya Mountains. In other areas, marine fossils have been found in rock above fossils of land animals or plants, giving proof to cycles of submergence and emergence of the land. By studying the occurrence of marine fossils in rock, it has been possible to determine the areas that were once covered by ancient seas.

The sudden appearance of animal fossils in a layer of rock where they do not appear in the underlying rock allows geologists to conclude that a migration had occurred. In this way, it can be deduced, for example, that a land bridge existed between Asia and North America 15 to 20 million years ago, because fossil remains of elephants, long residents of Asia, were found in the rock formed at this time in the North American continent, but not in earlier rock. By evidence of this type it was also found that North and South America were separated at least for the past 70 million years, and that the two land masses were connected by the Panamanian Isthmus only a few million years ago.

The evolution of species can be traced from fossils. A notable example is that of the horse, of which there is a fairly complete record in the North American rock during the past 50 million years. During this time horses increased from a small animal about the size of a dog to their present size, and changed from a five-toed to a one-toed or hoofed animal. Horses migrated to Asia and disappeared from the American continent during the last one million years and were reintroduced by explorers from Europe about 400 years ago.

Environmental conditions in the past can also be determined by studying fossil remains. As previously mentioned, marine fossils give proof that the now-emerged land area was once submerged beneath a shallow sea. Climatic conditions for a given period can also be ascertained from the type of fossils found in the rock. It may be assumed that palm trees now growing in semi-tropical or tropical regions were to be found in the same environment at least in the more recent geologic past. Thus, the occurrence of palm fossils in 25 million-year-old rock in the northern United States would lead one to conclude that the climate in the past in that region was much more moderate than it is at present. Coal deposits in such areas as Spitzbergen and in the Antarctic would indicate a climate permitting plant growth at some time in the past. It may well be that apparent climatic changes from the distant past to the present are the result of continental drift.

Fossils have served as an important factor in solving the mysteries of the rock record. In the early nineteenth century, William Smith of England was the first to note the importance of fossils in the correlation of rock from different areas. He found that where fossils occurred, the rock *unit* (see section 12.4) contained fossils that were unique to that rock unit. For example, shales at different levels in the rock strata might be quite similar physically but the included fossils differed, thereby indicating that the shales were formed at different times. In addition, by the presence of a particular fossil in a rock outcrop, Smith was able to distinguish whether this was the top or the bottom of the particular rock unit.

It was subsequently discovered that certain organisms occurred only during specific times of the earth's history. For example, trilobites were to be found only in the rocks of the Paleozoic era, dinosaurs mainly in the Mesozoic era, and large mammals in the Cenozoic era. Variations in the structure and size of these organisms made possible the more accurate pinpointing of the relative age of the rock.

12.4 THE GEOLOGIC TIME SCALE

The organization of a geologic time scale was begun in the early nineteenth century when it was found that the occurrence of unconformities made it possible to group rock strata into identifiable units. The grouping of types of fossils and breaks in their orderly evolution also provided a means of establishing rock sequences. Prominent unconformities, combined with important changes in the fossil record, led to divisions in the rock record which were designated as geologic *eras* (table 12.1).

Four major eras are now recognized and are distinguished on the basis of the stage of development of life in each. The Precambrian era represents about 80 percent of the earth's geologic history and includes that time from the beginning of the earth's history up to 600 million years ago. The Paleozoic era ("early life" or "old life"), extends from 600 million to 230 million years ago; the Mesozoic era ("middle life"), from 230 million to 63 million years ago; and Cenozoic era ("modern life"), from 63 million years to the present.

Within each era, unconformities and changes in the fossil record of lesser prominence than those used to distinguish eras are recognized and are used to subdivide the eras into time units called *periods*. The rock unit within a period is called a *system*, and the system and the period are generally identified by the same name. The name is taken from the locality in which the rock unit was first studied and recognized as a unit. This area is then considered a *type locality* and similar rock units in other areas may be correlated with it and classified as having been formed during the same geologic age.

Correlation of the rock units is accomplished by comparing the physical characteristics of the rock or by the use of *index fossils*. Index fossils have

Historical Geology

Table 12.1
Geologic Time Scale

Era—Millions of Years Ago	Period	Epoch	Duration in Millions of Years
Cenozoic 0–63	Quarternary	Recent (Holocene)	0.025
		Pleistocene	1–3
	Late Tertiary (Neogene)	Pliocene	12
		Miocene	12
	Early Tertiary (Paleogene)	Oligocene	11
		Eocene	22
		Paleocene	5
Mesozoic 63–230	Cretaceous		72
	Jurassic		46
	Triassic		49
Paleozoic 230–600	Permian		50
	Pennsylvanian		30
	Mississippian		35
	Devonian		60
	Silurian		20
	Ordovician		75
	Cambrian		100
Precambrian 600–4,500			

been useful for this purpose because these fossils are of animals or plants which existed during a defined geologic age and had a widespread geographic distribution. These methods were useful in correlating similar sequences of rock but did not serve to put this chronology into actual years. This was accomplished in the twentieth century by radioactive dating techniques.

12.5 THE PRECAMBRIAN ERA

The earth's history may be studied by starting at the present and unraveling events backward in time, or by considering the remote past and reconstructing the sequence of events as they occurred to the present. The latter approach may be most advantageous to the student (who is normally accustomed to going forward in time) and is the usual approach. Because of the limitation on space here, only a brief account of the events leading to the present day is possible, and the discussion will be restricted mainly to such events on the North American continent.

As has been previously stated, the Precambrian era represents about 80 percent of the earth's geologic history, and it is the portion of the history about which the least is known. Much of the rock record of this era is buried beneath great thicknesses of sediments of more recent origin, and many of the Precambrian features that do exist have been greatly altered from their original form. Precambrian rock is exposed in some of the great mountain ranges and in large continental masses called *shields*. Shields are continental blocks of the earth's crust, mostly of igneous and metamorphic rock, that have been stable for long periods of time. The shields are thought to form the core of the continents, with successive periods of volcanic activity and mountain building extending the margins of these continents.

The greatest exposure of Precambrian rock in the Northern Hemisphere occurs in Greenland, the northeastern and north central portion of Canada, extending south to the north central part of the United States. This area of about 3 million square miles is geologically known as the *Canadian shield*. Study of the area in the past 100 years resulted mainly from the interest in the mineral deposits, and progress was slow because of transportation problems. More recently the use of aerial photographs and the radioactive age-dating technique greatly accelerated study and correlation of rock units of the area. Prior to the use of radioactive age dating, correlation was extremely difficult in Precambrian rock, which contains few, if any, fossils.

The oldest rock in the Canadian shield occurs in an area lying from east and south of Hudson Bay to the Great Lakes region. Rock of similar age has also been found in the Northwest Territories around Slave Lake. These rocks have been dated at over 2.5 billion years old and are thought to be part of an original nucleus of the North American continent. Rock units found extending to the southwest of the Hudson Bay rock are somewhat younger, and this trend of decreasing age continues south, east, and to the west of the oldest rock. The occurrence of successively younger rock would appear to support the contention that the continents were gradually built up around a central core. However, orogenic activity in the western part of the continent 0.5 to 1 billion years ago tends to complicate this picture.

The scarcity of fossil evidence in Precambrian rock does not eliminate the possibility of life existing during the time these rock units were being formed. Fossils first appeared in abundance in Cambrian rock, and it is reasonable to assume that life existed prior to this time. Worm borings and some reeflike masses of calcareous algae found in late Precambrian rock, and carbon which possibly was associated with organisms in early Precambrian rock, is evidence used to conclude that life began at least two billion years ago on earth.

12.6 THE PALEOZOIC ERA

During the early Cambrian period, most of the central portion of the North American continent was land. The peripheral areas were invaded by shallow seas which covered marginal basins or geosynclines. Those geosynclines found

Historical Geology

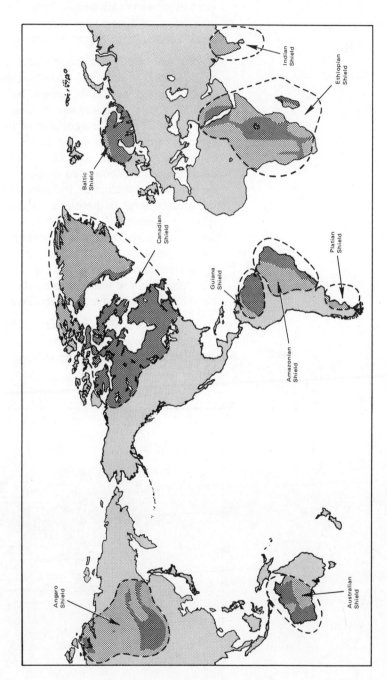

Figure 12.3 Map of the world showing location of prominent shields.

on the extreme eastern and western portion of the continental United States were associated with considerable volcanic activity along with sedimentation and are termed *eugeosynclines*, whereas those inland from the *eugeosynclines* characterized by sedimentation only are *miogeosynclines*.

By the late Ordovician period, at least half of the continent was covered by shallow seas. Most of the Canadian shield and a transcontinental arch of land which extended across Wisconsin, Nebraska, Kansas, and Wyoming and down through Utah, Arizona, and east through New Mexico and Texas were exposed. Uplift in the eastern part of the continent between the existing eugeosyncline and miogeosyncline along what was later to become the Blue Ridge of the Appalachian mountains occurred in the early Ordovician. Intense deformation, best observed in the New York-Vermont region, occurred in the eastern eugeosyncline in the late Ordovician, where earlier rock units were intruded by large batholiths in what is called the *Taconian orogeny* (mountain-forming process). Another phase of this same activity occurred late in the Silurian or early Devonian Period and is known as the Acadian orogeny. This second phase was more extensive than the Taconian orogeny and extended along the entire length of the eugeosyncline from Newfoundland to Alabama. The second phase was also accompanied by considerable volcanic activity in New England and the Maritime Provinces of Canada.

During the early Paleozoic era, geological changes continued to occur in the Western United States. Deposition continued in the geosyncline during the Ordovician and Silurian periods. In the late Devonian period, folding and

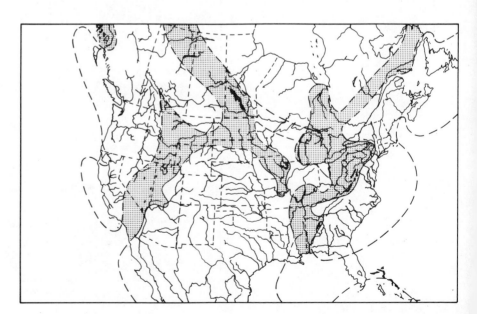

Figure 12.4   Stippled area shows possible submergence of portions of North America during middle Devonian period.

Historical Geology

uplift formed the Antler orogenic belt in mid Nevada and eastern Oregon and Washington, separating the western eugeosyncline and miogeosyncline.

A large percentage of the central United States was submerged during the Silurian and Devonian periods. The transcontinental arch extended from its previous position through Arkansas, Kentucky, and into Ohio and Indiana. These arches were not necessarily emerged land during the first part of the Paleozoic era, but they were sufficiently shallow to act as a barrier to the free movement of marine organisms, thus indicating the presence of the arches.

During the Mississippian period, the eastern part of the continent alternately emerged and submerged. Great swamps occupied the miogeosyncline in the Mississippian and Pennsylvanian periods, and huge quantities of organic material accumulated which was converted into coal. At the same time the Acadian Mountains were eroding and subsiding, permitting the seas to encroach upon the northeastern United States. In the late Pennsylvanian period, orogenic activity began in the eastern migeosyncline. This activity continued through the Permian period and possibly into the early Mesozoic era, and resulted in the formation of the Appalachian Mountain system.

Seas began to recede from the midcontinent by late Pennsylvanian, and by late Permian most of the continent had emerged. Several basins formed in the Kansas, New Mexico, Oklahoma, and western Texas region in which alternate filling and evaporation of sea water caused the precipitation of extensive salt deposits. Part of the far west was covered by water during the Permian period, especially eastern Nevada and western Utah, while further west in California a great deal of volcanic activity was occurring.

The beginning of the Paleozoic era is unique in that it was at this time in the earth's history that fossils of organic forms became abundant and complex. Organisms with well-developed digestive and reproductive systems were common and scattered over wide areas. Why this difference exists in the quantity of fossils found in Precambrian and Cambrian rock units is not completely understood.

Trilobites were the most abundant animals during the Cambrian, as indicated by the wealth of trilobite fossils found in Cambrian rock. These fossils have been invaluable in correlation of rock units formed during the Cambrian period. Many species of trilobite were recognized as the result of change through evolution during the entire Paleozoic era, providing a means of identifying rock units with the periods of this era. Brachiopods, a form of bivalve, became abundant during the Cambrian and have persisted until the present. Graptolites also appeared in the Cambrian, reached a peak during the Ordovician and declined during the Silurian. These organisms became the guide fossils of Ordovician rock units.

The first vertebrate fossil, a fishlike animal, was found in Ordovician rock and appears to have lived in fresh water. Fish, evolving from this early form, were not prominent until the Devonian period when they became numerous and existed in a variety of types.

Crinoids or "sea lilies" were animals that began their development during the Ordovician period and reached the peak of development in the Mississippian period. It was in the Silurian that land plants made an appearance. The land areas prior to this time were thought to be quite barren. Plants developed to a degree that tree ferns with trunks up to 1 meter in diameter were common during the Devonian period. Plant life became very abundant during the Mississippian and Pennsylvanian, forming the Appalachian coalbeds in the swamps of the Pennsylvanian period. More than 6,000 species of plants have been identified from the abundant fossil record in the coal beds formed at this time.

The amphibians made their appearance in the late Devonian and are considered the next on the evolutionary scale above the first fish. The amphibians were still closely tied to the water, laying their eggs and spending the early part of their life cycle in the water. The amphibians continued to evolve and increase in size, becoming 2 to 3 meters in length in the Pennsylvanian period.

The reptile, the first true land animal which evolved from the amphibian, appeared late in the Pennsylvanian period and became the forerunner of the dinosaurs of the Mesozoic era. The early types were only about a foot in length, giving little indication of the giants they were to become. The Pennsylvanian period is sometimes called the "age of cockroaches" because of their great abundance and variety during that period. Some species were as much as 8 centimeters in length, and the rock record has revealed that as many as 1,000 species existed during this time.

The Pennsylvanian period was well suited to the development of plant and animal life because climatic conditions were mild and humid. However, at the beginning of the Permian period conditions began to change. The climate became cooler and more arid on the North American continent and in certain other parts of the world. The swampy areas that existed during the Pennsylvanian period disappeared along with many species of plants which formed the coal beds. These plants were replaced by the conifers which developed in the Pennsylvanian and became well established during the Permian period.

Trilobites and crinoids, prominent during much of the Paleozoic era, were extinct by the close of that era. Reptiles became important, and diversified to the extent that some were herbivores, some carnivores, and still others insectivores. Mammals had not yet made an appearance by the close of the Paleozoic era. However, the therapsid, an offshoot of the reptiles, had certain mammalian characteristics and is thought to be an ancestor of the mammal. The therapsid appeared sometime during the middle of the Permian period.

12.7 THE MESOZOIC ERA

During the early Mesozoic era (Triassic period), the Appalachian Mountains, formed principally during the Permian period, were subjected to the forces of erosion. Debris resulting from this erosion was carried down both slopes to be deposited on the lowlands to the west and transported beyond the continental margins to the east. Faulting in the late Triassic resulted in the formation of fault troughs or *grabens* east of the Appalachian Mountains. Material eroding from the Appalachian highland accumulated in these subsiding troughs to form a variety of sedimentary rocks. No marine deposits or marine fossils representative of the Triassic period are found in the rock, indicating that no oceanic encroachments had occurred in the east. Some uplift in the late Triassic was accompanied by intrusion of basalt into the sedimentary deposits. This intrusive activity formed a series of sills, of which the Palisades sill on the western side of the Hudson River is a prominent example. This uplift was not as great as the Appalachian orogeny and is known as the *Palisades disturbance*.

In the western United States, the continental margin was the scene of volcanic activity. Volcanic deposits are found interbedded with other sedimentary deposits in Triassic rock in Nevada, which was then covered by a shallow sea. The west central United States was thought to be a flat, featureless plain sloping gently toward the site of the present Rocky Mountains system.

Little activity occurred in the eastern part of the United States during the Jurassic period. Mountain building had essentially ceased, and the area was subjected to extensive erosion which continued throughout the Mesozoic era.

The region now occupied by the Rocky Mountains was the site of the Rocky Mountain geosyncline during the Jurassic. This geosyncline was invaded by the Sundance Sea from the Arctic, covering the western Canada, Montana, Idaho, Wyoming, and Utah area. Marine deposits laid down in this sea were covered by nonmarine deposits formed after the sea retreated. These nonmarine deposits, known as the *Morrison formation*, are important because it is here that most of the fossils of American dinosaurs have been found.

Along the west coast, volcanic activity continued to pour volcanic material into the Jurassic seas occupying the California geosyncline. At the same time, widespread deformation of the crust and intrusion of batholiths formed fold mountains along the California-Nevada border and in Oregon and Washington. This *Nevadan disturbance*, as the orogeny was called, was responsible for forming the gold quartz veins of the "mother lode" belt in California.

The mountains formed by the Nevadan orogeny were well developed by the beginning of the Cretaceous period, and sediments eroded from the slopes were filling the California geosyncline from the east with sand and gravel. During the Cretaceous, the west side of the geosyncline was elevated, resulting in sediments being deposited in the California geosyncline from both sides.

By the middle Cretaceous, the continent was split in two by encroachment of the sea from the north and south Rocky Mountain geosyncline (fig. 12.5). In the late Cretaceous, the seas had retreated again and the geosyncline was subjected to deformation, identified as the Laramide orogeny. This activity, which included thrust faulting, folding and uplift of fault blocks, extended into the Cenozoic era and resulted in the formation of the Rocky Mountain system.

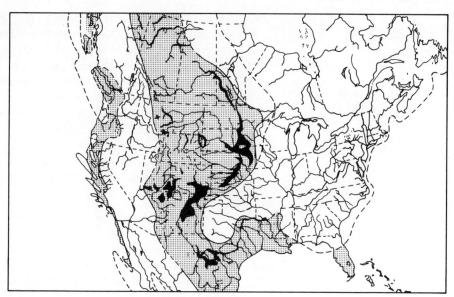

Figure 12.5 The North American continent was split in two by encroachment of the sea from north and south during middle Cretaceous period.

The entire system was not formed as a unit but rather as several units during the late Mesozoic-early Cenozoic eras.

The Mesozoic era, also known as the Age of Reptiles, was the era during which the reptiles reached the peak of their development. Dinosaurs ("terrible lizard") were the most spectacular of the reptiles, diversified to occupy the air and the sea, and to become the dominant animal on land. Some dinosaurs such as the *brontosaurus* grew to 30 meters in length and weighed over 50 tons. These animals were so ponderous they had to spend much of their lives in swamps where water helped to support their weight and sufficient vegetation was available to supply them with food. Dinosaurs lived on into the Cretaceous period, and it was during this period that *Tyrannosaurus rex*, a carnivore, roamed the earth. This dinosaur is considered to be the most ferocious animal ever to exist on earth, although the only basis for this is its apparently fierce appearance. The end of the Cretaceous period was also the end of the dinosaurs. Why the many forms became extinct is not completely understood, although changes in climate brought about by mountain-building activity have been suggested as a cause.

Definite evidence of mammals and birds have been found in Jurassic rock, although fossils of reptiles with mammalian characteristics occurred in the Triassic period. Early mammals were small, inconspicuous, capable of adapting to changing conditions, and thereby survived into the Cenozoic era.

Plant life also changed in form during the Mesozoic era. Conifers, which had their beginning in the Paleozoic era, were plentiful and continued into the

Cenozoic era. The petrified forest in Arizona is the fossil remains of conifers which lived during the Triassic period. Some of these trees were 30 meters or more in height, and their trunks were 3 meters in diameter. Ferns were less important during the Mesozoic era than previously, while seed plants or angiosperms evolved during the Cretaceous period to become the dominant plant during the Cenozoic era.

The seed plants may have been a factor in the more rapid development of birds and mammals during the late Cretaceous period, as the seeds provided a plentiful source of food.

12.8 THE CENOZOIC ERA

The Cenozoic era saw an end to the large-scale encroachments of the seas upon the North American continent. The land was invaded only along the coastal margins, with the exception of an embayment up the Mississippi River valley which reached the site of Cairo, Illinois. This marine encroachment had retreated by the end of the early Tertiary period to approximately the present coast line. The eastern coastal region was submerged beneath sea level during much of the early Cenozoic era, and Florida and the Gulf of Mexico coast were covered by the sea until late Cenozoic.

Much of the activity that occurred in the Cenozoic era was the result of large-scale epeirogenic uplift. The Appalachian Mountain system had not been active during most of the Mesozoic era and the early Tertiary period. Erosion had taken its toll and had leveled the system to a flat or broadly undulating plain called a *peneplain*. General uplift during the late Tertiary period raised the level of the old mountain system to approximately 1,200 meters elevation with gentle slopes to the east and west. Streams were rejuvenated, and downcutting through the newly formed slopes resulted in the present topography of the Appalachian Mountains. The tops of the resistant ridges are all at approximately the same elevation and are remnants of the old erosion surface.

The Laramide orogeny, which formed the Rocky Mountain system and had its beginning in the Mesozoic era, continued active until the end of the Eocene epoch. At this time, the mountain-building activity ceased, and erosion again became dominant. Fluvial material resulting from this erosion covered large areas of the Great Plains region, particularly during the late Tertiary period. By the end of the Oligocene epoch, erosion had leveled off the Rocky Mountains to a high peneplain of 600 to 900 meters elevation. Epeirogenic uplift began in the Rocky Mountain region at the beginning of the late Tertiary period and continued into the Pleistocene epoch. Stream erosion, which was renewed coincident with the uplift, etched the Rocky Mountain system into its present relief.

To the west of the Rocky Mountains and east of the California border, normal faulting during the Miocene and Pliocene epochs resulted in a series of uplifted mountain blocks, or *horsts*, separated by subsiding valleys, or *grabens*. This activity formed the Basin-Range Province, covering most of Nevada, western Utah, and southeastern California.

Geology

The west coast of the United States was most active geologically during the Cenozoic era. The Sierra Nevada range, a fault-block mountain about 550 kilometers long and 160 kilometers wide, was tilted westward to an elevation of almost 4,000 meters along its eastern edge. The western edge of the block is depressed about 6,000 meters below sea level in the Central Valley of California. Movement responsible for forming the Sierra Nevada Mountains began in the Miocene epoch, diminished during the Pliocene, and was renewed in the Pleistocene epoch.

North of the Sierra Nevada Mountains is the Cascade range, which extends through Oregon and Washington into Canada. Unlike the Sierra Nevada Mountains, which are composed of metamorphosed basement rock, the Cascade range is predominantly volcanic rock. Volcanism is this area was particularly active during the Pleistocene epoch, forming many cones, of which Mt. Mazama (which exploded to form Crater Lake), Lassen Peak, and Mt. Shasta are representative examples.

East of the Cascade Mountains, extensive volcanism formed the plateau basalts of the Columbia Plateau. Lava flows during the early Tertiary period covered much of the western part of the area. Great lava flows reoccurred in the Miocene and continued on into the Pleistocene to cover more than 250,000 square kilometers with a very fluid lava.

Folding and faulting in the late Cenozoic era resulted in the formation of the coastal ranges on the western margin of California, Oregon, and Washington. The great Central Valley basin, the Ventura basin and the Los Angeles basin are associated with this mountain-building activity. Great quantities of sediments of marine origin were deposited in these basins as they subsided during the Miocene and Pliocene epochs. The deposits were overlaid with nonmarine sediments during the Pleistocene. Major oil fields of California are located in these basins, where great quantities of oil had accumulated in the Miocene and Pliocene deposits.

Perhaps the dominant aspect of the Pleistocene epoch was the recurring ice ages which were responsible for shaping much of the landscape of Canada, the northern United States, and the higher elevations south of the glacial margins. Indirect changes to the landscape also occurred as a result of glacial action. These changes included the lowering of sea level an estimated 75 to 120 meters, which did much to shape the coastal areas. Lowering of the sea level may also have provided a land bridge over the Bering Strait from Asia to Alaska, which, would have permitted the first humans to migrate to the American continent 30,000 to 35,000 years ago. Drainage systems, particularly in the central United States, were altered to approximately the present pattern, while in the west, large lakes occurred where now the desert exists. The Great Lakes are a direct result of glacial action, and many of the scenic wonders of the great mountain ranges are the result of glacial erosion.

Earthquakes and volcanoes have been quite active on the North American continent during the Pleistocene, indicating that change continues unabated. The earth's crust is now subject to the same forces that were working in the past, although the change is slow and gradual and not always significant during one's lifetime.

The giant dinosaurs disappeared before the start of the Cenozoic era and were replaced by the mammals as the dominant form of life. With the development of the cereal grains and grasses, grazing animals made rapid advances. Horses, originally quite small, first appeared in the Eocene epoch and gradually evolved into their present form and size. Many of the hooved mammals followed this same evolutionary pattern of increase in size. The sabertoothed cat, as familiar to many as the dinosaurs, first appeared during the Oligocene epoch but became extinct during the Pleistocene, whereas many other members of the cat family survive to the present. Birds varied tremendously in form during the Cenozoic era, rapidly becoming better adapted to flight.

Although the first primates occurred in the Paleocene epoch, apelike animals did not appear until the Miocene, and manlike apes until the Pliocene. At what point man himself evolved is not yet clear, although the oldest remains of a manlike creature so far uncovered are about 1.75 million years old. Whatever his early beginning, man has undergone modification physically, mentally, and socially, and in the process has gone through a succession of cultures which continue to change even at present.

12.9 SUMMARY

Our knowledge of the history of the earth has become so extensive that only the briefest of reviews is possible in one short chapter. However, the intent here is to acquaint the student with a broad subject and not to train him for a career in paleontology.

To understand the earth's history it is necessary to know about the factors involved in its study. The measurement and understanding of relative and absolute time, and knowledge of the manner in which the rock record is studied, are necessary. Fossils of animals and plants have played an important part in unraveling the more recent parts of earth's history, especially the past half-billion years when plants and animals were abundant. Index fossils have been useful in identifying the era or period in which a rock unit was formed.

The geologic time scale permits the presentation of geologic history in an organized manner by correlating certain rock units with specific periods. The identity of the rock units are related back to a type locality, which is the area where the rock unit was first described.

The earth's history is often separated into eras, which are distinguished by prominent unconformities. The longest era, the Precambrian, is characterized by a lack of fossil material. This does not, however, signify that all rocks lacking fossils are Precambrian. An abundance of fossils first appeared in Paleozoic rock and include mostly the invertebrates, fish, and some primitive land plants. Reptiles, especially dinosaurs, were plentiful during the Mesozoic era, and mammals and seed plants became the dominant form of life in the Cenozoic era.

During these eras, extensive changes had occurred on the land surface. Regions that were once submerged beneath the sea have emerged to become land areas. Plains regions have been elevated into mountain systems and mountain systems have been eroded to become plains again. The cycles continue, with volcanism, mountain building, and erosion playing their important roles in modifying the landscape.

QUESTIONS

1. Describe the difference between relative and absolute time as applied to geologic time.
2. Briefly describe methods used to determine absolute geologic age of rock formations.
3. Briefly describe the nature of an unconformity. Make a diagram of an angular unconformity; disconformity; nonconformity.
4. In what forms may fossils appear?
5. Give a brief summary of the development of life during the Paleozoic Era; Mesozoic Era; Cenozoic Era.
6. Give a brief history of the geologic changes that have taken place in your locality during the past half billion years. Review other sources dealing more directly with these changes in your locality.

IV THE ATMOSPHERE

13 Introduction to the Atmosphere

Importance of weather to man and its prediction.
The structure of the atmosphere.
How the weather is produced.
Measuring atmospheric and weather conditions.
Cloud structure.

Throughout history, man's ability to live and work successfully has been in a large measure dependent on weather. Consequently, he has always attempted to predict the future course of weather in order to be better prepared for its severity and marked sudden changes. Permanent adverse changes in climate could bring ruin to the effective functioning of society and make portions of the earth's land area practically uninhabitable. Fortunately, the weather, unlike the reserves of some other natural resources, appears to be inexhaustible. Although there are fluctuations in rainfall, temperature, and other atmospheric elements from day to day, month to month, and even year to year, permanent changes in climate during the records of civilized man have been small. But even seasonal fluctuations have such profound influences on human livelihood and pursuit of happiness that no matter what our occupation we daily take keen interest in the weather.

13.1 INTRODUCTION TO METEOROLOGY AS A SCIENCE

In view of the fundamental influence of weather and climate on our food, clothing, and shelter, and therefore on our health and happiness, it is not surprising to find that from earliest times people have taken notice of seasonal variations. Early in human history, we find that people looked for signs or omens that would foretell the weather, particularly the approach of storms. In an attempt to court supernatural aid in warding off unfavorable weather, people made supplication to meteorological deities. Some of the weather signs observed over the ages contained an element of truth; others were mere superstitions, of which a surprisingly large number still persist. These are, however, giving way to a more scientific viewpoint and a better understanding of how weather phenomena functions.

Although people's observations of the weather began with the dawn of consciousness, meteorology, or weather science, is still a comparatively new member of the family of modern sciences. Accurate observations, truly representative of open-air conditions—that is unmodified by purely local or accidental influences—are not easy to obtain. Prior to the invention of meteorological instruments, it was impossible to accurately measure atmospheric pressure, temperature, humidity, wind velocity, and other weather elements. The barometer and thermometer developed in the seventeenth century permitted the gathering of more accurate weather records. However, daily weather service as it is now known could not be established until the telegraph was invented, and a widespread system of daily synoptic (simultaneous meteorological observations over broad areas) reporting was organized.

During the past century, the technological developments that improved weather observation systems have acted as stimulants to the rapid development of meteorology as a science. In this regard, space technology appears to be emerging as the current major stimulant. Rockets and satellites not only have provided new vehicles for carrying meteorological instruments, they have provided something more important—an opportunity for a new view of the atmosphere. With rockets it is possible to sample the atmosphere directly at altitudes never before attained. With satellites the atmosphere can be viewed from above and its motion and phenomena observed on a global basis rather than upward from discrete points on the surface. Further, it is now possible to measure the energy input from the sun without the filtering effect of the atmosphere.

The use of satellites in meteorological work led to two discoveries of significance in the 1960s. First was the demonstration that the cloud cover was highly organized on a global basis. Coherent cloud-cover systems were found to extend over thousands of miles and were related to other systems of equal dimensions. In this manner, the integrated character of the atmosphere on a global basis was clearly shown. Second, weather systems were directly identified by their cloud structure and so it was possible to identify and locate important atmospheric phenomena such as fronts, storms, hurricanes, and cloud fields and chart their course on a daily basis with extreme accuracy.

Cloud-cover observations by satellites have been almost continuous since 1962. The most spectacular observations have been those of tropical storms and hurricanes. For example, in 1964 there were 63 tropical storms in meteorological records. Meteorological satellites observed and tracked 46, and of these 18 were located and identified by meteorological satellites before they were noted in the forecast bulletins issued by forecasting centers. Of the tropical storms observed, 7 were hurricanes and 17 were typhoons.

Cloud pictures have also aided in the delineation of jetstreams and shear lines and have furnished information on cloud formation in areas between regular surface observation sites. The weather-satellite data have also been applied to nonmeteorological problems such as the location and movement of sea ice, the freeze-up and break-up of ice on major rivers and lakes, and the variation and extent of snow cover, swamps, and flooded areas. Such satellites have also been used in following changes in the detail of many geographical features.

13.2 THE ATMOSPHERE

Although the atmosphere appears to extend one to two earth radii (3,200 to 6,400 kilometers) above the surface of the earth, 90 percent of the gases exist within the first 30 kilometers, and it is within this zone that most of the meteorological activity occurs. The atmosphere is described and studied in terms of composition, temperature, and pressure, and the variation of these characteristics with altitude. This is insufficient, however, because of the complex mass motions resulting from heating and the chemical changes that result from the ionization of certain gases. These reactions produce phenomena in the atmosphere that are quite complex in nature. It is now known that the earth's atmosphere is very dynamic in that it is continually changing properties with time of day, season, year, and even with solar cycles.

Evidence exists which indicates that the present atmosphere differs from the primitive atmosphere that may have accumulated during the early formation and history of the earth. Some of the evidence for this is based upon a comparison of the abundances of elemental gases such as neon and nitrogen with nonvolatile elements of the earth, and on a comparison of these ratios with their cosmic abundances. One such comparison reveals that there is one ten-billionth the amount of neon with respect to silicon in the earth that might be expected from cosmic abundances of these elements. Consequently, it is believed that most of the original gases were lost early in the earth's evolution and a secondary atmosphere formed somewhat later. The most likely source of such a secondary atmosphere is volcanism, and this secondary atmosphere is quite different from the present one. The secondary atmosphere contained such gases as ammonia, methane, sulfur oxides, hydrogen sulfide, nitrogen oxides, carbon dioxide, and copious amounts of water vapor, but no elemental oxygen. The initial formation of atmospheric oxygen came from photodissociation of water and to a lesser extent of carbon dioxide. *Photodissociation* is the breaking down of the water molecule into hydrogen and oxygen by the action of sunlight. Subsequently, the major portion of oxygen came from the photosynthetic process of plant life that gradually evolved on the earth.

The present atmosphere is a mixture of gases, including nitrogen and oxygen as the major components and small amounts of argon, carbon dioxide, and other elemental gases (table 13.1). Water vapor is also present in the atmosphere in amounts up to 4 percent by volume, but it is generally not included as one of the atmospheric components. The combined weight of these gases in the atmosphere is about 1 kilogram per square centimeter of earth's surface at sea level. This represents a total atmospheric weight of 6×10^{15} tons. Most of the total atmospheric gases occur in the *troposphere,* the lower layer of the atmosphere.

Table 13.1
Percentages of Atmospheric Gases by Volume

Gas	Symbol	Percent of Volume
Nitrogen	N_2	78.08
Oxygen	O_2	20.95
Argon	A_2	0.93
Carbon Dioxide	CO_2	0.03
Other Gases (neon, helium, krypton, hydrogen, xenon, ozone, radon)		0.01

Even when apparently clear, the atmosphere contains an enormous number of tiny suspended liquid and solid particles called aerosols (dust particles). In addition to solid matter originating from such sources as volcanic eruptions, aerosols are comprised of windblown dust, meteoric dust from beyond the earth's atmosphere, microorganisms including pollen and plant spores, as well as contaminants introduced by man.

Very minute salt particles, particularly sodium chloride, are introduced into the atmosphere by salt spray from the sea. These particles are strongly *hygroscopic*: that is, they are believed to form condensation nuclei around which drops of water accumulate very readily because of the salt's affinity for water molecules. Water will also accumulate around the other solid particles, but not as readily. These particles are generally less than one micron (one-millionth of a meter) in diameter and may stay suspended in the atmosphere for days.

A manmade contribution to the atmosphere is the pollutants coming mainly from the combustion of various fuels. Toxic gases, vapor, and solids are discharged by this process, which together with the action of sunlight, contaminate the atmosphere with substances collectively called *smog*. In some areas, the smog concentration has reached levels so harmful to plants that the growth of certain crops is not feasible. The levels of concentration of smog are occasionally harmful to humans as well, requiring a decrease in smog-producing activities. These include a reduction in automobile traffic and a requirement that industries switch to fuels with less smog-producing properties.

The atmospheric components just discussed are those found mainly in the lower atmosphere or *troposphere*. The troposphere is one of the major distinct layers of the atmosphere and extends up to an elevation of approximately 15 kilometers above the equator and 8 kilometers above the poles. Within this layer, the gases shown in table 13.1 generally occur as a turbulent homogeneous mixture comprising about 75 percent of the total mass of the atmosphere. Temperatures within this layer decrease with elevation, since most of the heat comes from infrared (heat) radiation from the earth, thus heating the atmosphere from the bottom up. This results in temperatures in the troposphere decreasing from an average of 20°C for the total earth's surface down to −80°C at the *tropopause* (upper limit of the troposphere) above the equator, and −55°C above the polar regions. These differences in temperature at the tropopause are due to the variation in elevation above the earth's surface.

In the *stratosphere*—the layer which occurs above the troposphere and extends to an altitude of approximately 50 kilometers—temperatures increase with altitude. The temperature appears to be due to the presence of a layer of ozone (O_3) in the *stratopause*, or upper limit of the stratosphere. Ozone has the capacity to absorb energy coming from the sun and reradiate this energy as heat, thus causing the stratosphere to show a higher temperature at its upper boundary and a decreasing temperature downward toward the tropopause. This increased temperature with altitude is thought to contribute to stratification of the gases with very little vertical mixing or general turbulence in the stratosphere.

Above the stratosphere is a layer that is quite turbulent. This is the *mesosphere*, a layer extending up to an altitude of approximately 90 kilometers, in which the temperature decreases with an increase in altitude. The decrease in temperature continues to the *mesopause*—the upper layer of the mesosphere—where the lowest measured atmospheric temperature of −140°C is recorded.

Up to the altitude thus far discussed (90 kilometers), the atmospheric gases have been a fairly homogeneous mixture; thus, this portion of the atmosphere is called the *homosphere*. Above this elevation, the gases tend to separate according to their relative atomic weights and to stratify, with the heavier gases such as oxygen at the bottom of the zone, the helium layer above the oxygen, and hydrogen in the upper extremes of the atmosphere (fig. 13.1). This is the *heterosphere*, a zone which includes the *thermosphere* and the *ionosphere*.

The thermosphere lies above the mesopause and is a region in which temperatures increase with elevation. This increase in temperature results from the absorption of ultraviolet radiation by monatomic oxygen and causes temperatures in this zone to rise to as high as 1,275°C at an elevation of 700 kilometers. The heat implied by a temperature of 1,275°C cannot be felt because the density of gas atoms and charged particles is so low that transmission of energy by collision is very small. However, the energy levels of the individual particles is sufficiently high to be characteristic of such temperatures.

The Atmosphere

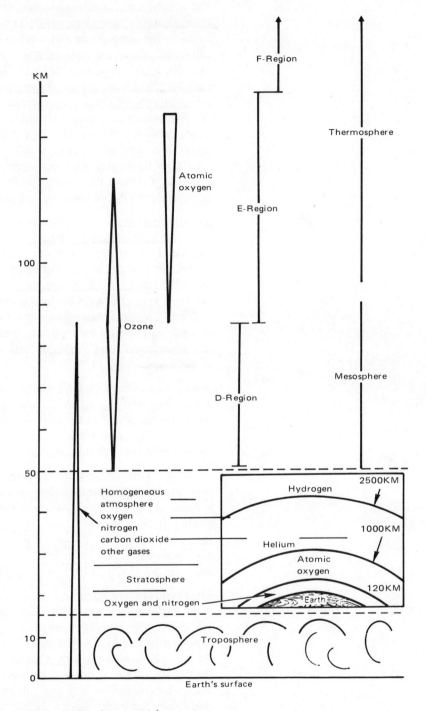

Figure 13.1 Profile of the earth's atmosphere.

The presence of monatomic oxygen results from the effect of solar radiation, particularly X-ray and shortwave ultraviolet radiation. Such electromagnetic energy causes molecular oxygen to be split into atomic oxygen at elevations as low as 50 kilometers. This form of oxygen increases until it is more abundant than molecular oxygen at approximately 200 kilometers. Such activity, beginning at 50 kilometers, overlaps the thermosphere and extends up to the limit of the atmosphere. The presence of free electrons and positively charged ions has caused this region to be called the ionosphere. The ionosphere is divided into a number of subregions identified basically according to electron density. These subregions are important because of the influence the free electrons and positively charged ions have on radio waves and radio communications (fig. 13.2).

The earth has a magnetic field (fig. 3.8) which extends from the earth's core to an indeterminant distance into space. The magnetic field has little influence upon the atmospheric gases except where free electrons and positively charged ions occur in the ionosphere. In this region, the magnetic field influences the motion of these charged particles. Beyond the ionosphere, the magnetic field or *magnetosphere* is influenced to a great extent by the solar winds. The magnetosphere extends approximately 10 to 12 earth radii in the direction of the sun and to a greater but unknown distance in the opposite direction. It is within the boundaries of the magnetosphere that the Van Allen radiation belts occur. These radiation belts represent concentrations of high-energy particles trapped in the magnetosphere.

13.3 WEATHER-PRODUCING MECHANISMS

Weather in the atmosphere is not the product of a single process but rather the result of several factors interacting in a dynamic fashion. Primary among these factors are heat and moisture, and these in turn give rise to secondary influences which contribute to the general pattern of weather.

Heat: The principal source of energy whereby the atmosphere and earth are heated is the sun. Only an extremely small amount (about 1 part in 2 billion) of the sun's total energy is intercepted by the earth, and this amount is measured as the *solar constant* (see section 3.3B). The solar constant has a value of 1.97 calories per square centimeter per minute for solar energy striking perpendicular to the upper surface of the atmosphere. Although the solar constant is not truly constant, it is useful in determining energy input into the atmosphere and on the earth's surface. The amount of energy that actually penetrates the atmosphere to reach the earth's surface is called *insolation*. Insolation varies with latitude and according to the time of the year and in this manner partly contributes to seasonal changes in temperature (see section 15.1).

Despite the passage of solar energy through the atmosphere, only a relatively small amount of this heat (14 percent) is absorbed by the atmosphere. The

The Atmosphere

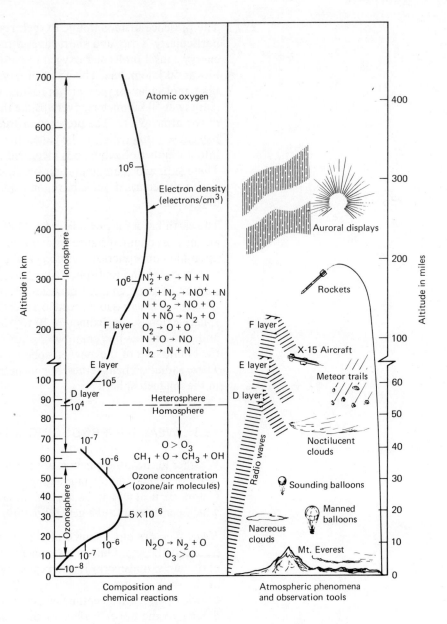

Figure 13.2 Vertical distribution of atmospheric properties and phenomena (above and at right).

source of most of the energy heating the atmosphere comes from the earth, which absorbs slightly more than half of the energy from the sun intercepted by the earth's surface (fig. 13.3). Most of the heat absorbed by the earth's surface is reconveyed back to the atmosphere by *radiation, conduction,* and *convection.*

Introduction to the Atmosphere

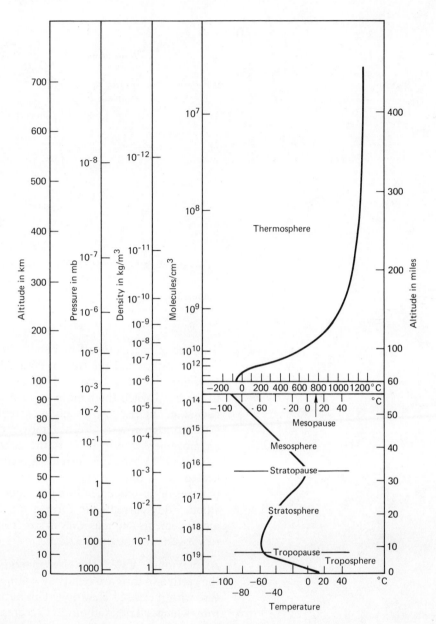

Figure 13.2 (Continued)

Part of the heat radiated by the earth is returned to the atmosphere as radiant energy. Radiant energy, in this case longwave or infrared radiation, is transmitted through space in waves at the speed of light. This energy heats in the same manner as a fire in the fireplace heats an individual several feet away. The heat is felt despite the fact that the temperature of the intervening space may be quite a bit cooler.

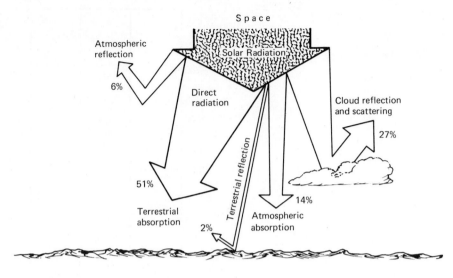

Figure 13.3 Distribution of solar energy as it reaches the earth's surface.

Transfer of energy (heat) by conduction occurs mainly at the interface of the earth's surface and the troposphere. The heat is conducted from molecule to molecule much the same as the heat from a kettle on the stove is transferred to the hand when the handle is grasped. In this way the lower atmosphere is heated, giving rise to convection currents whereby heated air, being less dense, is transported upward. Vertical movement of air is called *convection* (horizontal air movement due to temperature differences of adjacent air masses is called *advection*). Convection currents are considered the principal means by which heated air is transmitted to the upper atmosphere. In the process, complex patterns of air movement are generated in a manner not yet completely understood. Eddy currents are formed which involve both vertical and horizontal air movements and may vary in size from a few centimeters in diameter to large storm systems like the hurricane.

As previously stated, the atmospheric gases are heated for the most part by the earth and thus are relatively warm at the surface and decrease in temperature with increased elevation. This vertical change in temperature is the *lapse rate,* which averages about 2° C decrease per 300 meters elevation. The value has been determined by countless measurements of the vertical temperature gradient in many parts of the world.

The lapse rate represents the temperature of air which is not moving vertically. However, air masses do move both upward and downward, and this movement will modify temperatures depending upon the change in pressure of the moving air mass. When unsaturated air (air low in moisture) is heated from below and is forced to rise because of decreased density, the temperature of the air within this rising current will decrease at the rate of 3° C per 300 meters elevation.

This cooling rate of the unsaturated air is known as the *dry adiabatic lapse rate,* and the decrease in temperature is caused by the expansion of the air mass as it rises. The temperature of the air mass must decrease as it expands, because the energy required for expansion comes from the air mass itself.

The upward movement of saturated air results in condensation of the moisture. The heat released by the condensation process is absorbed by the air, and as a result the adiabatic cooling rate is decreased. The resulting rate of cooling, called the *moist adiabatic lapse rate,* varies from 1°C to slightly less than 3°C per 300 meters elevation. The variation in temperature in the moist adiabatic lapse rate results from the fact that at high air temperatures the moisture content of the air is high and a small decrease in temperature causes a relatively large amount of moisture to condense. Thus the amount of heat released due to condensation is greater at high temperatures than at low temperatures.

Occasionally, circumstances exist where the temperatures increase with altitude and a negative lapse rate occurs; this is called a *temperature inversion.* Temperature inversion may occur during calm, clear nights when ground cooling is rapid and there is a sudden cooling of the air at the ground surface, causing the ground-level air to become a good deal cooler than the overlying strata (fig. 13.4). The inversion is a climatological phenomenon found along the west coast of North and South America. Its presence up to 1,500 meters above the ground is responsible for trapping air contaminants and causing air pollution in industrial areas or regions of high population densities.

Moisture: In an unending exchange, water is transferred from ocean to atmosphere to land, and from land back to ocean again. The primary source of atmospheric moisture is the ocean, although some moisture is obtained from

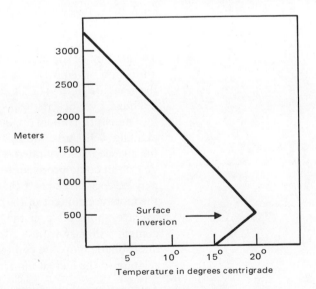

Figure 13.4 Relationship of temperature inversion to altitude.

water sources on land. By a distillation process that utilizes about one-third of the solar energy reaching the earth's surface, some 10 million billion gallons of water are evaporated each year. Approximately 20 percent of this comes from water sources on the land, and the balance from the sea. This atmospheric moisture in the form of vapor is transported over the land masses, where lifting and cooling cause condensation, and the moisture falls as rain or snow. Part of this moisture is returned to the atmosphere by evaporation. Ultimately, through runoff and seepage, the water is returned to the sea, completing the *hydrologic cycle* (fig. 13.5).

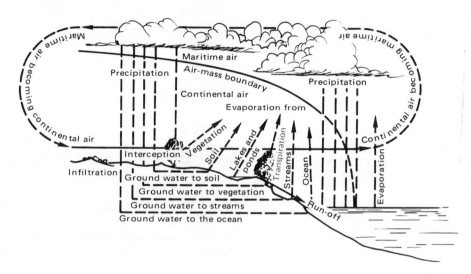

Figure 13.5 The hydrologic cycle.

The amount of moisture in the troposphere remains relatively constant and is a function of temperature. In the tropic regions, where temperatures are high, the moisture content of the air will reach 3 to 4%. The higher value may generally be found over deserts near the equator, where low rainfall is not the result of low moisture content in the atmosphere but rather to air temperatures remaining well above the *dew point*. The dew point is the temperature at which the air becomes saturated with moisture. Further cooling of the air below the dew point could produce clouds or rain. From this relationship, the conclusion may be correctly drawn that the lower the air temperature, the lower will be the moisture content of the air. In the polar regions, the moisture content will average about 0.5% in the winter, whereas in the middle latitudes moisture content will vary from 0.5% in the winter to 2.0% in the summer. Generally, atmospheric moisture content increases from the poles to the equator in direct relation to the increase in temperature.

Temperature of air masses and their moisture content play an important role in determining the relative stability and instability of the atmosphere. The degree of stability is important in weather forecasting. Generally speaking, when the atmosphere is stable, cloudless skies and clear weather prevail, whereas unstable air leads to the formation of clouds and possible precipitation. An expanded discussion of this topic is included in Chapter 14.

13.4 ATMOSPHERIC VARIABLES

The success of weather forecasting or of a weather research program is to a large extent dependent upon the ability to make accurate measurements of atmospheric variables. This data is of vital significance in the identification and analysis of air-mass movements (discussed in Chapter 14). The instruments used are varied both in kind and in degree of sophistication, from the simple dipstick used to measure rainfall to complex electronic equipment contained in modern weather satellites. Some of the more significant types of weather data and the instruments used to obtain this data will be discussed briefly.

Figure 13.6 Weather observation station with measuring instruments supplying data to a computer console. Courtesy of NOAA.

13.4A Temperature Measurement

Temperature is the measurement of heat accomplished by a thermometer, of which several forms are used. The most common thermometer is one that consists of a glass tube filled with a liquid such as mercury or alcohol which expands and contracts with a change in temperature. A scale (Fahrenheit, Celsius, or Kelvin) will indicate the temperature level as measured by the rise and fall of the liquid in the tube. Also certain conductors of electricity will vary in their resistance to current flow with changes in temperature. This change in resistance can be calibrated in terms of temperature. Metals vary in coefficients of expansion, and two laminated metal strips with different expansion rates will produce a torque that can be calibrated to changes in temperature.

The several types of temperature-measuring devices described are used in meteorological work. Aside from the temperature changes occurring with elevation discussed in section 13.3 the most usual measurements made are wet- and dry-bulb temperatures and daily maximum and minimum temperatures. The wet- and dry-bulb temperatures are obtained by means of a psychrometer and are used to measure humidity. The psychrometer consists of a pair of liquid-in-glass thermometers, one of which is used to measure temperature in the normal way, and the other, with the bulb covered by muslin cloth that is kept wet, is used to measure wet-bulb temperatures. Readings from these thermometers are made while air is rapidly moving past the wet bulb. This may be accomplished by whirling the psychrometer rapidly.

Maximum and minimum temperatures are obtained for specific periods, such as every 24 hours. Two special liquid-in-glass thermometers are used for this purpose. One has a constriction close to the bulb through which the liquid is forced when the temperature rises. When temperatures fall, the weight of the liquid is not sufficient to cause it to flow back into the bulb, and thus the maximum temperature reached is recorded. To reset the thermometer, it must be shaken to return the liquid to the bulb. The minimum-temperature thermometer has a dumbbell-shaped glass index in the liquid column. The index is slightly smaller than the bore of the thermometer, and the thermometer must lie nearly horizontal to function. The surface tension of the meniscus or surface of the liquid will move the index as the temperature decreases, but not as the temperature increases. Thus, the lowest temperature for the specified period will be recorded; the thermometer is reset by tipping it and allowing the index to return to the meniscus.

Placement of thermometers to obtain valid temperature readings is an all-important aspect of collecting weather data. Temperatures may vary as much as 15 to 20°C within a few feet of a concrete pavement. And of course the thermometer may accumulate excess radiant heat if placed in the direct sunlight. Thermometers provide the best information if placed about five feet above the ground surface in a shelter that shields the thermometer from the direct rays of the sun but at the same time permits free circulation of air.

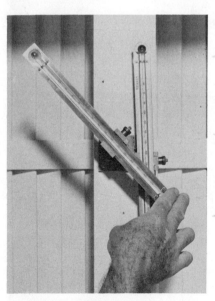

Figure 13.7 Maximum and minimum thermometers. Courtesy of NOAA.

13.4B Humidity Measurement

When a given volume of air contains all the water vapor it is capable of holding at the existing temperature, the air is said to be saturated. The quantity of water vapor in saturated air is dependent on the temperature of the air. (fig. 13.8.) The quantity of water vapor representing saturation increases as the temperature increases. If saturated air is cooled, the vapor condenses into droplets and precipitation can occur.

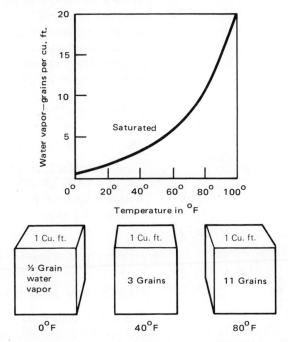

Figure 13.8 Air is capable of holding increased amounts of water vapor as the air temperature rises.

It is often desirable to know how close to saturation a parcel of air may be. This is expressed as *relative humidity*. Relative humidity is the ratio of water vapor actually in the air to the total amount of water vapor the air is capable of holding at that temperature; the ratio is stated in percent. For approximately every 20° F* increase in temperature, the capacity of air to hold water in the vapor phase is doubled. For example, if temperature change only is considered and the air is saturated (R.H. = 100%) at 40° F, and air temperature increases to 60° F, the relative humidity will drop to 50%.

---

*Fahreinheit is used in this instance, as tables for determining relative humidity are given in ° F and the R.H. equation makes use of ° F.

The temperature to which unsaturated air must be cooled at constant pressure in order for it to become completely saturated is the *dew point*. In the previous example, if the air at 60° F were cooled to 40° F, the amount of water vapor would be sufficient to accomplish saturation. Further cooling would produce dew on the ground or fog. The smaller the spread between the air temperature and the dew point, the higher will be the relative humidity.

As a general rule, clouds or low fog are considered likely when the air temperature is within 4° F of the dew point. When saturated air is cooled, the excess moisture begins to condense as fog, clouds, or precipitation. Thus, when air temperature near the earth's surface is near the dew point and is decreasing, as in the late afternoon, it often cools enough during the evening to cause night and morning fog. Similarly, warm, moist air moving over cold land or water surfaces may be cooled to its dew point, with subsequent fog formation. Under other conditions, the dew point may increase. For example, rain falling through an unsaturated layer will evaporate, causing the dew-point temperature to rise. This may continue until the dew point coincides with the existing air temperature, and fog will form.

Relative humidity may be determined by use of the psychrometer described in section 13.4A. The dry-bulb temperature and the difference between dry and wet-bulb temperatures are used to obtain relative humidity from tables or by using the following equation:

$$\text{R.H.} = 100 \, \frac{300 \, (T - W)}{T}$$

where T is the dry-bulb temperature and W is the wet-bulb temperature.

Relative humidity may also be measured by means of a *hygrometer,* which utilizes the ability of certain organic substances to expand and contract with changes in atmospheric moisture content. The human hair has been known since ancient times to have this property, and may be used with an appropriate linkage system and dial arrangement to give relative humidity. Electrical resistance and infrared hygrometers are also available; the former makes use of the capacity of certain salts to absorb moisture and conduct electric current, the latter the ability of water vapor to absorb certain wavelengths of light in the infrared portion of the spectrum.

13.4C Measurement of Atmospheric Pressure

Evangelista Torricelli made an important contribution to meteorology in 1643 when he discovered that he could weigh the atmosphere by balancing the weight of a column of mercury in an evacuated tube with the weight of the atmosphere on an open dish of mercury into which the open end of the tube had been placed. In figure 13.9, H represents the height of the column of mercury that is in balance with the weight of the atmosphere. At sea level, the average length of H is 760 millimeters, which means that the pressure exerted by a column of

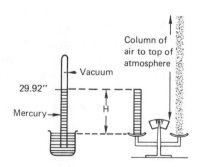

Figure 13.9 Principle of the barometer as developed by Torricelli is shown.

Introduction to the Atmosphere

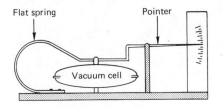

Figure 13.10 Principal parts of the aneroid barometer.

Figure 13.11 Rain gauge. Courtesy of NOAA.

mercury 760 millimeters high balances the pressure of the atmosphere. If water were used, the height of the column would be about 10 meters.

The weight of the air column above the earth's surface decreases with increased elevation. This decrease amounts to approximately 1 centimeter of mercury per 120 meters of elevation within the lower 1,200 meters of atmosphere. There are also horizontal variations in pressure, which, although not as great as the vertical change, are important in meteorology because it is these variations that are responsible for the wind. This indicates that atmospheric pressure is not only a function of height but also the temperature and density of the air.

The term "inches of mercury," although still popularly used, has been replaced by *millibar* as the unit of atmospheric pressure. One standard atmosphere is 1,013.2 millibars. This is equal under standard conditions to 29.92 inches of mercury, 760 millimeters of mercury, or 14.7 pounds per square inch.

Pressure-measuring instruments are of two general types—the mercurial and the aneroid barometers (fig. 13.10). The mercurial barometer is essentially as Torricelli developed it: a column of mercury sealed at the upper end and the open end in a reservoir of mercury. Changes in atmospheric pressure cause a change in the level of mercury at the sealed end, and a scale on the column provides a measure of atmospheric pressure. The essential feature of the aneroid barometer is a cell made of thin metal, corrugated to make it flexible. The cell, which is partially evacuated, responds to fluctuations in pressure by being compressed or expanded. This action is magnified by a system of levers coupled to a pointer on a dial marked with pressure readings.

13.4D Precipitation Measurement

"Precipitation" generally, is used in reference to rainfall. Actually, snow, sleet, hail, and drizzle as well as rain are included, although rain and snow are the major contributors to precipitation totals. Rainfall measurement is the oldest meteorological data to be collected—there are records of rainfall data collected by the Greeks 2,500 years ago. Rainfall patterns have a tremendous impact upon society, since rainfall influences agricultural enterprises, watershed management, flood-control activities, as well as all the activities inconvenienced by blizzards and rainfall of high intensities.

There are several uses for precipitation observations such as long-term records of annual rainfall. These records provide information on the fluctuation of rainfall during the year and over an interval of many years, and thus provide an estimate on the frequency with which rain may be expected. From such data it is also possible to determine rainfall patterns from a geographical viewpoint, enabling geographers to delineate desert, semiarid, and humid regions.

Observations of individual storms (rain or snow) are made to determine the general area covered by a storm as well as the amount and duration of the rainfall. From this information the general intensity of the precipitation may be calculated. This enables meteorologists, flood-control engineers, and others concerned to evaluate the storm in terms of water supplies and possible future flood danger.

Many types of rain gauges are used to measure precipitation, some of which are as simple as a collecting can and measuring stick. Basically the type used most frequently are variations of this simple type. One has a tipping mechanism which collects 0.01 inch of rain and empties automatically by tipping each time it is filled. The number of tips are counted by an electric counting device which permits a remote record to be made. Another system weighs the amount of water collected, and this weight is calibrated to inches of rainfall by various remote devices including a punch tape for computers. A simple yet accurate rain gauge may be made by using an eight-inch collecting container which funnels into a container with a diameter of 2.54 inches. The area ratio of the funnel to the container is 10:1, which permits magnifying the depth of the rainfall in the collecting container tenfold. A scale appropriately calibrated in fractions of an inch may then be used to accurately measure rainfall to 0.01 of an inch.

The use of remote recording equipment has enabled meteorologists to gather data from many remote areas which could not previously be visited on a frequent basis. This has greatly increased the accuracy of measuring total volume of rainfall for any one storm.

Snow is measured by several means to obtain data on snow depth, water content, and 24-hour snowfall. Snow depth is generally measured from calibrated stakes permanently placed in areas where the snow may be expected to remain undisturbed. This data gives an indication of the rate of melting or evaporation of the snow. The 24-hour snowfall data is measured at intervals at a site that can be swept clean after the observation is made. Water volume of the snow is obtained by weighing a given volume of the snow to determine its density or to obtain the equivalent in inches of water.

13.4E Wind Measurement

The wind moves both horizontally and vertically (usually in a 10:1 ratio), although only the horizontal direction is measured as one of the weather components. Both direction and speed measurements are made. Wind direction is determined by a *weather vane*—essentially the same method used by the ancient Greeks and a common ornament on houses throughout history. That weather vanes can be found on the ruins of buildings 2,500 years old indicates the importance people have placed upon the effect of wind upon their society. For example, wind direction is of significance in present-day agriculture, as it must be considered when crop dusting is to be practiced. The orientation of airport runways are to a large extent dictated by wind direction, and air pollutants are most annoying downwind from their source. Wind velocity also assumes importance because of the damage that can be inflicted by the wind, especially when the wind velocity reaches hurricane proportions.

Introduction to the Atmosphere

Figure 13.12 Anemometer. Courtesy of NOAA.

The weather vane still serves to show direction with the arrow (or pointer), indicating the direction from which the wind is blowing. Thus, wind blowing from the northeast is described as a northeast wind. Originally, wind direction data was given in terms of the points of the compass but the modern record is made in degrees of a circle with zero degrees indicating north. The *anemometer*, invented in the seventeenth century by Robert Hooke, is used to measure wind speed. The most common form of wind-speed indicator currently used is the rotating cup anemometer (fig. 13.12). It consists of three or more metal cups mounted on a freely rotating vertical shaft. The concave portion of the cups are all oriented in the same direction around the shaft and catch the wind, causing the anemometer to spin. The speed of rotation as an indication of wind velocity is recorded on a meter attached to the shaft or on a remote unit. The velocity is usually recorded in miles per hour or in knots, which is equivalent to 1.1508 miles per hour.

13.4F Measurement of Sunlight

Several meteorological values for sunlight are collected as a part of the weather-observation program. These include the length of time in minutes that the sun actually shines during the day. Such data may be collected by a *sunshine switch*, which consists of two photoelectric cells sensitive to sunlight at a preset intensity. The switch activates a counter when this level is reached and inactivates the counter when the sunlight intensity falls below the preset level. It is also possible to record the time of day that sunlight occurred as well as the total length of time.

Also of concern to man is the total amount of solar radiation of all intensities received at the earth's surface. This solar radiation influences evaporation rates from open bodies of water and from the soil, and influences photosynthesis in plants. Solar radiation is measured by a *pyranometer*, which yields data as a unit called the *langley*. The langley represents one gram calorie of energy per square centimeter of surface.

13.4G Remote Measurements

The atmospheric variables have been discussed in terms of their measurement on the land surfaces and usually at convenient stations on land. Now automated stations have been developed for recording weather data in remote areas on land and also at sea (see section 17.1).

The realization that upper atmospheric conditions must be understood to enable meteorologists to draw valid conclusions on weather conditions on the surface has been known for almost one hundred years. Attempts to probe the upper atmosphere with instruments were made in the nineteenth century with kites. While fairly successful, the kites were limited to altitudes of 3 to 4 kilometers. Most meteorological data on the upper atmosphere is now collected by means of a *radiosonde* carried aloft by a hydrogen or helium balloon. The

The Atmosphere

Figure 13.13   Radiosonde carried aloft by balloon. Courtesy of NOAA.

radiosonde is a lightweight radio transmitter which carries instruments for recording temperature, pressure, and humidity. Data on these variables are transmitted back to a receiving station as the balloon ascends, yielding a profile of atmospheric conditions with altitude. Wind direction and velocity may be determined by measuring the course and speed of the balloon, usually with radar or a radio direction finder. Atmospheric data may be obtained in this manner to altitudes of 30 kilometers, a great improvement over the kite.

Rockets are now used to collect atmospheric data above the level that can be reached by balloons. This is accomplished by launching rockets to altitudes of 70 to 75 kilometers, where radiosonde equipment is released to parachute back to earth. Data is transmitted as the equipment descends toward the earth in the same manner as the balloon devices.

Since 1960, satellites have been added to the instruments used in meteorological work. To date, the most dramatic application of the meteorological satellite observing system has been the identification and tracking of known meteoro-

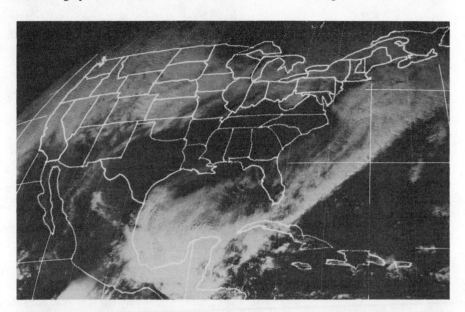

Figure 13.14 NOAA satellite photos taken over the United States on three successive days (above and page 292) reveal the changing cloud cover pattern. Courtesy of NOAA.

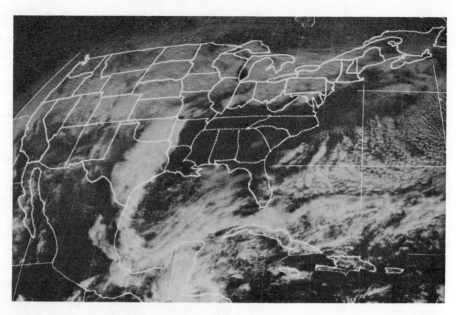

Figure 13.14 (Continued)

logical phenomena such as hurricanes, typhoons, cyclones, and frontal systems, as revealed by the global cloud cover (fig. 13.14). From more than 160 kilometers above the surface, weather satellites are able to detect the constantly changing pattern in cloud cover, which gives a clue to weather conditions on the surface. This technology, while as yet in its infancy, may eventually prove to be a major contribution in worldwide weather forecasting.

13.5 CLOUDS

Clouds, being composed of ice crystals or liquid water droplets, are a direct result of saturation-producing physical processes which take place in the atmosphere. An accurate description of the type and amount of cloud cover is an important aid to the meteorologist in analyzing the weather and making a forecast.

For identification purposes, meteorologists have divided cloud types into four groups, which are each further described in terms of their form and appearance. The four groups are: high clouds, with a base over 6 kilometers above the earth's surface; middle clouds (base 2 to 6 kilometers); low clouds (base below 2 kilometers); and clouds with accentuated vertical development. The high, middle, and low cloud groups are defined on the basis of the altitudes at which the bases of the cloud formations occur most frequently. These are shown in table 13.2. Clouds in the remaining group may have their bases in the low or middle levels but extend upward into the middle and high level.

Table 13.2
Approximate Heights of Cloud Groups at Different Latitudes

Cloud Group	Polar Latitudes	Temperate Latitudes	Tropical Latitudes
High	3–8 km (10,000–25,000 ft)	5–13 km (16,500–45,000 ft)	6–18 km (20,000–60,000 ft)
Middle	2–4 km (6500–13,000 ft)	2–7 km (6500–23,000 ft)	2–8 km (6500–25,000 ft)
Low	From the earth's surface to 2 km (6500 ft)	From the earth's surface to 2 km (6500 ft)	From the earth's surface to 2 km (6500 ft)

Within the high, middle, and low cloud groups are two main subdivisions: (1) clouds formed when localized vertical currents carry moist air upward to the condensation level, and (2) clouds formed when whole layers of air are cooled until condensation takes place. The clouds in the first category have a billowy appearance and are called *cumulus*, meaning accumulation or heap. Those in the second subdivision lie mostly in horizontal layers and because of their appearance are called *stratus*, meaning spread out.

In addition to the subdivision name, the word *nimbus*, meaning raincloud, is added to the names of those clouds that normally produce precipitation. Thus a horizontal strata of clouds from which rain is falling is called *nimbostratus*; a heavy and swelling cumulus that has grown into a thunderstorm cloud is referred to as a *cumulonimbus*. Clouds that are broken into fragments are usually identified by adding the prefix *fracto* to the classification name. For example, fragmentary cumulus are referred to as *fractocumulus*. The following cloud classification includes only the principal types most generally seen in the sky.

High Clouds

Cirrus: thin, featherlike clouds composed entirely of ice crystals.

Cirrocumulus: thin cloud, the individual elements of which appear as small, white flakes of patches of cotton, usually showing a brilliant and glittering quality suggestive of the presence of ice crystals.

Cirrostratus: thin, whitish cloud layers, appearing like a sheet or a veil; diffuse or sometimes partly striated or fibrous; due to their ice-crystal composition, these clouds are associated with halos—large, luminous circles or arcs of circles surrounding the sun or moon.

Middle Clouds

Altocumulus: white or gray-colored patches or layers of clouds, with the cloud elements having a rounded or rolled appearance.

Altostratus: a gray to bluish dense veil or layer of clouds having a fibrous appearance; the outline of the sun may show dimly through as though through frosted glass.

Nimbostratus: a low, shapeless thick cloud layer of dark gray color accompanied by continuous rain or snow.

Low Clouds

Stratus: a low, uniform, sheetlike cloud.

Stratocumulus: clouds having large globular masses or rolls, usually soft-appearing and gray to darker shading.

Clouds with Vertical Development

Cumulus: dense, dome-shaped, often isolated clouds, characterized by relatively flat bases with dark shading and by protuberances from the dome-shaped upper areas.

Cumulonumbus: towering cumulus clouds of large dimensions with califlowerlike tops often crowned with veils of thick cirrus.

Mammatocumulus: cumulus-type clouds whose lower surface form pockets or festoons; the cloud is usually indicative of severe turbulence.

13.5A Structure of Clouds

Clouds are composed of minute ice crystals or liquid water droplets. They are the direct result of moist air being cooled to the point where condensation takes place. Clouds with temperatures between $-10°$ C and $0°$ C are often composed largely of supercooled water droplets which can cause icing of aircraft. When the temperature is much below $-10°$ C the tendency is for the cloud to be composed mostly of ice crystals.

Cumuloform clouds always have some turbulence within them, because they are the result of local upward moving air currents. In contrast, predominantly stratified clouds have little if any turbulence within them, since there is little or no localized vertical motion. Sometimes, horizontal strata of clouds will partly change to cumuloform clouds as a result of the clouds being heated from below, or because the entire layer is being lifted by some means to the extent that the air becomes unstable. At times, the cumuloform clouds thus formed appear only as random puffs, at other times in groups or in lines. For example, lines of cumuloform clouds projecting upward out of a horizontal cloud deck sometimes mark the general location of a frontal zone (the boundary between warm and cold air masses). This may follow the orientation of a range of mountains below the clouds or the temperature difference between land and water at a seacoast.

13.6 SUMMARY

Man's activities are markedly influenced by weather, and therefore he has been preoccupied with the weather from his earliest history. Most early efforts in weather forecasting were steeped in superstition, much of which carries over

Introduction to the Atmosphere

Figure 13.15   Cloud genera. Courtesy of NOAA.

The Atmosphere

to the present. Not until the seventeenth century, when the thermometer and barometer were invented, were significant steps in weather forecasting accomplished. Since then a variety of instruments have furthered the advancement of meteorology, and the advent of rockets and satellites permit studying the atmospheric phenomena in depth and on a worldwide basis.

The atmosphere, composed of many gases, differs now in composition from the composition of the atmosphere when the earth was formed. A number of processes, including photodissociation and photosynthesis have contributed to the change, which leaves us now with a lower atmosphere—the homosphere—composed principally of nitrogen and oxygen. The lowest level of the homosphere is the troposphere, a turbulent layer of gases extending 8 to 15 kilometers above the earth's surface. This is overlaid by the stratosphere, a layer extending up to 50 kilometers, in which little vertical mixing of the gases occurs and in which temperatures decrease with altitude. Beyond the homosphere is the heterosphere, in which gases tend to separate and stratify according to their respective atomic weights. The lower portion of the heterosphere is the thermosphere, where the highest temperatures in the atmosphere have been detected. Above this is the ionosphere, a region in which free electrons and positively charged ions predominate. This layer actually overlaps down into the thermosphere. The motion of the charged particles is influenced by the earth's magnetic field.

A number of factors influence the weather, principal among which are heat and moisture. Heat as measured by temperature comes primarily from the sun and is transmitted through the atmosphere by radiation, conduction, and convection. The temperature of the lower atmosphere, where the weather is generated, decreases with altitude. Parcels of air are heated and caused to rise, or cooled and caused to fall, both of which create atmospheric disturbance. The amount of moisture in the atmosphere, originating mostly from the sea, is relatively constant and is a function of the temperature. Moisture may condense as a result of lowering temperature to form dew, fog, or rain, or it may be evaporated and become a constituent part of the atmosphere.

Successful weather forecasting requires the collection of data on a number of atmospheric variables, including temperature, humidity, pressure, precipitation, wind, sunlight, and cloud cover. Instruments used for the collection of these data vary tremendously in sophistication, from a simple measuring stick for measuring precipitation to earth orbiting satellites capable of photographing cloud cover over a large area. Now automated stations are used to collect and relay weather data from remote sites, thus improving the accuracy of weather forecasting.

QUESTIONS

1. What have been the advantages and benefits of using satellites in monitoring the earth's weather phenomena?

2. Trace the evolution of the atmosphere from the time of the earth's formation to the present.
3. Briefly describe the properties of the troposphere, the stratosphere, and the ionosphere.
4. How is the atmosphere heated?
5. Describe the path of water through the hydrologic cycle.
6. How does a maximum-minimum temperature thermometer function?
7. Define the following: relative humidity, dew point, sling psychrometer, millibar.

14 Air Movement and Air Masses

Forces responsible for atmospheric circulation.
Movement of air masses.
The development of weather along a front.
The development of storm systems.

The conditions of wind and weather occurring at any specified place and time are related to the large-scale general circulation in the atmosphere. The problem of formulating a theory of general circulation that is capable of accounting for the major features observed is one of the most difficult in meteorology. At present, there is no universally accepted general theory which takes into consideration all the observed properties of atmospheric circulation.

14.1 CIRCULATION OF THE ATMOSPHERE

There are a number of principles which are involved in general circulation and which deserve attention. In addition, there are certain portions of the circulation which appear capable of a fairly simple physical explanation.

14.1A Effect of Solar Radiation

The primary energy source responsible for atmospheric circulation is the sun (see section 14.3). On the whole, the energy balance of the earth and its atmosphere is essentially constant. However, individual portions of the atmosphere are subjected to a variable incoming and outgoing energy, which creates an imbalance leading to the formation of wind systems. The wind systems move large volumes of air in order to alleviate surpluses and deficits of heat which would otherwise result.

Air Movement and Air Masses

Incoming solar radiation is far from uniformly distributed over the surface of the earth. Because of the low angle of incidence of solar radiation in the polar regions, a given area in the polar latitudes receives far less solar radiation than an equal area closer to the equator. In order to determine the consequences of this inequity, it is advisable to disregard the rotation of the earth for the moment and see what would happen if the earth stood still but the sun followed its normal path across the sky. As a result of the greater heat income at the equator, the radiation heat gained is greater than the heat lost, while the reverse is true at the poles. Thus, if radiation were the only factor, the equatorial regions would become progressively warmer and the polar regions correspondingly colder. This is in fact not in accord with actual observations, indicating that other mechanisms must exist which equalize the heat surpluses and deficits. Such a transfer of heat energy can only be accomplished by actual transport of air from one latitude to another, thereby creating atmospheric circulation.

If such circulation occurred on a uniform, nonrotating earth, the heating at the equator would cause the air at the lower levels to expand upward. The same would take place at the poles, but to a much lesser degree. This would produce a low pressure in the upper levels over the poles and a relatively higher pressure at upper levels over the equator. Air in the upper levels would then begin to move from the equator (a zone of high pressure) toward the poles (a zone of low pressure). This motion would result in a rise in atmospheric pressure at sea level at the poles and a reduction at the equator, since sealevel air pressure is a measure of the weight of air above sea level. This pressure gradient would result in a flow of air toward the equator at lower levels and toward the poles at upper levels (fig. 14.1). However, since the earth does rotate, it is necessary to take into account the other forces that affect the air circulation over the entire globe.

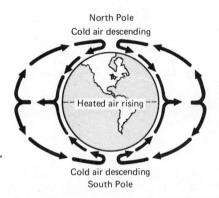

Figure 14.1 Air current circulation if the earth did not rotate.

The general effect of the earth's rotation upon the atmosphere serves to break up the previously described single circulation cell (moving air from pole to equator along the ground and from equator to pole in the upper atmosphere) into three cells (fig. 14.2). One cell extends from the equator to the horse latitudes (30° north latitude), another from that point to 60° north and south latitude, and the third from 60° latitude to the poles. Looking eastward at a vertical section through the northern hemisphere, one would observe counterclockwise circulation of the two extreme cells. These circulation patterns are direct in the sense that they conduct heat from a heat source to a cold area, while at the same time, they transform a small amount of the heat energy received into kinetic energy for motion. The direct cell to the south may be called the trade-wind cell, since the southward-moving lower portion of this cell is responsible for the steady northeast trade winds just north of the equator. The northern cell will be referred to as the polar-front cell.

In the two direct-circulation cells, strong westerly winds are continually being created at high levels. Along their boundaries with the middle cell, these strong westerly winds generate eddies with approximately vertical axes. Through the action of these eddies, the momentum of these high-level westerlies in the upper

The Atmosphere

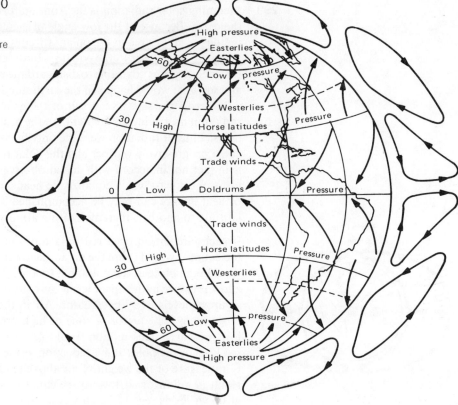

Figure 14.2 Wind patterns over rotating earth if surface were smooth and uniform.

portions of the direct cells is diffused toward the middle latitudes, and the upper air in these regions is thus dragged along eastward. The westerlies observed in the middle latitudes (middle cell) are thus frictionally driven by the surrounding direct cells.

To an observer looking eastward, the circulation of the middle cell is clockwise, opposite to the direct counterclockwise circulations of the trade-wind and polar-front cells. This middle cell serves as a necessary brake on the general circulation driven by the direct cells to the north and the south. Part of the relative momentum eastward generated aloft in the direct cells spills over into the middle cell, where the momentum is destroyed through slow southward displacement. The surface westerlies established in the middle cell serve the additional purpose of balancing the retarding force exerted on the earth's rotation by the easterlies in the direct cells.

14.1B Forces Acting on the Atmosphere

The forces that influence air movement over the surface of the earth are: (1) influence of earth's rotation, (2) Coriolis force, (3) gravity, (4) pressure gradient, (5) centrifugal force, and (6) friction.

Air Movement and Air Masses

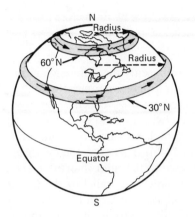

Figure 14.3 Wind at high latitudes moves faster (west to east) than winds closer to the equator.

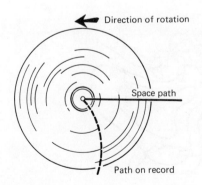

Figure 14.4 Dotted line indicates path of straight line drawn on rotating disk. This illustrates effect of Coriolis force on rotating earth.

1. *Influence of earth's rotation:* To understand the influence of the earth's rotation, it may be helpful to use the following analogy. If a weight attached to a piece of string is whirled around in a circle, the weight will achieve a given velocity. If the string is shortened it will be found that the speed of the weight increases generally in proportion to the decrease in the length of the string. If the string is shortened to one half its original length the speed of the weight is doubled.

Now consider a band of air around the earth at 30° north latitude (fig. 14.3) moving with the rotating earth. The air is at rest with respect to an observer standing on the earth at the same latitude. If the band of air is moved northward to 60°, the band will decrease in radius from 3,500 kilometers to 2,400 kilometers. The result would be similar to that of the weight on the string when the string was shortened. In a like manner, the band of air decreasing in radius as it moves northward will increase its speed of rotation. Instead of being at rest with respect to the earth, the wind would now appear to an observer at 60° latitude to be coming from the west. If, however, a ring of air at 60° north latitude originally at rest relative to the earth is displaced toward the equator, the ring of air will increase in radius and will decrease its speed of rotation. Thus, there will then appear to be a wind from the east relative to the surface. These relative winds would be of very high speed. For example, in moving a band of air from 30° to 60° the relative speed at 60° would be about 300 kilometers per hour. This, however, is not the case, since other factors such as friction severely limit this effect. While the wind activity described pertains to the Northern Hemisphere, the same forces are influencing the atmosphere in the same manner in the Southern Hemisphere.

2. *Coriolis force:* Another force acting on the atmosphere is an apparent force brought about by the rotating earth. This is the Coriolis force, which is also responsible for the deflection of ocean waters (see section 19.2). The Coriolis force causes the wind to be deflected to the right in the Northern Hemisphere and to the left in the Southern Hemisphere.

The Coriolis force can be illustrated by using the turntable of a record player. With a ruler held fixed above the rotating turntable, a chalk line may be drawn from the center to the edge along the ruler. To the person drawing the line, the chalk traveled in a straight line. If the turntable is stopped, the line on its surface would not be straight but curved, as indicated in figure 14.4. The Coriolis force has caused a deflection opposite to the direction of rotation. This force is strongest at the poles and decreases to zero at the equator. It also varies with wind speed, so that as the wind speed increases, the Coriolis force increases. Thus, as the wind moves northward away from the equator, it is deflected to the right (to the left in the Southern Hemisphere), and finally, is no longer moving northward but rather from west to east.

Under these circumstances, air piling up in the horse latitudes produces a high-pressure zone between the trade-wind cell and the middle cell. The air moving southward out of the horse latitudes is deflected to the right and becomes the *northeast trades* which, because of deflection, results in east winds at the equator.

Air moving north from the horse latitudes is deflected to the right to become *prevailing westerlies* (see fig. 14.2).

In the polar regions, a similar process is in operation with the southward-moving polar air being deflected to become an east wind. When this air meets the prevailing westerlies, a polar front is formed. Here the westerly current overlays the easterly polar air current at approximately 60° north. Figure 14.2 shows the general circulation of a uniformly rotating earth—that is, an earth assumed to have a relatively smooth surface everywhere composed of the same material.

3. *Gravity:* The force of gravity maintains the envelope of air in close proximity to the earth. Gravity is responsible for the layered distribution of the gases so that more dense air lies below lighter air.

4. *Pressure Gradient:* The pressure gradient represents the difference in pressure in the atmosphere due to differences in atmospheric temperatures. Points of similar atmospheric pressure are connected by lines called *isobars* on a weather chart (fig. 14.5). The movement of air (wind) would be perpendicular to the direction of the isobars from a point of high pressure to a point of low pressure if the Coriolis force did not exist (fig. 14.6). However, the presence of the Coriolis force causes the wind to be deflected to the right (Northern Hemisphere) and ideally would result in the wind blowing parallel to the isobars. When the wind is blowing parallel to the isobars, the Coriolis force is just balanced by the pressure-gradient force (fig. 14.6). Winds blowing approximately parallel to isobars that are essentially parallel and without curvature are known as *geostrophic* winds and found mostly at high altitudes.

5. *Centrifugal force:* The movement of air along a curving path is also influenced by centrifugal force, which in essence opposes the deflection resulting from Coriolis force. The resulting deflection is, therefore, somewhat less than that

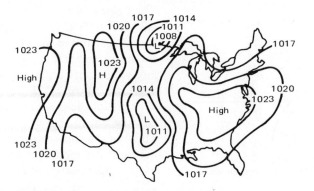

Figure 14.5   Lines (isobars) connecting points of equal pressure indicate areas of high and low pressure on a weather map.

Air Movement and Air Masses

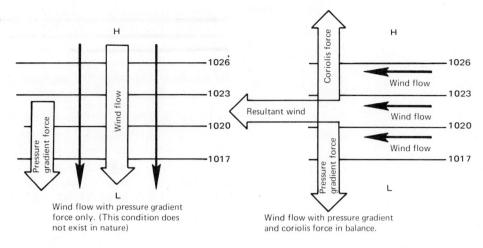

Figure 14.6 Pressure gradient causes wind to flow from high pressure area to low pressure area. Coriolis force deflects wind countering the pressure gradient and causes wind to flow parallel to isobars.

caused by Coriolis force alone, making centrifugal force a factor in determining the direction of a geostrophic wind.

Centrifugal force also acts on air moving in a curved path in such a way as to increase the speed of the wind in high-pressure areas and decrease it in low-pressure areas.

6. *Friction:* Friction tends to retard air movement, and the effect of friction is most strongly felt near the ground. Friction will modify the direction and the velocity of the wind, depending upon the nature of the terrain. The least influence due to friction occurs over the ocean; it increases over the land areas, with the greater effect over the mountainous regions. The influence of friction is also felt above the ground surface as a result of turbulence, but this effect decreases rapidly 600 meters above the surface and may be considered negligible from that point upward.

14.1C   Effect of Land and Sea

Consideration has been given thus far to air circulation on the basis of the earth as a smooth, uniformly shaped globe. However, complications in the general circulation are brought about by the irregular distribution of the oceans and continents, the relative effectiveness of different surfaces in transferring heat to the atmosphere, the daily variation in temperature, the seasonal changes, and the topographic features of the land surfaces.

The variable factors lead to the establishment of quasi-permanent regions of high pressure or *highs* that rotate in a clockwise direction in the northern hemisphere and are called anticyclones. Regions of low pressure or cyclones are called *lows*. The name *centers of action* is sometimes applied to these semipermanent or seasonal lows and highs because their counterclockwise or clockwise circulations control to some degree the general movements and conflicts of air masses in their vicinities. In addition, the lows and highs influence both the position and the movements of various segments of the polar front and the development and motion of the traveling cyclones that are largely responsible for most temperate-zone bad weather.

In figure 14.7, the flow of air around high-pressure and low-pressure areas at the surface are illustrated. Low pressure is to the left and high pressure is to the right of a person who has his back to the wind. A traveling anticyclone or "high" usually indicates the approach of a period of good weather with drier, cooler air and cloudless skies. On the other hand, the approaching cyclone or low-pressure area brings cloudy, rainy weather.

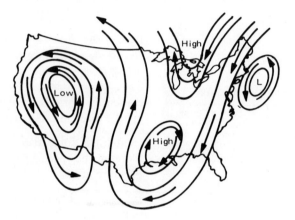

Figure 14.7 Direction of wind flow around high and low pressure areas.

14.1D Local Wind Movement

Local conditions of land formation and heat transference frequently lead to drastic modifications in wind direction and speed close to the earth's surface.

Convection currents, one of the main causes of local variations in wind activity, may sometimes result in the bumpiness experienced by pilots of small aircraft. The vertical circulation that occurs is caused by the uneven heating of the atmosphere due to the fact that some kinds of ground surface are more effective than others in heating the air directly above them. For example, rocks and sand, plowed land, and barren ground in general radiate more heat than bodies of water or land covered by vegetation.

The convection effect is particularly noticeable where land is adjacent to a body of water. During clear weather, the land surface heats up and warms the air above it and the warm air rises. The water temperature does not change significantly, so the relatively cooler air above the water moves toward the land, causing the warm air to rise and producing an onshore wind or sea breeze (fig. 14.8a). At night the land cools rapidly, and since the water does not lose heat as readily, it is relatively warmer and the above process is reversed. The surface air tends to blow offshore, producing a land breeze (fig. 14.8b).

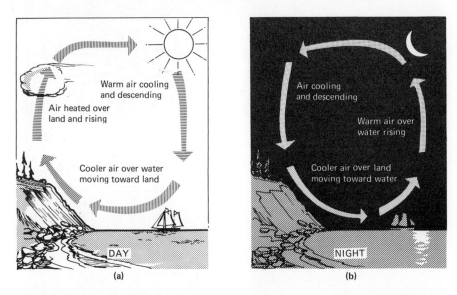

Figure 14.8 Surface winds flow (a) toward land from sea during the day and (b) away from land at night.

Small eddy currents are created when the wind blows over rough surfaces, and once the eddy currents are formed they are moved along by the wind. Under these circumstances, if the wind velocity is measured by a wind-speed indicator, it will be noted that there are brief periods of low velocity and high velocity, with irregular intervals between the gusts of wind.

Eddies or rotating pockets of air are produced and remain near the obstruction when wind speeds are less than 30 kilometers per hour (fig. 14.9). If the wind speed exceeds 30 kilometers per hour, the flow may be broken up into irregular eddies which are moved along some distance downstream from the obstruction.

A similar and much-disturbed wind condition occurs when the wind blows over large obstructions such as mountain ridges. In such cases, the wind blowing up the slope on the windward side is relatively smooth, whereas wind on the leeward side spills rapidly downslope setting up strong downdrafts and causing the air to be very turbulent (fig. 14.10). The downslope movement of the wind is analogous to water flowing downstream over a rough stream bed. The downdraft turbulent conditions are generally several thousand feet deep and represent a hazard to small aircraft approaching mountains on the leeward side.

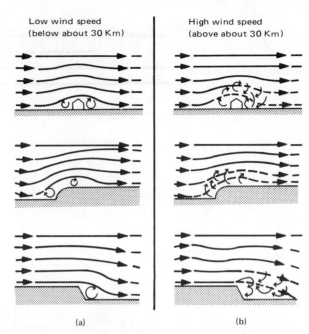

Figure 14.9 Air turbulence remains near obstruction when wind speed is (a) less than 30 kilometers per hour and moves downwind from obstruction when wind speed is (b) more than 30 kilometers per hour.

It is not unusual for winds over mountain ridges and through passes to be considerably stronger than surrounding winds, and as a result they are usually more turbulent. These turbulent winds generally flow in the direction of the pass or valley, even though the surrounding winds are moving in a different direction.

Mountain and valley breezes also result from temperature differentials. In the daytime, the air next to the mountain surface is heated by contact with the surface, while air some distance above the surface is not affected. The warmer air at the surface tends to move upward to higher levels and is called a *valley wind,*

Figure 14.10 Effect of large obstruction on wind flow.

which is quite shallow. At night the air may be cooled along the slopes of the mountain, causing it to increase in density and flow downslope producing a *mountain breeze*. Such a wind may be several hundred meters in depth and flow at velocities in excess of 80 kilometers per hour in extreme cases.

When air flows downslope from a higher elevation, its temperature is raised by adiabatic compression. The rise in temperature results in a lowering of relative humidity, bringing about what are called *Foehn winds*. Foehn winds are quite deep and occur frequently in the western United States. In Montana and Wyoming the *Chinook* is a well-known phenomenon, and in southern California the *Santa Ana* is known particularly for its hot, high-speed winds. The Santa Ana winds originate from a high-pressure area in the Great Basin region, which due to the resulting pressure gradient generates northeasterly or easterly winds. The winds blow from the interior of southern California offshore in the Los Angeles area, in some instances with considerable force. The winds are usually accompanied by hot, clear weather.

14.2 AIR MASSES

During World War I, interruption of ocean weather reports led Norwegian scientists to intensify studies of air currents in an effort to develop improved methods of weather forecasting. The studies focused attention on the fact that most weather changes are related to the boundaries between air currents which have different conditions of temperature and humidity. The more or less continuous conflict between warm, moist currents, usually from south or west, and cold, dry currents from the east or north in the Northern Hemisphere so resembled the tides of battle in Europe that the Norwegians applied the name *front* to the boundary between different *air masses*. This concept let to a great step forward in meteorology—the evolution of the polar-front theory and the air-mass method of weather analysis, which systematized and simplified the picture of atmospheric formations most frequently responsible for weather changes.

The meteorologist views an air mass as a large body of air which is fairly uniform with respect to temperature and moisture. That is, layer for layer, the air in one area of the air mass has about the same characteristics as other areas of the same mass (fig. 14.11).

When a large portion of the atmosphere comes to rest, or moves slowly over land or sea areas having fairly uniform surface conditions, the air will take on the characteristics of the underlying surface with respect to temperature and moisture. The formation of a cold air mass in winter can be used as an example. Warm air from the sea moves over the North American continent into northern Canada and adjacent Arctic regions and stagnates. The stagnation of the air over these regions is an ideal condition for the formation of large, very dry, cold bodies of air, because during the winter season the area is covered with

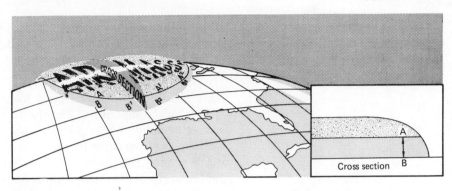

Figure 14.11 Horizontal layers in an air mass are generally uniform with respect to temperature and moisture.

snow and ice. Later, as the air moves out of this region under the action of the pressure gradients, the air will tend to maintain its characteristics, although some modification will occur during its travel.

The region where the air mass acquired its identifying characteristics is called its *source region*. The depth to which an air mass becomes modified by the source region depends upon the length of time the air remains in the source region and the difference between the original temperature of the air and that of the underlying surface. When the air is colder than the surface temperature, heating from below results; vertical currents are produced which carry heat and moisture aloft, thereby modifying the air mass to a great depth. On the other hand, when the air is warmer than the surface, cooling from below takes place. Thermal currents do not develop under these conditions, and the air is modified to a lesser height.

There are two general source regions for air masses—the tropical region and the snow- or ice-covered polar region. The latter region is often further subdivided into the Arctic region and the polar region. The arctic region is the colder and extends from the pole outward to the limit of the perpetual frost area. The polar region extends from this area to a point where the mean temperature of the warmest month exceeds 10° C. In addition to the general source regions, there are two types of surfaces in the source regions—continental and maritime. Although the characteristics of the individual air masses vary, in general the air from tropical regions is warm and the air from polar regions is cold. Again in general, continental air is dry and maritime air is moist. Where two air masses meet from different source regions—such as cold, dense polar air and warm, light tropical air—then instead of simply mixing, a definite surface of discontinuity will appear. This boundary between cold and warm air is called a front.

The source regions have been found to be convenient criteria for identifying the air masses, using the following notations:

A—Arctic air masses

P—Polar air masses
T—Tropical air masses

Added to this is a prefix *m* or *c* indicating whether the air mass has maritime or continental characteristics (fig. 14.12).

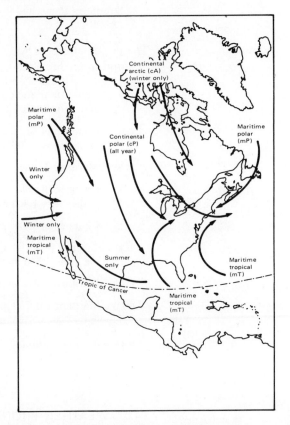

Figure 14.12 Typical paths of air masses affecting North America.

A third letter is added to indicate whether the air is colder (*k*) or warmer (*w*) than the surface over which it is moving. The *k*-type air mass will be warmed from below and convection currents will form. Characteristics of this type of air mass are:

1. Turbulence up to about 10,000 feet
2. Unstable lapse rate (nearly dry adiabatic)
3. Good visibility except in showers and dust storms
4. Cumuliform clouds such as cumulus and cumulonimbus
5. Showers, thunderstorms, hail, sleet, and snow flurries

The *w*-type of air mass being cooled from below exhibits different characteristics. This type tends to maintain its original properties and is modified only in the lower few thousand feet as it moves. Some characteristics of this type are:

1. Smooth air above friction level
2. Stable lapse rate
3. Poor visibility due to smoke and dust held at lower levels
4. Stratiform clouds and fog
5. Drizzle

The principal North American air masses by names and symbols and a brief description are included below.

cAw   Continental arctic air warmer than the surface over which it lies; stable in about the lower 1,500 meters.

cAk   Continental Arctic air colder than the surface over which it is passing; unstable in the lower 1,500 meters.

mAk   Maritime Arctic air colder than the surface over which it is passing; unstable.

cPw   Continental polar air warmer than the surface over which it is passing; stable in the lower 1,500 meters.

cPk   Continental polar air colder than the surface over which it is passing; unstable.

mPw   Maritime polar air warmer than the surface over which it is passing; stable in the lower 1,500 meters.

mPk   Maritime polar air colder than the surface over which it is passing; unstable.

mTw   Maritime tropical air warmer than the surface over which it is passing; stable in the lower 1,500 meters.

mTk   Maritime tropical air colder than the surface over which it lies or is passing; unstable.

14.3 WEATHER FRONTS

In the previous section it was stated that when two contrasting air masses meet, a rather sharp transition zone or front is formed between them. At a discontinuity of this type, the cold air mass, being heavier, will tend to underlie the warmer, lighter air mass.

The formation of a discontinuity surface between two air masses may be illustrated by placing two unlike fluids such as water and oil in a container. If the fluids are allowed to remain undisturbed for some time, the water, being the heavier of the two liquids, will underlie the oil. A definite boundary surface will be visible between them, and if the liquids remain without relative motion the boundary will be horizontal. This is analogous to the boundary surface separating cold polar air and warm tropical air. However, in the atmosphere the fluids involved are ordinarily in relative motion, thus bringing forces into play which cause the boundary surface to become a sloping one. This boundary

Air Movement and Air Masses

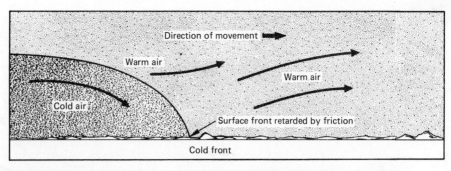

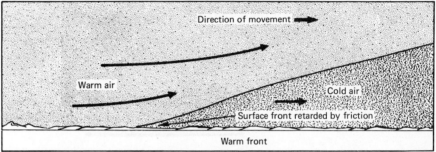

Figure 14.13 Relationship of cold front and warm front. The vertical dimensions are exaggerated.

surface or front slopes in such a way that the cold, heavy air underlies the warm, light air in the form of a flat wedge (fig. 14.13).

The leading edge of an advancing cold air mass is called a *cold front*, and an advancing warm air mass is called a *warm front*. Stated another way, the front is called a cold front when cold air is overtaking and replacing warm air, and a warm front when the reverse is happening. If the front is not moving, it is called a stationary front.

In the Northern Hemisphere, the wind near the surface will invariably shift in a clockwise direction as a front passes. From figure 14.14 it may be seen that an observer standing at point *A* in the warm air will notice a clockwise shift in the wind as the cold front passes. Similarly, an observer at point *B* will also notice a clockwise shift in the wind as the front passes him.

Figure 14.14 also illustrates the point that a front is usually found along a low-pressure trough, that is, the pressure will usually be higher on both sides of the front than at the front itself. Thus, when a front approaches a station, the pressure will usually decrease and then begin to rise after the front passes. The pressure rise in many cases may be quite pronounced.

Any clouding and precipitation occurring along a front is dependent upon several factors. If the air that is being lifted is unstable, cumuliform clouds will usually be produced. The precipitation will be of the showery type and there

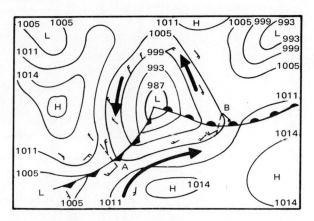

Figure 14.14 Air movement and pressure relationships in vicinity of a weather front.

will be turbulence in the clouds. On the other hand, if the lifted warm air is stable, it will form stratified clouds with steady precipitation and not much turbulence within the clouds.

Frontal weather also depends on the amount of moisture available. If the air is very dry to a great height, clouds will not form. If the air is very moist, the tendency will be toward cloudy weather.

The third factor that determines the weather along a front is the slope. A cold front will tend to have a steep front, especially near the ground, because the surface friction will tend to hold back the air near the surface and create a bulge in the frontal surface. In a warm-front surface, friction tends to hold back the receding cold air near the surface, resulting in a very flat slope to the frontal surface.

Flat frontal surfaces tend to give extensive cloudiness with large areas of precipitation. Steep frontal surfaces tend to produce narrow bands of cloudy weather with precipitation localized along the immediate front. Fronts with flat slopes occur when the temperature difference between the air masses is large and the wind velocity difference between two air masses is small. Fronts usually have steep slopes when the temperature difference is small and there is a strong wind discontinuity. Also, fronts tend to be steeper in high latitudes than in the lower latitudes.

Not all of the important weather activity occurs along fronts, there being large areas of low clouds and unstable weather conditions far removed from a front. Also the weather associated with fronts may be variable.

14.3A Cold Fronts

Cold-front surfaces have slopes ranging from 1/50 to 1/150. This means that the front will reach an elevation of $1\frac{1}{2}$ kilometers above ground surface at a distance of from 80 to 240 kilometers from the locale where the front intersects the surface.

In the Northern Hemisphere, strong cold fronts are usually oriented in a northeast to southwest direction and move toward the east and southeast. They are usually followed by cooler and drier weather, often by severe cold spells in winter, and occasionally by windstorms. The sequence of events with the passage of a cold front is as follows: First, an increase in the southerly winds in the warm air lying ahead of the front will be observed. Then, high cumuliform clouds such as altocumulus will appear on the horizon in the direction from which the front is approaching, and the barometric pressure will decrease (fig. 14.15). Next, the clouds will lower and rain will begin as the clouds change to cumulonimbus; the rain will increase in intensity as the front nears the point of observation. As the front passes, the wind will shift to a westerly or northerly direction, and the pressure will rise sharply. This type of cold-front passage will normally be followed by fairly rapid clearing, falling temperature, and dew point, but the actual cloud conditions in the cold air will depend on the moisture conditions and the stability of the cold air. The sequence of events may in some cases vary from that described because of other factors involved in the development of weather.

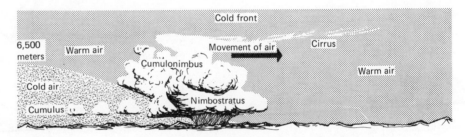

Figure 14.15   Cloud formation developing along a cold front.

For example, with the slow-moving cold front there is a general upgliding motion of the warm air over the frontal surface, which results in the formation of a rather broad postfrontal cloud pattern in the warm air lying above the cold layer (fig. 14.16). If this warm air is stable, stratiform clouds will form. If the warm air is conditionally unstable, cumulonimbus clouds and frequently

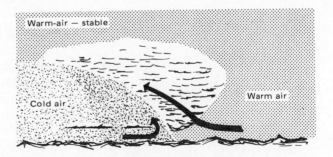

Figure 14.16   A slow moving cold front with stable warm air

thunderstorms may develop (fig. 14.17). In the fast-moving cold fronts, there is a combined downward-moving air above and below the frontal surface and upward-moving air in the area ahead of the front as shown in figure 14.18.

Friction retards the front near the ground and causes the frontal surface to be rather steep. This steep frontal slope results in a narrower band of weather concentrated along the forward edge of the cold front. If the warm air is stable, an overcast sky may occur for some distance ahead of the front, accompanied by general rain. In this case, the low ceiling, poor visibilities, and heaviest pre-

Figure 14.17   A slow moving cold front with unstable warm air.

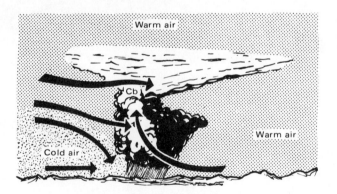

Figure 14.18   A fast moving cold front with unstable warm air.

cipitation will usually be limited to the immediate zone of the front. If the warm air is moist and conditionally unstable, scattered thunderstorms and showers may form along a front or ahead of it. This line, referred to as a *squall line,* is often characterized by an almost impenetrable wall of turbulent clouds building to 12,000 meters or higher (fig. 14.19). Behind the fast-moving cold front there is usually rapid clearing, gusty turbulent surface winds, and colder temperatures.

Air Movement and Air Masses

Figure 14.19 A squall line. Courtesy of NOAA.

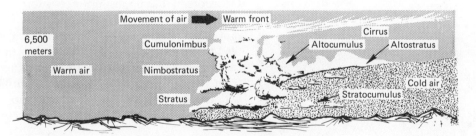

Figure 14.20 A warm front.

14.3B Warm Fronts

As previously stated, when warmer air is replacing colder air, a *warm front* is formed. Warm-front slopes are on the order of 1/50 to 1/200, with an average value of 1/100. A rise of 1½ kilometer over a distance of 150 kilometers is a

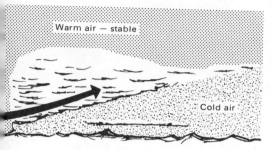

Figure 14.21 A warm front with stable warm air.

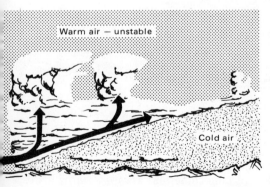

Figure 14.22 A warm front with unstable warm air.

relatively gentle slope. With the warm front, the warm air gradually moves up over the sloping frontal surface and will usually form a broad prefrontal cloud system (fig. 14.20). The clouds will be of the stratiform type if the air is stable, and, as illustrated in figure 14.21, the sequence of cloud types encountered in the cold air as one moves toward the warm front is cirrus, cirrostratus, altostratus, nimbostratus. If, however, the warm air is unstable (fig. 14.22), cumulonimbus or altocumulus clouds and frequently thunderstorms will be found ahead of the warm front.

When the air being lifted is stable, the precipitation that falls from the cloud system ahead of the warm front increases gradually and often continues until the system has passed. However, with unstable air, cumulonimbus clouds develop and the warm front precipitation becomes spotty in character. The rate of movement of warm fronts is usually about half that of cold fronts. Warm fronts are seldom as well marked as cold fronts, because the warm rain falling through the cold layer modifies the cold front considerably.

The widespread precipitation area ahead of a warm front often causes very low ceilings and visibilities, because the falling rain raises the humidity of the cold air to near saturation and brings about foggy conditions. The frontal zone itself may have zero visibility over a wide area. If the cold air has below freezing temperatures, the precipitation may take the form of freezing rain.

Very cold air underneath a warm front resists displacement and may force the warm air to move over a thinning wedge of cold air with waves in the upper surface. This gives the effect of secondary upper warm fronts and may cause parallel bands of precipitation at unusual distances ahead of the surface warm front.

14.3C Stationary Fronts

Occasionally the opposing forces exerted by adjacent air masses of different densities are such that the frontal surface between them shows little or no movement. Under such circumstances, it is usually found that surface winds tend to blow parallel to the front rather than toward or away from it as is the case with cold or warm fronts. Since neither air mass is replacing the other, the front is referred to as a *stationary front*. The weather conditions associated with a stationary front are similar to those found with a warm front, but are usually less intense and may persist for several days.

14.3D Occluded Fronts

Before a discussion of occluded fronts can be undertaken, it is necessary to to become acquainted with the development of lows that form along fronts. The occluded front occurs during the development of the low which meteorologists call a *wave cyclone*.

When a warm air mass lies adjacent to a cold air mass and either of these is caused to be accelerated along a part of the front, there is a tendency for a wave motion to be set up along the front. The wave travels along the front much like a sea swell, but on a front the wave is in a horizontal plane. In this phase, the front is said to be stable. However these waves usually become unstable after a time, and this instability is analogous to the formation of whitecaps on the ocean or breakers on the beach. The unstable frontal wave develops as the two air masses adjacent to the frontal surface tend to whirl together and the pressure of the wave becomes less than in the surrounding area, thereby forming a region of low pressure.

When the low forms, that part of the original front lying to the west of the center becomes a cold front with the cold air moving in from the north or northwest, while that part of the front lying to the east of the center begins to move northward as a warm front. In this unstable type of wave motion, the cold front tends to catch up with the warm front. As they run together, an occlusion or occluded front is formed.

Lows do not continue to exist indefinitely but develop through a series of stages and then die. Some develop into very active mature storm areas, which, after occlusion, gradually die out. Others develop only partly before giving way to other forces. The life history of a typical low is shown in figure 14.23. In (A) the front is shown extending along an approximately straight line. This is followed by the formation of a small wave (B) on the front with an associated precipitation area. In (C) wave development has progressed to the point where there is a definite circulation, a warm sector, a well-defined crest with warm and cold fronts on either side and a typical precipitation area. The warm sector becomes more narrow (D) as the cold front overtakes the warm front. The occlusion process takes place (E) as the low reaches maximum development and the warm sector is being rapidly pinched off. As the warm sector is being eliminated (F), the low begins to die out and is represented only by a whirl of cold air that is rapidly dissipating in strength. Figure 14.23 (B), (C), and (D) represent the stable phase of the wave, (E) and (F) represent the unstable phase of the wave, and (G) and (H) represent dissipation.

Figure 14.24 shows a plane and two vertical sections of a typical low. The shaded portions show the precipitation areas, the largest of which is ahead of the warm front, but there is also a narrow band of precipitation along the cold front. The warm front is preceded by considerable cloudiness of various types and by fairly large areas of precipitation closer to the front. After the passage of the warm front, there is an area where the sky is relatively clear and the air is warm. As the cold front approaches, clouds again begin to appear. There is then a brief period of precipitation during and shortly after the passage of the cold front, and the weather becomes colder. Frequently, showers occur for a time after the passage of the cold front.

The upper figure represents a vertical section through the atmosphere a short distance north of the low center. It shows cold air everywhere at the surface, with the warm air and cloud systems aloft. Rain is falling through the cold air from the clouds in the warm layer. The lower figure shows a vertical section

318

The Atmosphere

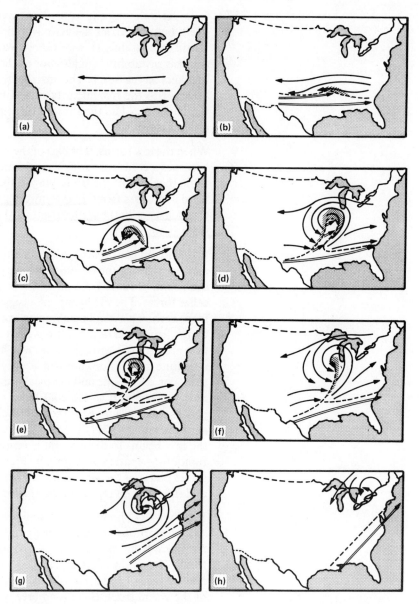

Figure 14.23 The "life cycle" of a low pressure area forming along a front is pictured. The front is indicated by broken line.

some distance south of the center with typical sectional views of the warm and cold fronts. While the above description represents a typical low, it must be understood that deviations from this model can and do occur.

Having examined the life cycle of the lows, it is now possible to discuss the

Air Movement and Air Masses

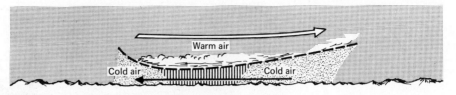

Cross section along line "A – A"

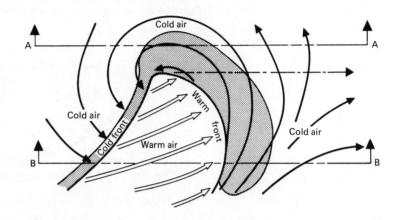

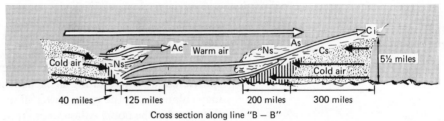

Cross section along line "B – B"

Figure 14.24 The center panel shows a top view of a low-pressure system with cross-sections at points (a) and (b) shown in upper and lower panels.

occluded front, which represents one phase of the low-pressure system. The structure of the occluded front depends upon the temperature difference between the cold air in advance of the system and the cold air to the rear of the system. If the air in advance is colder, the cold front which overtakes the warm front will move up over the colder air in the form of an *upper cold front*. In this case, most of the weather is in advance of the surface front. The surface front is called a *warm-front type occlusion* (fig. 14.25).

When the air behind the system is colder, it will push in under the cool air in advance of the system and produce a *cold-front type occlusion* (fig. 14.26). In this case, most of the weather will occur near or behind the surface front.

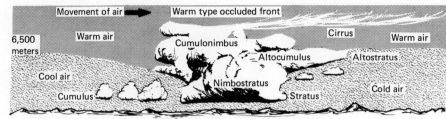

Figure 14.25 A warm-type occluded front.

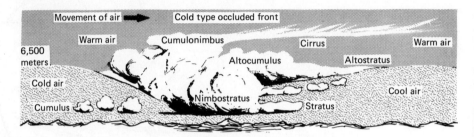

Figure 14.26 A cold-type occluded front.

14.4 STORMS

Previously, the cyclone or low-pressure state had been described as an atmospheric condition resulting in inclement or stormy weather. Such conditions may culminate in the most powerful and destructive of storms—the hurricane (if formed in the north Atlantic) or typhoon (if formed in the Pacific) (fig. 14.27). The hurricane or typhoon, also called a tropical cyclone, forms at low latitudes, usually between 5 and 20° latitude. Midlatitude (30 to 70°) cyclones are more numerous and are milder than the tropical cyclone. Midlatitude cyclones are responsible for most of the rain or snow storms that occur in the temperate zones; however, the activity may result in a tornado, a fierce storm covering a relatively small area.

The hurricane or typhoon is a nearly circular storm with extremely low pressure at the center and accompanied by high winds, dense clouds, and heavy precipitation. The diameter of the storm is usually 480 to 720 kilometers and appears as a tremendous atmospheric eddy or whirlpool bordered by altostratus clouds at lower elevations and cirrus and cirrostratus at high elevations. These clouds merge into dense overcast as the storm approaches. The disturbance is classified as a tropical storm until wind velocities of 120 kilometers per hour are attained at which point the storm has reached hurricane status. Wind velocities of 120 to 200 kilometers per hour are usual, with winds increasing in strength from the perimeter toward the center, where gusts of more than

Air Movement and Air Masses

Figure 14.27 Hurricane Agnes. June 19, 1972. Courtesy of NOAA.

300 kilometers per hour have been recorded (fig. 14.28). Around the center of the hurricane, the winds revolve in a gigantic spiraling vortex which surrounds the *eye* of the storm. The eye is 15 to 30 kilometers in diameter, surrounded by clouds up to 10,000 meters; it is a region of relative calm, with little or no wind and a few scattered clouds. Low humidity and no rainfall at the center are in marked contrast to the high humidity and intense rainfall measured in the storm. Amounts of up to 30 centimeters in 24 hours have been recorded during a hurricane; a record 190 centimeters in a 24-hour period was dumped on Baguio in the Philippines during a typhoon.

Following their formation in the tropics, the storms move westward or northwestward through the trade-wind belt to a latitude of 30 to 35°, then change to a northeasterly direction. The change in direction indicates the effect of the easterlies and the westerlies on these storms as they travel toward the poles, where their energies are normally dissipated.

There are six regions in the world which produce tropical cyclones (fig. 14.29):

1. West Indies, Gulf of Mexico, and the Caribbean
2. Western north Pacific including the Philippines, China Sea, and Japanese Islands
3. Arabian Sea and Bay of Bengal
4. Eastern Pacific coastal region off Mexico and Central America
5. South Indian Ocean off Madagascar
6. Western south Pacific near Samoa and Fiji Islands and eastern Australia

Hurricanes and typhoons are always born over the ocean, never over a land mass, although as these storms move they often cross over the continental

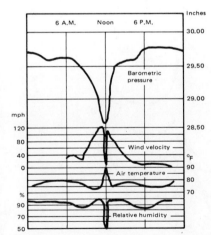

Figure 14.28 Changes in relative humidity, air temperature, wind velocity, and barometric pressure during passage of a hurricane.

The Atmosphere

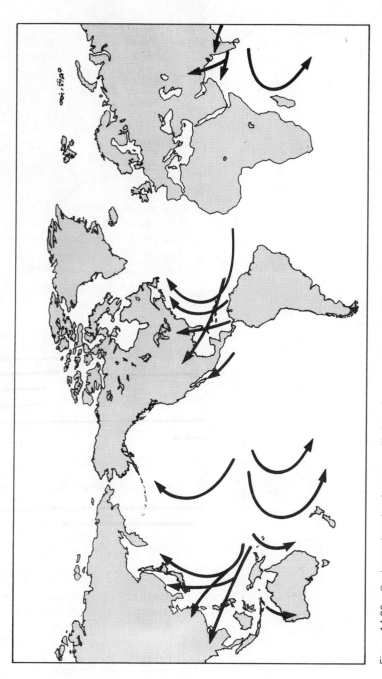

Figure 14.29 Regions where hurricanes are likely to form, and the paths followed.

margins. Hurricanes formed in the Gulf of Mexico and Caribbean Sea travel northwest, often passing over Cuba and other Caribbean islands as they move toward the Florida coast. Occasionally, the hurricane will turn northeastward and not strike the land mass, but many times it will pass over the continent and cause great damage as it travels northward. At times, a hurricane will travel westward into the Gulf of Mexico, then turn north to cross the Texas-Louisiana coast. These storms usually do much damage as they pass over the continental area.

Destruction along the coast by storm waves is perhaps one of the more serious effects of tropical cyclones. For example, in 1900, a hurricane struck Galveston, Texas, causing a sudden rise in sea level, inundating the low coastal city, and drowning approximately 6,000 people. It has also been reported that in 1737, a severe cyclonic storm hit the mouth of the Hooghly River on the Bay of Bengal, where a twelve-meter wave generated by the storm killed 300,000 people. Waves of such magnitude are usually the result of water being piled up and confined by the force of the wind within the central portion of the storm. The force of a tropical cyclone can also be illustrated by the destruction on the island of Barbados in 1780. The storm tore stone buildings loose from their foundations, destroyed forts, carried small buildings a hundred feet from their original locations, and stripped the bark from trees. More than 6,000 people were killed.

The best protection at the present time against such cyclonic storms is warning of its advance. With sufficient warning, the populace can be evacuated to higher ground and buildings can be boarded up for protection against high winds and flying objects. With the advent of weather satellites, sighting and tracking of these storms can be accomplished with a greater degree of accuracy than ever before. Some research is now being conducted to find methods of destroying the killer storms before they reach the populated coastal regions of continents.

Tornadoes, sometimes called cyclones, are perhaps the most destructive of all storms, even surpassing hurricanes in fury, although they affect a much smaller area. Tornadoes are typically a weather phenomenon of the United States, where 90 percent of the world's tornadoes occur. Of the remaining percentage, a substantial proportion occur in Australia, and a few arise in tropical and subtropical regions of the world.

Tornadoes originate during hot, humid, oppressive weather conditions. Dark, dense cumulus clouds prevail, and thunderstorms are in the offing. A furiously rotating portion of the overcast may form a funnel-shaped cloud spiraling at high velocity upward in a counterclockwise direction (Northern Hemisphere) and touch down on the surface, causing dust and debris to be carried skyward. The size of the funnel at the lower end may be 100 to 500 meters in diameter and, because of its meanderings, may cut a swath 1.5 kilometers wide for a distance of 15 to 65 kilometers at a speed of 30 to 80 kilometers per hour. Velocities within the funnel have never been directly measured, but they are estimated to reach 800 kilometers per hour. The exact mechanism that produces these violent storms is not completely understood, other than that they originate in cumulus clouds along cold fronts. In the central United States, where cold continental polar air lifts warm moist tropical air and forms a cold front, the conditions are favorable for the formation of tornadoes (fig. 14.30).

Figure 14.30 The adolescence and maturity of a tornado was photographed near Enid, Oklahoma, June 5, 1966. Thin light colored funnel cloud touches down (1), darkens as dirt and debris flow up into the violent winds (2), thickens and grows coarser (3), as it reaches a rope-shaped, frightening maturity (4).

The destructive characteristic of the storm is brought about by the high winds and extremely low air pressure within the funnel. If the storm passes over a closed structure, the pressure differential inside and outside will cause the structure to explode. The destruction in the narrow path of the tornado may be complete, while structures on either side of the path will be unaffected. The strong winds associated with tornadoes have been known to pick up large animals and even automobiles. Occasionally the funnel will touch the ground, then hop a short distance before touching down again. The best protection against tornadoes is a storm cellar which is constructed completely below ground.

Figure 14.31 shows the frequency of tornadoes on a monthly basis for the United States. Note that the summer months are the months of highest frequency. These are also the months during which the greatest temperature differences

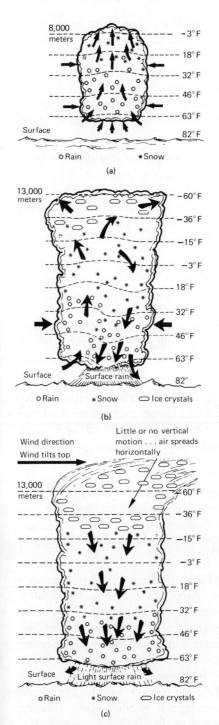

Figure 14.32 Idealized cross-section of a thunderstorm cell in cumulus stage (a); in mature stage (b); in dissipating stage (c).

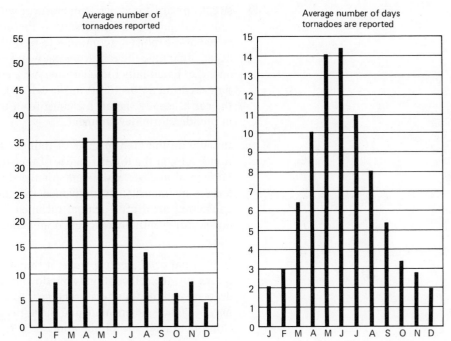

Figure 14.31 Tornado frequency in the United States from 1916 to 1960 by months.

occur between the continental polar air masses and the tropical marine air masses that meet over the central part of the country.

By far the most common type of storm, although much less destructive than those already described, is the *thunderstorm*. It is estimated that in excess of 40,000 may occur in one day throughout the world, with 2,000 in progress at any one time. Thunderstorms form within a warm, moist air mass that is heated from below, thus causing an unstable condition and convection currents. These currents form separate units of convection circulation called thunderstorm cells. Within the cells, strong turbulence is associated with the vertical currents and is accompanied by thunder, lightning, heavy rain, and occasional hail. The cell boundaries are characterized by smooth, cloudy air.

For convenience of description, the development of a thunderstorm has been divided into three stages. The first of these, the *cumulus stage,* begins with the development of a cumulus cloud or cell within which there is a strong updraft extending the cloud from approximately ground level to above 8,000 meters. The updraft may vary in strength and speed, but speeds up to 1,000 meters per minute are not unusual (*A* in fig. 14.32). A certain amount of horizontal inflow of air through the sides of the cloud occurs. This *entrainment*, as it is called, brings about a mixing of air from the surrounding environment with the air in the cloud column.

During this early stage, the air within the cell is warmer than the surrounding air at comparable elevations. The moisture in the air tends to condense as it is swept upward to cooler temperatures with raindrops forming at lower elevations and snow and ice crystals at the upper reaches of the cell. The upward sweep of the moist air continues to feed the cumulus cloud as it develops, while at the same time the raindrops, snowflakes, and ice crystals increase in number and size. Eventually these become large enough to fall to the earth, at the same time creating a downdraft in the cell (*B*). The arrival of the first raindrops on the earth usually signals a change from the cumulus stage to the *mature stage* in thunderstorm development.

Initially, in the mature stage of thunderstorm activity, the downdraft is to be found only in the lower reaches of the storm cell, but gradually it increases in horizontal and vertical extent for the duration of this stage (*B*). Rainfall increases in intensity, accompanied by thunder and lightning. Because of the conflicting air drafts, general turbulence is encountered in the cell, with the downdraft eventually reaching dominance and indicating the end of the mature stage.

In the final or *dissipating stage* of thunderstorm activity the updraft gradually ceases and the cell contains only downdrafts (*C*). Rain, thunder, and lightning decrease and the thunderstorm comes to an end. At this point, the thundercloud may be 20,000 meters high, and the upper air currents produce the anvil-shaped cloud characteristic of the thunderstorm's final stage. Temperatures within the cell assume values close to the environment, and wind direction and speed becomes normal for that area.

Lightning during a thunderstorm occurs as a result of a change in electrical charge within a thundercloud. The updraft of a large convection cloud will transport electrically charged particles from near the earth to the upper portions of the cloud. Tiny ice crystals growing in size will fall and separate the charges, in the process leaving the upper part of the cloud positively charged and the lower part negatively charged. This induces a positive charge on the ground, reversing the fair-weather electrical field which is normally negative (fig. 14.33). Eventually, the field strength between the positive and negatively charged portions of the cloud reaches the *breakdown potential*. At this point, the air can no longer insulate the charges and a surge of current in the form of lightning is generated.

Measurements have shown that the center of negative charge is located between $0°C$ and $-10°C$, while the upper positive-charge center is near the $-10°C$ level. As the thunderstorm progresses through the mature stage, a small region of positive charge also develops in the downdraft associated with the heaviest rain. Between the upper positive charge center and the negative-charge center is usually the region where the first lightning develops. This is called the *lightning hearth region* and is apparently associated with the existence of both liquid water particles and ice or snow crystals. The exact mechanism which produces the large electrical potential is, however, not completely understood. Thunder is thought to result from the sudden expansion of the air due to the heat generated by the electrical discharge of lightning.

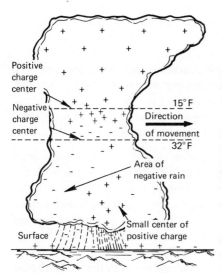

Figure 14.33 Location of electric charges inside a typical thunderstorm cell.

Figure 14.34   Electrical discharge and thunderhead typical of a thunderstorm.

It is estimated that the total potential required to produce a lightning stroke 3,000 meters long would be 20 to 30 million volts. The current may be as large as 60,000 amperes and in extreme cases may be in excess of 100,000 amperes.

14.5   SUMMARY

Weather is essentially the result of a number of factors affecting the circulation of the atmosphere and modifying movements within the general circulation pattern. The primary source of energy for this movement comes from solar radiation. Solar radiation is not uniformly distributed over the entire earth because of the earth's spherical shape. This results in warmer temperatures over the equator and progressively colder temperatures toward the poles. The resulting temperature differential causes the atmosphere to circulate in a complex manner. Additional modification of this circulation is caused by such factors as rotation of the earth, Coriolis force, gravity, pressure gradient, centrifugal force, and friction. In addition, the nonuniform distribution of the land masses and variation in land topography, as well as land-sea relationships, bring about more localized changes in weather conditions.

Air movement is generally described in terms of movement of air masses. The boundaries between air masses are known as weather fronts, and most weather changes are related to the occurrence of these fronts. Several major types of air masses are recognized, based upon the regions in which the air mass originated. Thus an Arctic air mass originates in the perpetual frost region of the Arctic circle, and the polar air masses from this area to a point where average temperatures for the warmest month exceeds 10° C. Tropic air originates in the warm regions. The air masses are further defined on the basis of whether the source region is continental or maritime in nature.

Fronts are described as the boundary between several air masses, and the nature of the front is a major factor in weather conditions. Cold air replacing warm air represents a cold front; a warm front is the reverse. The moisture available and the slope of the front are factors which contribute to variations in weather conditions along a front. Occasionally, the front formed by two air masses displays little or no movement and is known as a stationary front. An occluded front is related to a low-pressure system or wave cyclone. The structure of the occluded front is dependent upon the difference in temperature between the cold air in advance of the system and the cold air to the rear of the system.

Low-pressure areas in the atmosphere usually result in inclement weather and occasionally in destructive storms such as hurricanes. Hurricanes (typhoons in the Pacific region) are circular storms with extremely low pressures at the center. The hurricane may be 480 to 720 kilometers in diameter and have winds with velocities in excess of 120 kilometers per hour. High precipitation rates are also associated with hurricanes, as well as storm waves, which may create havoc along coastal areas.

Tornadoes, occurring principally in the United States, are small cyclones that rotate at velocities up to 800 kilometers per hour. A great deal of damage results where the funnel-shaped storm touches down on the earth's surface. As in the case of the hurricane, the center of the tornado is a region of extremely low pressure.

Thunderstorms, probably the most common form of storm, are generally milder than the hurricane and tornado. Thunder, lightning, and some rain are associated with these storms. Wind and lightning may cause small damage, but not to the degree experienced with hurricanes and tornadoes.

QUESTIONS

1. What conditions exist that cause different parts of the earth to be subject to differing amounts of heat energy from solar radiation?
2. Briefly describe the forces responsible for influencing air movement over the earth's surface.
3. Describe air movement around high and low pressure areas in the northern and southern hemispheres.
4. What causes onshore and offshore breezes to blow?
5. What factors control clouding and precipitation along a weather front?
6. What weather events are related to the passage of a cold front in the northern hemisphere?
7. What is an occluded front and how does it develop?
8. Describe the characteristics of a hurricane.

15 Climate and Life on the Earth

The world pattern of climate.
Causes of seasonal variation.
Classification of climate.
Relationship of climate and vegetation.
Relationship of climate and soils.
Relationship of climate and geology.
Relationship of climate and society.
Weather forecasting.

Climate is frequently defined as the average weather that one may experience at any one place over an extended period of time. However, such a statement is unsatisfactory in that it does not define "average weather," nor is such a definition readily available. Since climate is recognized as changeable, some effort has been made to standardize statements with respect to these changes. This has been accomplished to some degree by the World Meteorological Organization by establishing "period means" for a specific span of years. The "normal" is the mean value of weather data obtained during a stated 30-year period. The "standard normal" 30-year period most recently completed included 1931 to 1960. The current standard normal period runs from 1961 to 1990. The "norms" of these periods are useful in comparing and determining trends in the climate.

The world pattern of climate today depends primarily upon definite aspects of atmospheric behavior related to surface conditions, such as the relative proportion of ocean cover, or the sizes and shapes of continents and their positions with respect to each other, to the poles, and the equator. The distribution of plains, plateaus, and mountains upon the continents are also factors in the development of climate. The world pattern of climate during any part of the geologic past was, in general, related to exactly the same factors that influence climate today. The amount of atmosphere has not changed appreciably, nor has its average temperature, viscosity, composition (within the past half billion years), or other significant physical characteristics. The speed and direction of the earth's rotation, the rate of escape of heat from the earth's interior, the amount of solar radiation received by the earth, and similar fundamental climatic factors have either remained constant or have varied by only inconsequential amounts. The most variable factors affecting climate have been those relating to continental size, position, and elevations. It is thus reasonable to regard these as the most probable causes of such past climatic changes as is indicated by geologic evidence.

The earth has experienced several brief periods of crustal unrest during its long history, each of which has been accompanied by evidence of glaciation on various parts of the surface and by intense aridity in other parts. Between these revolutionary periods have been vastly longer periods of quiet and climatic monotony.

Man appeared on the scene during a revolutionary geologic period and has experienced glacial climates. In the geologically recent portion of his experience, during which he has progressed through a cultural development culminating in such scientific advances as the introduction of modern instrumental observation, he has witnessed a slow amelioration of extreme glacial climate, but he has at no time experienced the normal climate indicated by most of the geologic record.

Man has observed that climatic conditions fluctuate rather widely from time to time at a given place. In seeking to understand such natural phenomena, he has been tempted to explain such fluctuations on the basis of recurring cycles. However, except for seasonal variations, no definite proof has been advanced to contradict the opinion that all such relatively short-term climatic changes are nothing more than matters of chance. The world pattern of climate today is the product of climatic variations and not the expression of recurring mean or normal conditions. The extent of desert climate will not be the same next year as this. The humid margin of the desert is the product of an ever-changing distribution of extreme aridity. The time may come when such changes may be well enough understood to be of definite forecast and economic value, but it is likely that such information will be the result of long-term, patient research.

Interest in changes of climate of geologic proportions will remain intellectual. There is satisfaction in learning the secrets of earth's history, even though the investigations are based almost entirely upon evidence that accumulated long before the appearance of man on the earth, while all very long-term forecasts relate to a time in the future when he may no longer be present to verify or contradict them.

15.1 THE SEASONS

Climate and Life on the Earth

The seasons of the year—spring, summer, fall, and winter—follow one another in endless cycle. These seasonal variations are due to the earth's axial inclination of 66.5° from the ecliptic and are not the result of varying distances from the sun during the year. The earth's elliptical orbit brings it closer to the sun at perihelion in January (147 million kilometers), and further away at aphelion in July (152 million kilometers). This results in approximately 7 percent more solar radiation being received in January than in July by the earth as a whole, but it is not an important factor in climate.

The earth's axial inclination causes the Northern Hemisphere to be subjected to direct solar radiation at one time of the year and the Southern Hemisphere at another (fig. 15.1). The sun will be directly overhead at 23.5° north latitude

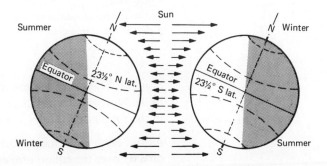

Figure 15.1   Because of the earth's axial inclination to the ecliptic, the sun's rays are concentrated alternately north of the equator and then south of the equator. The alternating concentration of solar radiation results in the changing seasons.

about June 21 (*summer solstice*) in the Northern Hemisphere, resulting in this latitude receiving maximum solar radiation or *insolation*. In the southern latitudes, the surface receives the sun's rays at an angle. The amount of insolation received for a given area in the Northern Hemisphere would therefore be distributed over a greater area in the Southern Hemisphere and would result in less heat per unit of area (fig. 15.2). In addition, the direct rays in the Northern Hemisphere travel through less thickness of atmosphere than do the slanting rays south of the equator. Thus, there is less radiant energy lost by atmospheric absorption and reflection in the north than in the south (fig. 15.2). The length of the day during which any area of the earth's surface is subjected to insolation is also a factor in the rhythm of the seasons. During the summer months in the Northern Hemisphere, days are longer than in the Southern Hemisphere, permitting a longer exposure to solar radiation (fig. 15.1). This allows an increased amount of heat to be delivered in a day. The diagram reveals that at this time of year the North Pole has 24 hours of daylight and the South Pole has 24 hours

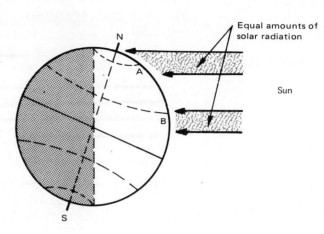

Figure 15.2 The sun's energy is most concentrated where it strikes the earth vertically (b) and less concentrated where the energy strikes at an angle (a). The same amount of energy is distributed over a greater area at (a) and is more concentrated at (b).

of night. This combination of factors results in generally warmer temperatures in the Northern Hemisphere associated with summer, and colder temperatures in the southern latitudes associated with winter.

As the earth continues in orbit, the sun will appear to move south until it is directly over the equator on approximately September 22. This is known as the *autumnal equinox*, at which time the days and nights are of equal length at all latitudes. By December 22 (*winter solstice*), the sun will be directly overhead at 23.5° south latitude, at which time cold weather is experienced in the Northern Hemisphere and warm weather in the Southern Hemisphere, or the reverse of what was described as occurring six months previously. Approximately three months later, on March 21, the sun will have again reached a position directly over the equator, but now the sun appears to be traveling northward. This is the *vernal equinox*, at which time the days and nights are again of equal length. By the following June 21, the sun will again be directly overhead at 23.5° north latitude and the earth will have completed its annual cycle of seasons.

The occurrence of maximum insolation and day length, coincident with minimum atmospheric absorption and reflection of solar energy during June in the Northern Hemisphere (December in the Southern Hemisphere), would lead to the expectation that the highest average temperatures would be experienced at that time. This does not occur, because time is required for the atmosphere and land surface to accumulate heat, resulting in a seasonal lag of about a month. For this reason, maximum temperatures are generally encountered in July (January in the Southern Hemisphere). By the same token, minimum temperatures also generally lag the period of minimum insolation by approximately a month. Thus, January, on the average, becomes the coldest month north of the equator, and July the coldest south of the equator.

It must be recognized that in addition to the normal seasonal cycle there are other factors which modify climate. Such modifications may be seen even along the same latitude and within short distances. Variations in temperature are shown in figures 15.3 and 15.4, where isotherms representing average temperatures for January and July are not parallel to the equator. If the earth were a uniformly smooth sphere of similar surface material, then temperatures might be expected to grade uniformly from equator to pole. The fact that the earth is not such a sphere causes climate to deviate from a regular pattern of distribution.

15.2 CLIMATES OF THE WORLD

The distinction between climate and weather is more or less artificial, since the climate of any place is merely a buildup of the day-to-day weather over a long period, while, conversely, weather is the day-to-day breakdown of the climate. It seems to be a useful distinction, however, and there will continue to be meteorologists concentrating on the daily weather and climatologists concerned with the more long-term outlook.

Climate may be classified on the basis of the pattern resulting from the interaction of moisture and dryness, heat and cold. The interaction of these characteristics in the production of a climate pattern in turn creates a pattern of natural vegetation, and both together have been the principal forces—though not the only ones—in creating the general pattern of soils that cover the earth's land surface.

The forces responsible for the world's climate are those related to the general circulation of the atmosphere discussed in Chapter 14. These forces result in eight major climatic divisions. Five of these climates are in regions where the prevailing temperatures range from hot to warm to cool, indicating that moisture, not temperature, is the chief factor affecting plant growth. These five categories are therefore classed on the basis of moisture as: *very humid, humid, subhumid, semiarid, and arid.* The three remaining divisions occur where the prevailing temperature is cold enough so that temperature rather than moisture is the chief factor affecting plant growth. These divisions are: the cold regions called *taiga*, after the dominant vegetation; the very cold regions called *tundra*, also after the dominant vegetation; and the regions of *perpetual frost.*

The temperature and moisture characteristics will of course influence the type of vegetation that results, and some systems of classifying climate make use of vegetative types of describing each category. However, such a system may not be completely satisfactory, since the categories are based on the result rather than the cause of climate. Nevertheless, vegetative types do serve a useful purpose and may be correlated with the categories listed above and in table 15.1.

Very humid climates appear in equatorial regions, particularly on east coasts and on high middle latitudes on the west coasts. Arid climates appear on the west coasts of continents in low middle latitudes and extend as lobes into con-

The Atmosphere

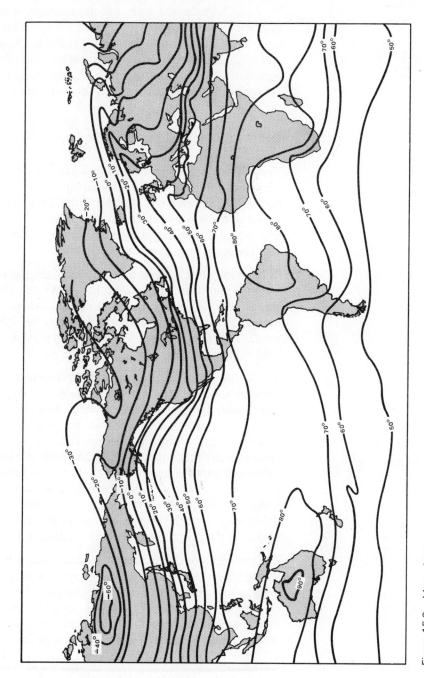

Figure 15.3 Mean January temperatures.

Climate and Life on the Earth

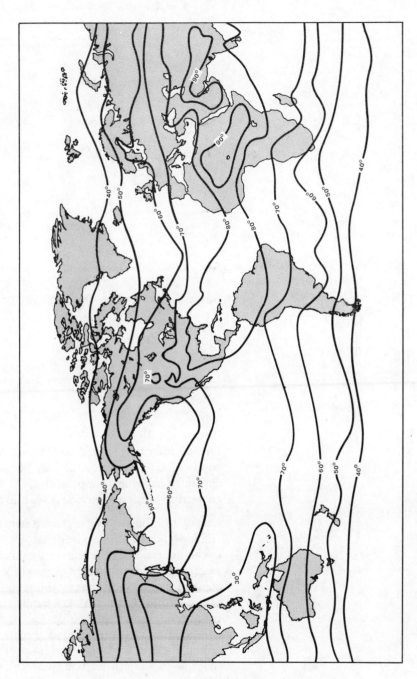

Figure 15.4 Mean July temperatures.

Table 15.1
Precipitation, Temperatures, and Vegetative Types Generally Found in
Each of the Climatic Categories

Climatic Type	Annual Rainfall in Centimeters	Temperature Ranges	Vegetation
Very humid	More than 200	Above 18°C all year	Tropical rainforest
Humid	100–200	Coldest monthly average temperature above 0°C	Broadleaf forest; deciduous
Subhumid	50–100	Coldest monthly average temperature below 0°C; warmest monthly average above 10°C	Tall grass or prairie
Semiarid	25–50	Coldest monthly average below 0°C; warmest monthly average above 10°C	Short grass or steppe
Arid	Less than 25	Wide range of temperatures	Desert shrub, sparse vegetation
Taiga	100–200	Long, cold winters, short summers	Coniferous forests found from 50° to 65° north latitude
Tundra		Warmest monthly average below 10°C above 0°C; short summer	Mosses and lichens
Perpetual Frost		Ice cap	No vegetation

tinental interiors. Humid, subhumid, and semiarid climates make up broad bands which lie between the very humid and arid regions. The three climates characterized by low temperatures, i.e., taiga, tundra, and perpetual frost, form concentric bands around the poles and appear on higher mountain slopes.

Irregularities in the shape of continents and in the distribution of mountains and lowlands result in some departure from the generalized pattern. However, as can be seen from the map of world climates (fig. 15.5), the general arrangement as delineated on the generalized continent is preserved. Very humid climates appear in the equatorial region and include the East Indies and part of the Philippines, the Malay Peninsula and the coastal parts of Burma and Indochina, the west coast of India, the coasts of Ghana and Nigeria in Africa, most of Central America, and the northwest coast of South America. Very humid climates are also found on continental west coasts in high middle latitudes as in western Europe, western North America (Olympic peninsula in Washington), and southwestern Chile.

Arid climates are found on the west coasts of continents in low middle latitudes

Climate and Life on the Earth

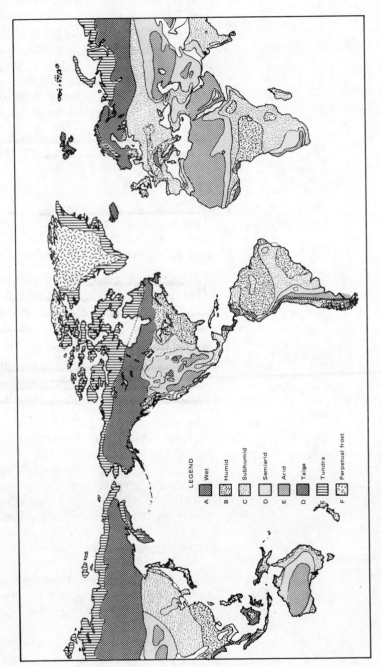

Figure 15.5 Distribution of principal climates of the world.

and inland to continental interiors. The largest arid regions occur in Eurasia and Africa, and include the Sahara, the Arabian deserts, the deserts of Iran and Turkestan, the Thar desert of India, and the Gobi desert of Mongolia. The arid regions of North America are restricted largely to northern Mexico, southern California, Nevada, Arizona and Utah. They are much less extensive than the Old World deserts, but they occupy identical positions in the climatic pattern. Arid regions in the Southern Hemisphere include the Atacama and Patagonian deserts in South America, the Kalahari desert of South Africa, and the Great Australian desert.

Semiarid climates occupy broad belts around the deserts and are most extensive in Eurasia and North Africa. Other large areas are found in western North America, Australia, and South Africa. The semiarid part of South America is relatively small.

The humid climates occupy extensive areas in the tropics and middle latitudes. They are found in the equatorial regions of South America and Africa, the eastern and northwestern United States, central and western Europe, China and eastern Australia, and southeastern South America.

The subhumid climates occur between the humid and the semiarid climates. They are found chiefly in the central United States and Canada, northeastern Argentina and interior South America, south and central Africa, eastern Australia, north China, and peninsular India, and in a belt from central Germany eastward into Siberia.

Extensive areas of taiga and tundra climates are found in Canada and Siberia. The climate of perpetual frost appears in the interior of Greenland and other islands and occupies the entire continent of Antarctica.

The world map in figure 15.5 represents the average position of the climatic regions. Actually the boundaries shift from year to year as the regions themselves expand and contract. These shifts are important to society because of the influence climate has on the basic industry of agriculture. However, it is the mean position of the climatic regions which is most strongly related to the distribution of natural vegetation and soils and the development of minor landforms; these ramifications of the climatic patterns will now be considered.

15.2A Climate and Vegetation

Since very early times, it has been recognized that there is a close relationship between climate and vegetation, and many terms have come to be applied interchangeably to both of them. For example, the term "desert" calls to mind a region which is excessively dry and is characterized by sparse vegetation peculiarly adapted to arid conditions. "Tundra" applies to those cold subarctic lands, frozen much of the year, where only mosses and lichens grow.

The close identification of climate and vegetation is the consequence of thousands of centuries of plant differentiation and adaptation. Since plants first appeared upon the earth, they have been subjected to the influence of climate.

Through the elimination of nonadapted species and through the frequent origin of new forms, many different types of plants have become adapted to widely different climatic conditions. For example, plants capable of withstanding prolonged drought have developed. Cacti have extensive root systems for drawing moisture from a wide area and spiny leaves which decrease moisture loss by transpiration. Short grasses such as buffalo grass have a low growing habit that makes them more drought resistant than the taller, more luxuriant, prairie grasses.

Because of this adaptation, each major climatic region has a dominant vegetation group made up of several plant species, each of which is adjusted to the climate of that region.

The dominant plants are not the only plant species found in a particular climatic region, as other species do exist. However, the members of the dominant group are known as the *climax vegetation* for the region. Locally, the climax group may not exist because of soil conditions or because fire has temporarily destroyed the climax vegetation and there has not been sufficient time for the dominant species to become reestablished. The climax vegetation for the immediate circumstances may exist, but this is not the true climax—only the climax due to climate is the true climax for the region.

A comparison of the world climate map (fig. 15.5) and the world vegetation map (fig. 15.6) reveals the correspondence between climate and vegetative patterns. The following is a brief description of the major vegetative types:

1. *Rain forest:* characterized by tall trees with interlocking crowns but lacking an understory (undergrowth) due to shading by trees. In the tropics, rain-forest regions are characterized by high precipitation and high temperatures throughout the year. A short dry season will result in a lighter tropical forest in which trees are not as tall as in the true rain forest, but a dense understory of shrubs, brush, and smaller trees exists. In some tropical areas where climatic conditions grade to subhumid, a scrub and thorn forest may occur that is composed of small trees and thorny vegetation more widely spaced than the trees in a tropical forest.

2. *Broadleaf forest:* characterized by trees usually of a deciduous nature, with heavy undergrowth found in regions of strong seasonal characteristics. Some conifers may be mixed with the broadleaf trees. If a long dry season alternates with a cool wet season the forest will grade into *chaparral*, which is composed of scrub oak and a variety of shrubs typical of the Mediterranean and coastal southern California areas.

3. *Tall grass or prairie:* luxuriant grasslands typical of subhumid climate. Between this vegetative type and the forested areas may lie the *savanna*, which is a grassland interspersed with trees. This may occur where total rainfall is high but the rainfall pattern assumes a markedly seasonal aspect with a long dry season.

4. *Short grass or steppe:* characteristic of semiarid regions where short grasses such as gramma grass and buffalo grass are dominant.

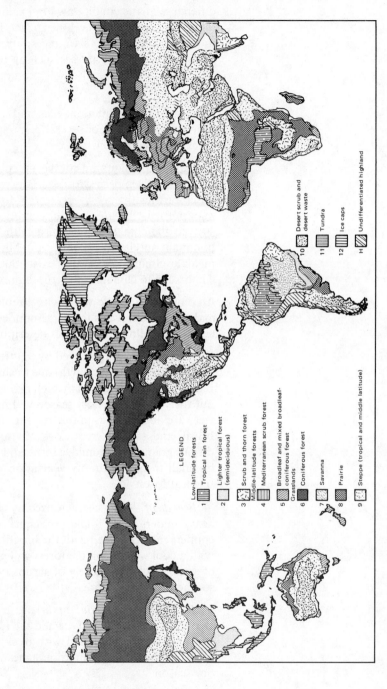

Figure 15.6 World distribution of the principal vegetative formations.

5. *Desert:* populated by a variety of plants adapted to arid conditions. Usually plants have a short life cycle which permits them to survive on the low rainfall levels that exist, or the plants have the ability to remain dormant during protracted dry periods. The plants such as the cactus are efficient users of water.

6. *Taiga:* consists of tall coniferous stands, mainly spruce and fir, with heavy crowns. Understory is composed of dense fern and mosses and not as many small trees as are found in the deciduous forests.

7. *Tundra:* characterized by the presence of mosses, lichens, and sedges. Usually a treeless expanse in the Arctic areas.

8. *Perpetual frost:* no vegetation because the region is covered by ice and snow.

15.2B Climate and Soils

Numerous studies have clearly revealed that the major differences between soils are due to the effects of climate operating through soil-forming processes. It is true that locally, the nature of the underlying rock, peculiarities in drainage conditions, or unusual vegetative conditions may outweigh climatic factors in influencing soil character, but the distribution of the major soil groups can best be interpreted in climatic terms. This is partly due to the direct influence of climate on soil formation and partly to the fact that soils are strongly influenced by vegetation, which in turn is related to climate.

The recognition of the importance of climate in soil genesis led to the investigation of the relationship between climatic conditions and the distribution of soils of various kinds. In the nineteenth century, it was realized that the amount of precipitation as compared with the amount of evaporation was of critical significance in soil formation. This indicated that there existed a climatic boundary along which the effect of precipitation was balanced by the effect of evaporation. This boundary, which lies between the moist and dry subhumid climates, divides the humid soils from the arid soils. The humid soils, formed under climates in which the precipitation exceeds the evaporation, are leached soils—that is, the predominantly downward movement of water has removed material from the top zone into the lower zones. Arid soils, on the other hand, are characterized by carbonate accumulation at or within a few feet of the surface, resulting not only from the comparatively small amount of downward percolation of water, but also from the upward movement of water that evaporates from the soil and deposits basic salts.

The effective precipitation (precipitation minus evaporation) has an effect not only on the amount of leaching, but also on the acidity, the nitrogen content, and the amount of fine clay minerals or colloids present in the soil. Increased effective precipitation is associated with increased acidity and nitrogen content, concentration of clay minerals, and decreased carbonate accumulation.

In addition to precipitation, temperature is an important factor in soil distribution. A great variety of chemical changes can occur only under high temperatures. Thus, with an increase in temperature, there is an increase in chemical weathering. In addition, the rate of accumulation of organic matter in the soil tends to increase with increase in temperature. This relationship is frequently masked by the more important influence which vegetation has on organic matter content. Grasses yield more humus than does a forest cover, and maximum organic matter content is therefore found in warm subhumid areas.

Vegetation is also of great significance in its effect on the mineral content of the soil. In the northern coniferous forests, leaf fall constitutes the principal source of humus, and this kind of leaf litter, being low in mineral constituents, leads to the formation of a highly acid, peaty surface layer. In a deciduous forest, the leaf fall is higher in mineral matter, and the mineral compounds are added from plants covering the forest floor, so that a more nearly neutral humus is formed. The highest mineral accretions to the soil from plants are found in grasslands, where plants are high in mineral content and where their disintegration both below and above the surface insures the dissemination of this mineral matter throughout the upper portion of the soil mass.

In studying soils and relating soil conditions to such fields as agriculture, it is convenient to use a classification system of some sort. Prior to the mid 1960s, such a system was based upon the influence of geologic and climatological processes on soil formation. At the present time, a new soil classification system has been inaugurated which focuses on the soil properties themselves rather than on the means by which they are formed. Since this new system cannot be easily related to climatological processes, the system will not be discussed here. However, in figure 15.7, the major soil groups are shown based upon the old classification system so that a comparison with figures 15.5 and 15.6 will show a correlation between climate, vegetation, and soils.

15.2C Climate and Geology

How does climate influence the geological processes? This is illustrated by the variation in the landscape from one climatic region to another. These variations are a reflection of the effectiveness of the forces that strip or denude the surface. Streams, waves, ground water, frost, winds, and glaciers are agents which participate in the leveling of the ground surface. In opposition to these, one must place the forces that thrust up rock, such as tectonic forces and volcanic forces. These are the agents responsible for the formation of positive relief features.

The agents of erosion act in a variety of ways in molding the landscape (see Chapter 11). It is necessary only to consider briefly those forces that are important because of their relationship to climate. (Thus, wave action may be eliminated from consideration because climate does not primarily effect the tensity of waves in promoting erosion.)

Virtually all erosion occurs through the action of running water, through mass movement, as a result of wind, or in association with glacial movement. Figure

Climate and Life on the Earth

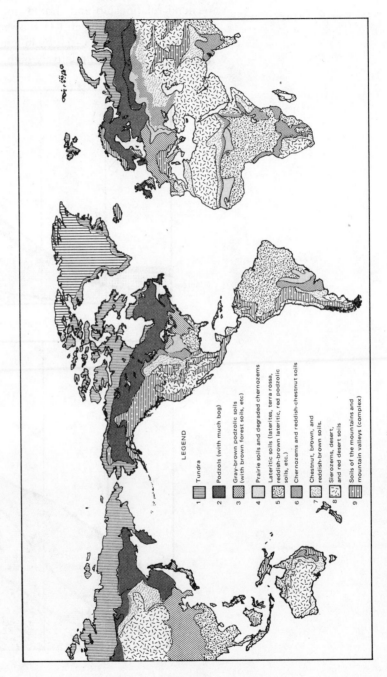

Figure 15.7 World distribution of principal soil groups.

15.8 brings out how the effectiveness of these agents varies with differences in temperature and precipitation. The diagrams are meant only to present the general relationship and should not be interpreted in a strictly quantitative manner. The diagram in (A) shows the variation in the effectiveness of the erosional agents with variation in temperature where the effective precipitation remains constant and relatively high, approximately equal to that found in the

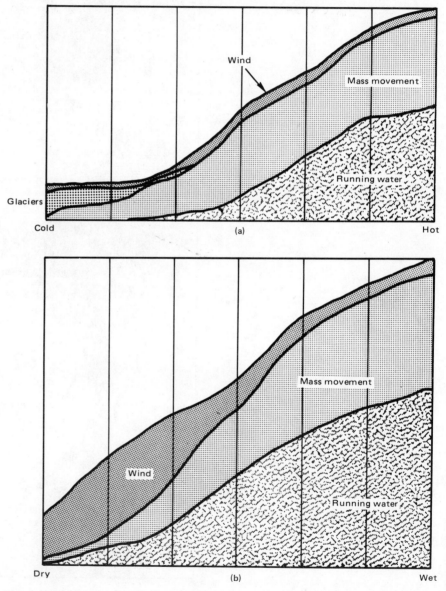

Figure 15.8 Variation in effectiveness of various erosional agents due to temperature (a) and precipitation (b).

southeastern United States. Running water becomes less important as the temperature decreases, since a decrease in temperature results in a longer frost period. Mass movement remains an important mechanism throughout the entire thermal range, except perhaps for the very lowest portion of the scale. At the warm end of the scale, soil slumping is the principal mass-movement type; at the cold end of the range, solifluction or mud flow becomes significant; for intermediate values, soil creep, as related to frost and ice action, becomes significant. It will be noted that wind has relatively little influence where effective precipitation is high. This is due to the fact that under hot and warm conditions, with no deficiency in rainfall, the ground surface is normally well protected by vegetation; whereas with very low temperatures, frozen ground and snow cover both aid in protecting the surface against wind erosion. Glacial action becomes significant only at temperature values below freezing, although some glaciers are encountered in areas where the temperature is high. Overall, the diagram illustrates that there is a decrease in the rate of erosion from areas of high temperature to areas of low temperature.

Differences in the rate of erosion and in the importance of the agents of erosion with variations in precipitation, if temperature is constant, may be seen in (*B*). The variation is illustrated with temperatures constant near the center of the mesothermal range. It is evident that erosion by running water will decrease as precipitation decreases. Similarly, mass movement will become less effective, because water is an important element in the mass-movement mechanism, and in arid regions mass movement is restricted to rock slides. The effectiveness of wind, however, varies inversely with precipitation, as the diagram indicates. As the effective precipitation decreases, so, in general, does the density of the natural vegetation. The exposure of the soil makes it possible for wind to work with maximum efficiency. Glaciers do not enter in the diagram because mesothermal conditions are too warm for the existence of glaciers.

The variation in total erosion as effective precipitation changes (*A*) does not parallel the variation resulting from temperature differences (*B*). More erosion will occur in dry climates than in cold climates, and more in wet climates than in hot climates. However, it would be erroneous to conclude from this that precipitation efficiency is more significant than temperature in influencing the erosional processes.

Low values of either temperature or effective precipitation result in low rates of erosion. This accounts for the general similarity in local relief in polar and arid regions. The fact that erosional forces are not particularly active results in a rugged landscape, as can be seen in figures 15.9 and 15.10.

Although polar and arid regions appear similar in broad outline, there are numerous minor differences. Wind, which is far more significant in desert than in polar regions, is responsible in arid lands for the formation of dunes, the scouring of dry basins, the carving of solid rock by means of sand blast, and the development of desert pavement. In polar climates, the wind can seldom attack the surface with any degree of effectiveness because the ground is usually

Figure 15.9 Landscape typical of desert regions.

frozen or covered by snow, and the wind is not provided with sand particles to use as blasting tools. On the other hand, solifluction and glacial action, so characteristic of tundra and areas of perpetual frost, are absent in the arid regions. Such land forms as cirques and moraines which occur in cold climates are not found in dry regions unless as a remnant from a previous glacial era.

While the deserts and polar regions are characterized by angular landscapes, the warm humid regions of the earth are typically regions of gently rolling slopes well covered with vegetation (fig. 15.11). An extensive network of streams and the gradual movement of soil particles downhill soon obliterate surface breaks caused by faults and sharp folds. The vegetation protects the land against the scouring action of the winds, and the temperatures are too high for the formation of glaciers. Thus, regions of warm and hot climates display a natural landscape markedly different from that of arid or polar regions.

15.2D Climate and Society

Landscapes are the integrated combinations of natural and cultural features that characterize the surface of the earth. Landforms, surface waters, vegetation, animal life, soils, and rock are foremost among the natural elements of the landscape. Buildings, roads, crops, domesticated animals, railroads, and canals are a few of the many cultural elements. In the interactions between climate, vegetation, soils, erosion, and the molding of the land surface, climate

Climate and Life on the Earth

Figure 15.10 High rugged mountain terrain where erosional activity is influenced by low temperature is similar to polar conditions.

plays a dominant role. It is therefore understandable that the natural landscape, solely the product of natural forces in the sense that man's influence has not been felt, is closely bound to climate. The cultural landscape is related to climate to a somewhat lesser degree. The elements of the cultural landscape such as buildings, roads, and the distribution of crops are strongly influenced by heritage, yet they remain the outward manifestations of an economy and social organization which is in delicate balance with the natural environment. So, while a wide variety of cultural landscapes may exist under any one set of natural conditions, each displays certain characteristics that are in harmony with the natural landscape and hence with the climatic complex.

In the cultural landscape, the climate is a passive rather than an active agent, setting limits beyond which certain human activities cannot reasonably be pursued. Climate influences the crops grown and the techniques used to cultivate them. Climate also influences human activities indirectly through the medium of the natural landscape. Thus cultivation of crops is limited for the most part to the subhumid and humid regions where suitable soils and natural

Figure 15.11 Landscape typical of humid regions.

vegetation make it possible to practice agriculture. Though climates vary to a very limited extent from year to year, their mean positions change very slowly. Hence the average conditions and normal fluctuations are operative over long periods of time. The result is that the cultural as well as the natural landscapes tend more and more to reflect climatic conditions, although the cultural landscape forms themselves may vary within one climatic region because of the different cultural backgrounds of the people inhabiting it.

The world landscape pattern, then, whether or not the landscapes are altered by man, constitutes the integration of the world patterns formed by vegetation, soils, landforms, and land utilization. The study of climate makes it possible to have a better understanding of these patterns and is indispensable in determining how man can most intelligently utilize the resources and environment that nature has provided.

15.3 WEATHER FORECASTING

Climate and weather are basic natural resources, but they must be understood if they are to be turned to good advantage. No one is in a better position to realize this than the farmer. As civilization has become more complex, the dependence on intimate and accurate knowledge of climate and weather has increased. Today, this knowledge is so indispensable that every civilized country has an elaborate weather service. In the United States, this weather service functions twenty-four hours a day and endeavors to bring up-to-date information to every individual in the land who needs it.

15.3A Uses of Meteorological Service

Weather information has innumerable uses and applies directly or indirectly to almost every human activity. The Weather Bureau provides meteorological information which contributes to the success of business and commercial activities as well as to the agriculture industry.

Power companies in large cities plan to meet peak loads in heating and lighting on the basis of forecasts of temperature and unusual cloudiness. Without advanced knowledge of these conditions, uneconomical operation and perhaps power failure would result. Hydroelectric companies regulate their use of water power in the light of forecasts of rainfall, even going to the expense of starting steam auxiliary plants when a long period of dry weather is in prospect.

In winter, when a great mass of cold air is about to engulf the country, the forecast leads to precautionary activities in almost every field. Water pipes and valuable shrubs are wrapped for protection. Automobile radiators are filled with antifreeze, and wholesalers of antifreeze products rush their advance shipments to areas that will be affected. Retailers increase their advertisements for winter articles. Fuel dealers increase their supply of coal and oil and ration their sales to customers if there is likely to be a shortage. Farmers drive their stock to shelter. In some parts of the country, the first forecast of cold weather in the fall is the signal for preparation of the winter's meat supply and the completion of fall harvesting. Other precautionary measures affecting almost every walk of life in one way or another follow the publication of forecasts of hurricanes and floods.

From the foregoing, it can be seen that weather is potentially one of the greatest upsetters and interrupters of human plans and activities. It must constantly be taken into account in the calculations of industry and agriculture if a smooth operation is to be achieved. Because so much depends upon it, the pressure in modern times is for a more elaborate, refined, and accurate weather service.

15.3B How Forecasts Are Made

The primary function of weather forecasting is to provide weather information as accurately as possible and for as far into the future as practicable. In order to accomplish this, it is necessary for the meteorologist to know intimately what weather changes have taken place in the past few days. When he has become well acquainted with the sequence of events, and with weather phenomena as shown on the principal weather charts along with other material prepared on a regular basis for weather forecasting, he can project the present conditions into the future. Information from hundreds of observations made all over the world, and more recently, information from meteorological satellites, provide the needed data for the forecast. This is presented in a synoptic weather report.

A synoptic weather report is a concise synopsis or summary, usually in a simple code for brevity, describing weather conditions in a locality at a certain time. It is important to understand the role of the synoptic report in modern meteorology, since the report as represented on the weather map is the basis for understanding weather changes.

A single isolated weather observation tells little of the general state of the atmosphere or of the changes that are about to take place in it. Experienced outdoor observers such as farmers and mariners may recognize the approach of a storm from a single observation of clouds or the appearance of the sky, but such an observation does not enable them to describe the weather in detail or to describe it for a large area each day. As stated in the previous chapter, the moisture that falls as rain is usually transported by a large body of air from some distant ocean. These vast bodies of air, technically known as air masses, come together from widely separated regions and by their contrasting characteristics produce changes in the weather. Clouds and rain usually occur along the boundaries of air masses and are the result of their overlapping. In order to have a comprehensive understanding of weather and climate, it is necessary to view a major portion of the atmosphere as a whole, such as a polar hemisphere or the hemisphere encompassing the region under consideration.

The state of the atmosphere at any instant may be likened to a great jigsaw puzzle in which one local weather observation is a single piece. The synoptic weather reports from well-distributed observing points make up the numerous pieces of the weather puzzle. Many simultaneous observations are necessary to reveal the total pattern. When these observations are pieced together in the form of a daily weather map, the complete view of the weather as seen by observers on the ground is available. The weather map is an important aspect of meteorology and is the foundation of the modern weather service.

The surface weather map provides a composite display of the surface weather observations made by weather stations. These maps are usually compiled from simultaneous observations taken four times a day (0130, 0739, 1330, 1930 EST). In constructing the weather map, the data reported by each station are entered around the station's location shown on the map. When all the reported data are entered on the map, it is possible to see the broad picture of the weather conditions existing at the time of observation over a large geographical area. It must be stressed that the map does not directly represent what the weather conditions will be like at some time in the future but only represents the actual conditions at the time of observation.

Figure 15.13 shows a section of an actual surface map and the placement of the data around the station. The symbols used are standardized and their arrange-

Climate and Life on the Earth

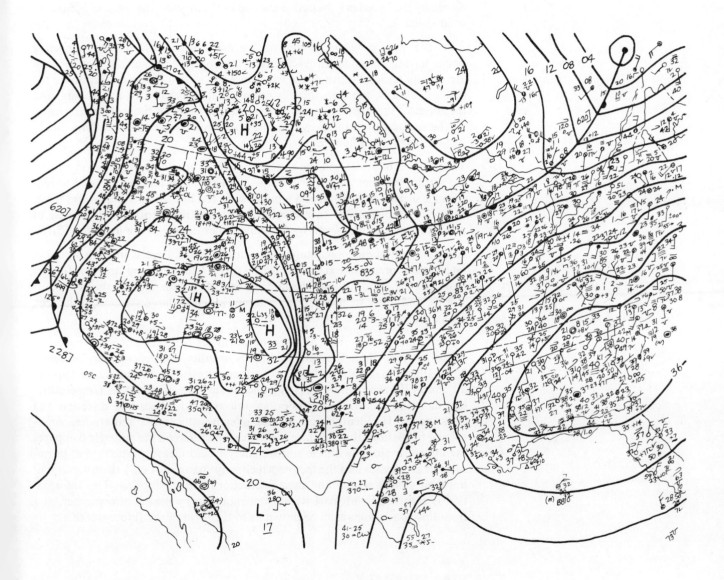

Figure 15.12 Section of a weather map showing isobars and placement of weather data around each station.

ment follows a pattern that is essentially uniform around the world. The surface weather map also provides information concerning winds at the surface. The wind direction at each station is shown by an arrow with the head pointed at the station. The arrow points in the direction toward which the wind is blowing, and the wind is identified by the direction from which it blows. Thus, a wind blowing from the northwest is a northwest wind. The speed of the wind is shown by the "feathers" placed at the end of the arrow—the number indicating wind speed.

After the basic data are plotted on the map, isobars are drawn connecting points having the same sea-level pressure. Isobars on the surface weather map are usually representative of the wind flow one to two thousand feet above the surface. That is, the flow there is approximately parallel to the isobars, and the speed of the wind increases as the distance between the isobars decreases. At the surface, the wind flow is usually at a small angle to the isobars due to surface frictional influences.

The position of a front on the earth's surface is represented by a heavy line on the surface weather map. For example, a heavy solid blue line is used to represent a surface cold front, and a heavy solid red line for a surface warm front. Where it is not feasible to use colored lines for the fronts, as in the case of printed maps and maps sent by facsimile, other notations are used (fig. 15.13).

Areas of continuous precipitation are shown by solid green shading; areas of intermittent precipitation are shown by green hatching; and areas of fog are indicated by solid yellow shading. Areas of thunderstorm, showers, dust storms, and other such phenomena are indicated by large symbols in the area where the phenomena are occurring.

The synoptic reports from all over the world are collected at certain meteorological centers where they are studied by weather forecasters. The forecaster has available to him the surface weather map, pressure-trend charts, and upper-air charts showing winds aloft, significant surfaces of pressure, temperature, and moisture at various altitudes, as well as wind speeds and direction and cloud cover. He identifies air masses, sketches the fronts and studies atmospheric formations that produce the weather pictured on the map. He compares the upper-air charts with the surface analysis and measures the rate of movement of air masses and the factors which may change not only their movement but their characteristics and their interactions with each other. This information is compared with past weather phenomena by means of computer, and from this conclusions may be drawn as to the form of the future weather.

15.4 SUMMARY

The characteristics of the atmosphere and those factors influencing it have remained essentially unchanged, at least within the past half billion years. These influences resulted in a fairly uniform climate during this period of the

Climate and Life on the Earth

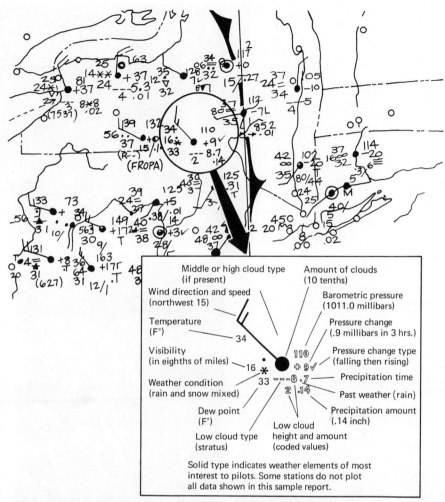

Figure 15.13 Section of surface map, with signs and symbols commonly used shown in inset.

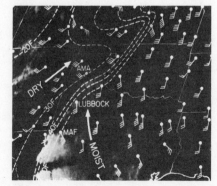

Figure 15.14 Monitoring of weather conditions over a large area permits detection of potentially severe storm conditions, such as tornadoes, in early stages of formation.

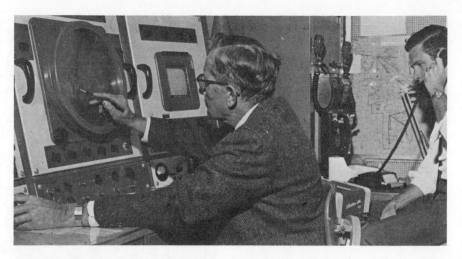

 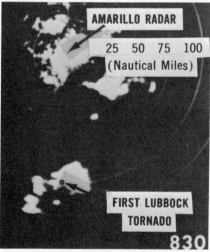

Figure 15.15 Local offices of the National Weather Service, once alerted to a severe storm, can monitor the atmosphere over their area for signs of dangerous weather activity.

earth's history. There have been brief occasions when the earth has experienced unusual climatic conditions of glacial activity and intense aridity between long periods of climatic quiescence. Man appeared on earth and developed to his present state during the latest period of unusual climate.

The earth's axial inclination to the ecliptic is the cause of the cycle of seasons. This inclination of the earth's axis has results in variations in rate of insolation at different times of year, as well as a change in day length—or the length of time a portion of the earth is subjected to solar radiation. The angle at which the sun's rays enter the atmosphere will influence the amount of radiation that is absorbed by the atmosphere. In effect, the atmosphere acts as an insulator of greater or lesser efficiency, depending upon the angle at which the sun's rays enter the atmosphere.

The general circulation of the atmosphere is in part responsible for the climate, as is the type of surface the air masses move over and the temperature of the air masses themselves. Moisture and temperature are the principal factors used in classifying climate and dividing climatic types into eight classes. Those based mainly on moisture are: very humid, humid, subhumid, semiarid, and arid. Taiga, tundra, and perpetual frost are classes in which temperature rather than moisture is the chief factor in determining climate.

The moisture and temperature characteristics have a direct bearing on the vegetation found in each climatic class, and it is possible to correlate the principal types of vegetation with climate. Climate will result in the development of climax vegetation, which is the dominant vegetative group in a climatic zone.

Climate and vegetation together are mainly responsible for the development of soils on the earth's surface. High precipitation will result in leached soils wherein material is removed from the upper layers of soil by the downward percolation of water. In dry climates, carbonate accumulation in the surface soils results from the rapid evaporation of moisture from the soil. Vegetation will not only influence the mineral content of the soil but also the type and amount of organic matter present. Forest litter results in different types of soils than those found under grassland conditions.

The landscape varies from one climatic zone to another, indicating the influence climate has on the geologic processes. Flowing streams, wave action, groundwater activity, frost, winds, and glaciers are geologic agents that are directly affected by climate. The effectiveness of these agents varies with differences in moisture and temperature.

Society, too, is affected by climate. The natural landscape, a product of natural forces including climate, will influence man's mode of living, his housing, the manner in which he provides his food, and his cultural development.

Weather forecasting has become an important aspect of civilization because many of society's activities are dependent upon weather. Successful forecasting, however, is dependent on the knowledge of the activities of the entire atmosphere—a capability not yet completely available. Temperature, pressure, precipitation, wind activity, cloud cover, movement of air masses, and location of fronts are the types of information accumulated for a synoptic weather report. These reports are used to make up a surface weather map which gives weather conditions as they exist at the time the data were collected. From this and other meteorological data, the meteorologists develop a forecast.

The state of the art is as yet imperfect, but with the use of new equipment such as computers, meteorological satellites, and weather buoys, comprehensive weather data may be available in the future for accurate short- and long-range forecasts.

QUESTIONS

1. Describe the cycle of seasonal changes and why they occur.
2. Describe briefly the eight recognized climatic types and give a location for each of the types.
3. What is meant by "climax vegetation" and what local factors may result in the absence of such a vegetative community?
4. In what way does climate influence the development of soils?
5. How do the geological erosional agents vary with differences in temperature and rainfall?
6. How is a weather forecast made up?

16 Air Pollution and Weather Modification

Societal causes of air pollution.
What are the various forms of air pollution?
Influence of air pollution on humans, plants, animals, and inanimate materials.
Movement of smog.
Methods of controlling air pollution.
History of attempts to modify weather.
Methods used to modify weather.
Problems in modifying weather.

Economic expansion and technological progress are hallmarks of a modern society in which the standard of living is constantly being improved. But those trends, particularly in the twentieth century, coupled with an increasingly urban population, have had many unplanned and unwanted side effects on health and welfare. Air pollution is one such side effect—a very serious one whose impact is being felt by millions of people all over the world. At the same time, efforts are being made to modify weather for our benefit. This is a somewhat more positive aspect of our activity in the atmosphere as compared to the negative effect of air pollution.

16.1 AIR POLLUTION AND SOCIETY

Is air pollution a new phenomenon—an outgrowth only of twentieth-century industrialized society? Evidently not, for prior to the arrival of the white man, Indians called the San Gabriel Valley near Los Angeles the "valley of the smokes." This was due to the accumulation of smoke caused by cooking fires and periodic forest fires beneath the inversion layer. As early as the fourteenth century, Londoners complained of the noxious smoke and odors in their city; later, the coming industrial revolution added to the problem.

Initiated by the industrial revolution, magnified by the proliferation of the internal combustion engine, and multiplied by the technological explosion of the last several generations, the increase in the pollution of the air we breathe has been alarming. Pollution has continued to increase at an even faster rate during the past several decades until it has become one of the most serious environmental problems of urbanized society. The abundant good air accepted as normal in the past is being polluted, just as are the waters of the world, both through a lack of early recognition of the problem and development of effective controls.

Moreover, almost every technological advance which contributes to the world's progress also adds to the burden of contaminants which the atmosphere and waters must carry. This applies to the growth in population, in urbanization, in industrialization, and in living standards, as well as to increased per-capita use of power and transportation. Even advances in flight technology, as exemplified by jet aircraft, rockets, and missiles, are new sources of pollution. In fact, some concern is being expressed about the discharge of beryllium in rocket launchings.

Meteorological factors and pollution-emission rates determine air-quality levels or the degree of contamination. In the short term, variations caused by atmospheric motion determine local contamination concentration. But in the long view, the importance of weather variability diminishes and it is the characteristics of the emissions themselves that determine the level of contamination. There are also important interactions between pollution emissions and meteorological conditions. For example, the effects of sunlight, humidity, and rainfall can change the character, the amount, and distribution of airborne pollutants. Therefore, before one can take effective steps to restore and preserve the air quality, it is necessary to understand the interrelationships of weather characteristics and pollutant emissions on a regional and global scale.

The sources of air pollution are numerous and varied. They include such activities as the burning of fuel to produce heat and power, the use of motor vehicles, the burning of refuse, and the manufacture and use of such staples of modern life as steel and other metals, paper, and chemicals. No city is without some source of air pollution; in thousands of cities there are enough sources emitting enough gases and solid particles into the atmosphere to cause a community air-pollution problem.

The solution to the problem is by no means simple. Modern civilization would collapse if all activities contributing to air pollution were suddenly stopped. However, it is not necessary to take such drastic action. There are acceptable

16.2 TYPES OF AIR POLLUTION

Pollution of the atmosphere was a natural phenomenon long before man appeared on earth. Forest fires, volcanic action, and windstorms contributed to the gaseous and particulate matter normally not a part of the atmosphere. In due time, man's appearance added to the problem to a minor degree because of smoke from his fires. Gradually, as civilization developed, the variety of gases and solids being exhausted into the atmosphere increased until now the

Figure 16.1 (Top) A view of Southern California's San Gabriel Mountains on a clear day. (Bottom) The same view of the San Gabriels obscured by smog.

amounts are at a noxious and occasionally at a toxic level. Basically, the two most offensive and dangerous air contaminants are sulfur dioxide and photochemical smog. Both are the result of burning fossil fuels and the interaction of the atmosphere with the by-products of the combustion process.

Sulfur dioxide is a by-product of coal burning. With the increased scarcity of wood as a fuel in medieval Europe, coal became the main source of heat and energy after its introduction into Europe by Marco Polo. With the advent of the industrial revolution, the burning of coal increased and with it the attendant increase in air pollution. Even in the seventeenth century it was suggested that the coal industry be banned in England. But because coal had become the principal source of heat for the Englishman's hearth and the prime factor in the economy, the suggestion was largely ignored. The problem persisted, however, and in 1905, the combination of smoke and fog prompted a Londoner, Dr. Harold Des Voeux, to coin the term *smog*. It was the unusual concentration of such smog that resulted in illness and death in the Meuse Valley, Belgium, in 1930; in Donora, Pennsylvania, in 1948; and in London in 1952. In each case the outpouring of smoke from coal fires, combined with weather conditions that prevented the dispersal of the smoke, created a killer smog.

Sulfur dioxide, although noxious, is not particularly toxic in the amounts usually emitted by coal-burning furnaces. It is the combination of sulfur

Figure 16.2 High density traffic is the major source of smog in many large cities. Courtesy of Los Angeles County Air Pollution Control District.

dioxide with other air contaminants that produces lethal effects. Sulfur dioxide in combination with moisture in the atmosphere will form sulfuric acid, which may exist as a fine vapor or be absorbed by solids in the smoke such as the *fly ash*. The external portions of the body are not particularly vulnerable to attack by these aerosols. But the delicate membranes of the eyes, nose, and respiratory tract, including the lungs, are susceptible, and here is where the damage is accomplished.

Figure 16.3 Smoke from milling operations contributes to air pollution.

Photochemical smog is not truly a smog in that it is not a combination of smoke and fog. Rather, it is a by-product of automobile exhaust and the evaporation of gasoline from the carburetor and gas tank. The action of sunlight converts hydrocarbons from exhaust emissions and fumes into a yellowish brown haze. Gasoline evaporation provides about 20 percent of the hydrocarbons released into the atmosphere, crankcase blow-by about 30 percent and tailpipe exhaust about 50 percent. High-compression engines tend to vent more of the unburned fuel than low-compression engines. High-octane gasolines contain volatile components that evaporate more readily than ordinary gasolines. In all, motor vehicles contribute about 70 percent of the hydrocarbon pollution in a city like Los Angeles. Although first identified in Los Angeles, photochemical smog may be found in every large city which has great numbers of gasoline-driven vehicles and a fairly good percentage of sunny days.

Gasoline is composed primarily of carbon and hydrogen, which, when burned in the presence of oxygen, ideally produce water and carbon dioxide. Since combustion is almost never complete, carbon monoxide is also produced. Oxygen comes from the atmosphere, which also contains approximately 78 percent nitrogen. At the high temperatures developed during combustion in an internal combustion engine, nitrogen dioxide is formed and exhausted with the other emissions. A number of additives are included in the fuel to improve its performance, and they contribute to a small degree of air pollution. Such additives include rust inhibitors, antioxidants, metal deactivators and detergents. Tetraethyl lead is added to reduce engine knock, but it accumulates on engine parts as lead oxide. To prevent this from occurring, compounds such as ethylene dichloride and ethylene dibromide are added to the gasoline. All of these compounds, either as added to the gasoline or altered in the combustion process, ultimately end up in the atmosphere as pollutants.

Once in the atmosphere, some of these compounds are modified by sunlight, or more specifically, by ultraviolet radiation. Such shortwave radiation will split nitrogen dioxide into nitric oxide and atomic oxygen. The atomic oxygen in turn will react with molecular oxygen to produce ozone. Certain other compounds formed as a result of reactions of nitrogen dioxide and nitric oxide include formaldehyde, peroxyacetylnitrate (PAN) and arolein.

16.3 EFFECTS OF AIR POLLUTION

Knowledge of the harmful effects of air pollution on human health may be gained in three principal ways. First, by statistical studies of health records to search for correlations that might exist between illnesses or death and exposure

Figure 16.4 Los Angeles City Hall on a clear day (see figures 16.5 and 16.6). Courtesy of Los Angeles County Air Pollution Control District.

Figure 16.5 Smog engulfs the Los Angeles Civic Center when the base of the temperature inversion is approximately 465 meters above ground level. The inversion layer—a layer of warm air above a strata of cool air near the ground—prohibits the natural dispersion of air contaminants into the upper atmosphere. Courtesy of Los Angeles County Air Pollution Control District.

to air pollution; second, on relationships that may exist between illness or death and factors that influence a person's exposure to air pollution, such as his place of residence and his need to travel in areas subject to pollution; third, laboratory studies on the ways in which exposure to individual pollutants or combinations of pollutants affect humans and animals.

Through these types of investigations, a substantial amount of information on the health hazards of air pollution has been accumulated. This evidence indicates that exposure to ordinary levels of air pollution impairs the health of many people and is associated with the occurrence and worsening of chronic respiratory diseases. Air pollution is often a contributory factor in the premature death of aged or ailing persons. Among the specific diseases associated with air pollution are asthma, chronic bronchitis, emphysema, and lung cancer. There is evidence that exposure to air pollution increases a person's susceptibility to upper respiratory infections, including the common cold. In England, where much of the population lives in air-polluted urban centers, 21 percent of the men between 40 and 59 years old suffer from chronic bronchitis, a condition causing about 10 percent of the deaths in England.

Further appraisals are being made between relationships of air pollution to specific diseases. For example, efforts are being made to assess the impact of

Figure 16.6 Smog is trapped by a temperature inversion at approximately 100 meters above the ground. The upper portion of Los Angeles City Hall is visible in the clear air above the base of the temperature inversion. The inversion is present over the Los Angeles Basin approximately 320 days of the year. Courtesy of Los Angeles County Air Pollution Control District.

relatively brief exposures to high levels of air pollution resulting from meteorological conditions that prevent dispersion of the contaminants. Such a condition, which may occur in a community for three or four days at a time, was responsible for the Meuse Valley, Donora, and London disasters mentioned above. The difference in air-pollution levels may not be generally noticeable. But in terms of the effect on health, the consequences may range from an increase in the incidence of colds to an increase in deaths of older people or of those susceptible to respiratory disorders.

There are also a number of insidious ways in which air pollution may impair human health. For example, concentrations of up to 10 ppm (parts per million) carbon monoxide may be tolerated without visible effect. A smoggy urban area may have concentrations up to 30 ppm in the atmosphere, which is about the maximum level at which no physical impairment will occur. Above this level, a person may experience lack of energy, reduced endurance, and a decline in reaction rate to external stimuli. A heavy smoker in a smoggy situation may approach dangerous levels of up to 100 ppm carbon monoxide. In traffic, when stopped behind other cars at a red light, carbon monoxide levels may reach 350 to 400 ppm. Such levels may certainly affect a driver's ability to cope with a difficult traffic emergency. This and many other factors, such as the subtle but progressive deterioration of the respiratory function by inhaled gases and particles and their influence on other physiological processes, need to be brought into focus before the full impact of air pollution on the human population can be understood.

Damage to plants from smog has for decades been quite evident and a cause for concern. Sulfur dioxide appears to be one of the most damaging materials in the atmosphere and the damage therefrom will range from minor to severe, depending on the concentration. Plants vary in their ability to resist damage. Thus, in areas like the Los Angeles basin, some of the leafy vegetables can no longer be grown. Occasionally, the concentration of pollutants such as sulfur dioxide becomes so great that all plants are destroyed. High sulfur dioxide emissions resulted in the destruction of a forested area near Ducktown, Tennessee, early in the twentieth century. A copper smelter in that area discharged up to forty tons of sulfur dioxide per day into the atmosphere. Almost all the vegetation on over a twenty-thousand acre area downwind from the smelter was destroyed. Despite the fact that controls were instituted that converted the waste sulfur dioxide into profitable sulfuric acid, the region is still virtually barren. Damage to plants from sulfur dioxide comes about as a result of the contaminant entering the plant leaf through the stomates, which are small openings on the underside of the leaves. The sulfur dioxide will damage internal plant cells, resulting in dead tissue, which detracts from the appearance and quality of the vegetation.

Photochemical smog also damages plant tissue. Ozone will attack the upper surface of the leaf and destroy the palisade layer, causing brown spots to appear. This has resulted in damage to Ponderosa pine near Los Angeles and white pine near Oak Ridge, Tenneessee, and along highways in Connecticut. Orange trees are susceptible to PAN, which along with other photochemical smog components, has reduced production of oranges in the Los Angeles area.

Even those things considered inert substances are not immune to the deleterious effect of smog. Sandstone (containing carbonates), limestone, and marble, used in the construction of buildings, are attacked by carbon dioxide and sulfur-bearing pollutants. Steel is affected by sulfuric acid mists formed in the atmosphere from sulfur dioxide. Fabrics of various types, rubber, paper, and leather will generally deteriorate as a result of contact with smog. Nor can one additional, very important, and costly effect of smog be overlooked. This is the dirtying of one's possessions—home, car, clothes, etc.

The effects of air pollution on the economy as a whole are enormous. As stated above, air pollution soils and damages buildings and other structures as well as clothing and home furnishings and thus adds to the expense for cleaning and replacement. It contributes to urban decay and the depression of property values, and causes injury to crops and livestock in rural areas. In addition, it reduces visibility, thus increasing the risks of accidents on the highway and in the air.

Economic losses in the United States resulting from air pollution are estimated to be $11,000,000,000 annually, which is approximately $55 per person. The real cost is even higher, as the estimate does not include medical care for persons who have respiratory diseases associated with air pollution, nor does it include resultant factors such as lost earnings and reduced productivity. A precise method of evaluating total losses from air pollution is not yet available, but it is necessary before an overall impact of air pollution on the economy can be determined.

16.4 DISPERSION OF SMOG

Air pollution knows no boundary lines. Air flows freely from one community to another, and the pollutants along with it. Movement of air over low-relief terrain is not difficult to define, but most pollution sources are located in or near variable terrain. Rough terrain, combined with the fact that the atmosphere is rarely steady, makes it difficult to describe the dispersion of smog over a wide area.

Some progress in describing and predicting the movement and occurrence of smog in local areas has been made. Computerized maps of the buildup and movement of smog in the general Los Angeles area have been developed by Dr. Joseph Behar. These maps show smog concentrations for different times of the day in the various parts of the Los Angeles basin, and the data can be used to predict smog levels for several days in advance. Dispersion on a large scale applicable to the spread of pollutants from one urban area to another hundreds or thousands of miles away cannot as yet be accurately described.

Much of the knowledge that does exist on large-scale movement of pollutants from the Northern to the Southern Hemisphere and between troposphere and stratosphere has come from observation of the spread of nuclear debris. At high altitudes, the information is scanty, and it is not clear what role general atmospheric circulation plays in the transport of air pollutants.

The stratosphere can play a major role as a storage reservoir for pollutants, which in the troposphere are removed by precipitation. The stratosphere, although constituting only about 20 percent of the atmosphere, can hold over 90 percent of the burden of pollutants initially injected into it. This storage capacity, and its implications for the cleanliness of ground-level air and the mechanisms by which such pollutants are reintroduced into the troposphere, are only beginning to be explored.

Figure 16.7 The photo of exhaust emissions from a truck is used by a pollution control officer as evidence to issue a citation. Courtesy of Los Angeles County Air Pollution Control District.

16.5 CONTROL OF AIR POLLUTION

Control of air pollution is focused in a large measure on two of the most important aspects of the air-pollution problem—motor-vehicle pollution and sulfur dioxide pollution from fuel combustion.

Of the two, motor vehicle pollution is the greatest source of air pollution in some communities. Some control techniques are available although by no means perfected. Starting with the 1968 car models, crankcase blow-by has been reduced by recycling the emissions back to the carburetor. This was followed by a requirement for a vapor-tight gas-tank cap. One of the more difficult problems is control of tailpipe emissions. These can be controlled to some extent by one of several methods, each of which has disadvantages. The methods include use of afterburners, decreasing compression ratios, and using leaner and more carefully measured fuel-air ratios. Another suggested method of control is to modify the gasoline by reducing additives that end up as pollution components, and by making the gas less volatile. This would require some engine modifications to permit the use of such fuels and devices, as well as additional maintenance, something the average motorist generally neglects on his car.

A reduction in permissible horsepower may be a partial solution, at least until acceptable alternatives are available. One of these alternatives may be replacing the internal combustion engine with steam- or electric-powered cars, but this also presents problems requiring solutions before the alternative is acceptable.

Sulfur dioxide pollution, arising chiefly from the combustion of oil and coal containing sulfur, is a widespread and serious problem. In some large cities, sulfur dioxide pollution consistently reaches high levels and represents an urgent problem requiring control, for two principal reasons: first, because sulfur dioxide is one of the pollutants most injurious to health and the most damaging to crops and property; second, because increasing demands for electric power, most of which is produced by burning sulfur-containing fuels, threatens to result in a sixfold increase in sulfur dioxide pollution by the year 2000.

Sulfur dioxide pollution can be reduced by using low-sulfur fuels or natural gas, which is essentially sulfur free. If these types of fuel are not available, then one of two approaches may be used to solve the problem. One is to remove the sulfur from fuels before they are burned. Methods of accomplishing this for oils are available, but they are not generally used for the grade of oil burned at electric power plants. Coal does not lend itself to the removal of sulfur very readily, and only a small fraction can be removed with present methods.

The second approach to the sulfur dioxide problem is to remove the sulfur dioxide from the combustion gases before they escape into the atmosphere. A number of devices and techniques are available for removing a variety of pollutants, including sulfur dioxide, from flue gases. These devices are expensive and are an indication that air-pollution control is costly. However the cost of control may be far less than the price paid for damage from polluted air. It is felt that overall savings realized by eliminating air pollution would be more than sufficient to pay for cleaning the air.

Figure 16.8 Uniformed inspector issuing a citation for an excessively smoking incinerator. Courtesy of Los Angeles County Air Pollution Control District.

16.6 WEATHER MODIFICATION

"We do not know whether it will be practically feasible for man to control the weather or favorably alter the climate in which he lives. Certainly some of man's acts already have an effect, usually unfortunate, upon rainfall and runoff and hence upon creation of new arid regions." Thus spoke Dr. Vannevar Bush in 1957. He went on to say that for man to successfully alter climate and weather for his own purposes, it will require a better understanding of meteorological processes than is now available.

Weather control or modification refers to the artifically produced changes in the composition, motion, or dynamics of the atmosphere. It covers a rather wide spectrum of activities, some of which, such as control of precipitation, are already recognized. Rainmaking is one of the oldest attempts at weather modification, and in the modern sense includes altering the timing, distribution, and amounts of precipitation. Currently, studies are underway to determine methods of hurricane, tornado, and hail suppression. Success in these endeavors depends, as stated by Dr. Bush, upon a better understanding of the total action of the atmosphere.

16.7 HISTORY OF WEATHER MODIFICATION

Man's attempt to control or in some way modify weather goes back in time long before recorded history. From the very beginning of civilization, man has had to depend on rainfall to quench his thirst and water his crops. The distribution of rainfall over the earth's surface has been a vital factor in designating the geographic areas where man could live, what he could grow and eat, and what kind of life he could lead.

In the beginning, man attempted to influence rainfall by hope and faith, magic and ritual. Eventually, witch doctors, medicine men, and charlatans all tried their hand at weather control or rainmaking. Rain dances performed by the Hopi Indians in Arizona still attract interest, but more as an example of a cultural exercise than for the purpose of rainmaking. African rain queens continue to perform mysterious rites to increase rainfall.

In Europe, noise was used for many years as a means of stimulating rainfall, probably stemming from Plutarch's (first century A.D.) observation on the occurrence of rain after great battles. Some centuries later, when the development of gunpowder produced great noisemakers such as muskets and cannons, even more stimulus was lent to the argument that noise and concussion could induce condensation in the atmosphere.

Gunfire and church bells were used for many years in Europe for the purpose of stimulating rainfall by means of noise, but it was not until the late nineteenth century that such tests were made in the United States under the direction of Robert Dyrenforth, a special agent of the Department of Agriculture. He advocated the use of noise and suggested a series of public demonstrations, held in Texas in the summer of 1891. As might be expected, the results proved inconclusive, but they did spark many sporadic attempts, mostly by charlatans, to profit from people's dependency on adequate rainfall.

Another popular early belief focused around the idea that conflagrations such as forest fires caused convection currents which in turn produced rain. Some evidence, based on a few observations of showers developing from cumulus clouds formed over large forest fires, favored this idea. The most responsible advocate of this idea was the American meteorologist James P. Espy, who, in his *Philosophy of Storms* published in 1841, discussed the importance of convection currents in the formation of clouds and precipitation. Three years earlier in a previous publication, Espy recommended that farmers save brush and waste timber to burn in dry weather in order to stimulate rainfall. He suggested to Congress that 40-acre blocks of timber in the West, spaced every 20 miles along a 700-mile, north-south line be ignited simultaneously to produce a rainstorm that would travel to the Atlantic coast. Congress never appropriated the funds and the test was never made.

With the invention of the airplane, dozens of new schemes for increasing rainfall were conceived. These consisted of dropping into a cloud any number of different substances, ranging from secret chemicals to electrified sand and soap flakes. There were also suggestions made for the use of extremely cold substances such as liquid carbon dioxide and even liquid air. The intent was to produce a sudden cooling that would result in condensation and precipitation. These early attempts to use such materials were done on a haphazard basis with virtually no knowledge of the basic principles of precipitation. Because the main objective in these early trials was to produce clouds by cooling rather than to operate on existing supercooled clouds, the efforts failed.

By the end of the nineteenth century some knowledge of the nature of the atmosphere had been gained but despite the schemes mentioned above it was not until after World War I that real serious attempts were made to modify weather by application of physical principles.

One of the first of the modern scientific attempts at weather modification was made in 1930 by August W. Veraart in Holland. He directed four experiments from planes using dry ice alone and in mixtures with supercooled water ice. Veraart made such extravagent claims of success that he incurred the scorn of Dutch scientists and his results received no consideration elsewhere.

The real groundwork for modern experiments in weather modification was made in 1935 by the Swedish meteorologist Tor Bergeron, who advanced the theory that rain in appreciable amounts could be released principally by the presence of ice crystals formed in or transported through water clouds. In 1939, the German physicist Walter Findeisen reemphasized that the coexistence of ice crystals and supercooled water droplets in the proper proportion is a necessary condition for precipitation from a cloud. This theory, known as the Bergeron-Findeisen theory, forms the basis of present-day techniques of seeding supercooled clouds.

Weather modification received added stimulus from the work performed by Dr. Irving Langmuir and Dr. Vincent J. Schaefer in 1944 when these scientists applied their knowledge to the problem of precipitation. Their research took them to Mount Washington in New Hampshire, where they became interested in aircraft icing and its relation to the growth of cloud particles. The explanation of why snow fell from some supercooled stratus clouds and not from others prompted Langmuir and Schaefer to search for an answer. Using an ordinary home freezer, Schaefer accidentally discovered that dry ice caused the formation of ice crystals in miniature supercooled clouds and thereby triggered the precipitation process, as the Bergeron-Findeisen theory suggested.

Schaefer's fortunate accident was a perfect example of serendipity—the art of profiting from fortunate accidents or unexpected occurrences. This has often resulted in new ideas and concepts in science. The discovery was immediately seized upon to design some experiments with natural clouds in the atmosphere and a series of cloud-seeding flights. As the dry ice pellets dropped, Schaefer noted that the supercooled water droplets in the clouds were changed to ice crystals and some of the ice crystals grew large enough to fall out of the clouds. Soon thereafter, Dr. Bernard Vonnegut discovered that microscopic silver iodide crystals proved more efficient as freezing nuclei than those ordinarily found in the atmosphere. Their greater effectiveness was due to the ability of silver iodide to form ice crystals at warmer temperatures than natural nuclei.

During the years that followed, a number of projects were sponsored, both by government and the private sector, to further the knowledge of the rainmaking process. Commercial operators generally use restraint in making positive claims, since there is still controversy as to the degree of their effectiveness. In 1962, the Rand Corporation made a study of weather modification efforts and concluded that an understanding of atmospheric processes was vital to the success of weather modification. The advent of the meteorological satellite in the 1960s has greatly increased the ability to gain knowledge of atmospheric dynamics. As a result, the study of the atmosphere in relation to weather modification has been intensified not only in the United States, but also in such technologically advanced nations as Australia, Japan, the Soviet Union, and others. The practical application of weather modification techniques must be undertaken with great care because of the uncertainty of the results (see section 16.9). Weather modification is not strictly a scientific problem but has far-reaching social and legal implications that are not yet completely understood.

16.8 METHODS OF WEATHER MODIFICATION

Up to now, the seeding of clouds has proved to be the most promising means of modifying weather, but discussed below are a number of other possililities whereby control may be exercised or modification accomplished.

1. *Clearing of supercooled stratus and fog:* The effects of seeding by dry ice and by silver iodide were first demonstrated upon supercooled stratiform clouds. More recently, attempts have been made to develop operational methods for clearing supercooled fog at airports. Tests were made at several airports in the United States by dispersing varying amounts of crushed dry ice just over the top of the supercooled fog. These initial methods were somewhat crude, since little data was available on the optimum amounts of dry ice necessary, the most practical size particles to use, or on the most appropriate method of dispersal. In the late 1960s this type of information gradually became available, and some success in fog dispersal by this method was attained. Tests of this type have also been conducted in the Soviet Union with some success. At Orly Airport in Paris, compressed propane gas was released, which cooled upon expansion and formed ice crystals, thus dissipating the supercooled fog. Cooling warm fog is much more difficult and no really satisfactory methods have been proposed.

2. *Change in amounts of precipitation by seeding:* Despite the many tests that have been made, controversy still exists as to whether precipitation can be enhanced by seeding. A statistical study on the results of commercial cloud-seeding operations using silver iodide yielded ambiguous results. A positive statement that precipitation was modestly increased or redistributed could not be made because the possibility of some unknown source of bias or a systematic error in the commercial seeding operation. It should be emphasized that the problem is an extremely complex one. There is great variability in cloud types and in the ways in which precipitation can be caused. The theoretical knowledge of how seeding nuclei are introduced into clouds from ground-based generators and how precipitation may be effected thereby is still rudimentary. Present indications, if taken at face value, are that local precipitation can be increased in many situations on the order of 10 percent by seeding. These positive results are obtained in the cases where rain should have fallen anyway, without seeding. There is no evidence that seeding can induce rain to fall when normally there would be none. Thus, seeding appears to be of limited value in relieving drought conditions. A limited number of tests have also been conducted on changing the form of precipitation—that is, from freezing rain to snow. Such knowledge would be useful in reducing damage to transmission lines by ice or reducing the hazard of icy roads. Little progress has been made in this line of research.

3. *Increased precipitation by forced convection:* Suggestions have been made that precipitation in some local areas could be increased by changes on the earth's surface which would promote greater absorption of heat and also greater transfer of heat and water vapor to the atmosphere. This would stimulate convection currents, hopefully in sufficient amounts to increase cloudiness and precipitation downwind. While some experiments have been suggested, no field tests have been made to evaluate these proposals.

Another method of creating convection currents, one which has given some indication of success in limited trials, utilizes the cloud-seeding process. It has been suggested that latent heat released by the increased condensation of moisture into water droplets caused by seeding in turn causes uplift and cloud formation. It may be that some of the observed increase in precipitation by cloud seeding resulted from enhanced convection activity rather than directly by nucleation of drops.

4. *Lightning suppression:* Lightning (see chapter 14.4) is generally most damaging in forested areas, since it is responsible for the major portion of the forest fires. For this reason, studies have been conducted by the United States Forest Service in an attempt to reduce the incidence of lightning during thunderstorms. Some tests in recent years have indicated that seeding can alter cloud-to-ground lightning from thunder clouds. However, the results were not too encouraging and the frequency of forest fires resulting from lightning were not significantly diminished. Another suggestion, not yet tested on a large scale, was to introduce metal strips into thunder clouds to decrease the electric field potential. The results of the limited application of this experiment have not been encouraging.

5. *Hail suppression:* Hail is of concern in agricultural areas because of the costly damage to crops it can cause. Studies of suppression of hail by seeding have been conducted in a number of countries, but most results have been rather inconclusive. Soviet scientists reported some success in hail suppression by seeding hail-producing clouds with silver iodide by means of artillery shells. The work could not be evaluated on a statistical basis and has not been duplicated elsewhere. In the Argentine, a similar experiment conducted on a randomized basis (to permit statistical evaluation) indicated that hail suppression is partially successful when the test is carried on in the vicinity of a cold front. However, the same tests in warm-front areas resulted in an increase in hail activity. Taken together, the data indicated that no real influence on hail suppression resulted from seeding hail-producing clouds. Work on this problem in the United States has been conducted in a rather piecemeal fashion and thus far has produced no results.

6. *Moderating severe storms, tornadoes, and hurricanes:* Some attempts have been made to reduce the intensity of major storms. Hurricanes have been seeded with the intent of producing a warming in the outer zone of the eye wall by releasing latent heat of fusion. This would alter the pressure and wind distribution by a movement of the rotating winds away from the eye of the storm and cause a consequent reduction in wind velocity. Results of these experiments have thus far been inconclusive, but it has been found that because of the tremendous energies involved, much broader basic research into storm mechanisms is required.

7. *Modifying the microclimate of plants:* The problems here are those largely concerned with methods of preventing frost, suppressing evaporation, and reducing the effects of wind. Some practical methods have long been in use, such smudging (now undesirable in many areas because of smog problem) and the

use of wind machines for stimulating the circulation of air. Thus far, there has been only a limited application of modern knowledge of micrometeorology to take best advantage of these weather-modifying techniques. More research on the improvement in the use of these techniques would be desirable.

8. *Modifying the weather and climate of large areas:* The interaction between climatic changes and altered land-use patterns is well known. Historically, there are many examples: The development of the dust-bowl region in the United States resulted from the loss of grassland cover. There has been a similar occurrence in the grassland areas of the Soviet Union. Reduction of forested areas in the eastern United States brought about more temperate winters. Intensive agriculture practiced by the Romans in Libya transformed that land into desert. In addition, the destruction of the Lebanon cedar forests resulted in the development of desert, with a consequent change in the economic pattern of that area.

The changes in vegetative cover and subsequent change in climate have come about as a result of man's folly and a misunderstanding of the long-term climatic effects resulting from a change in the vegetative cover of the land. All of these transformations have been destructive, and it is not known if the trend can be reversed. However, a systematic exploration of the possibilities for modification of climate is now possible through the large-scale examination of the atmosphere. Some restorations of former climates are perhaps possible, but the costs may be prohibitive. Benefits could be realized if a greater understanding of atmospheric dynamics resulted in preventing further deterioration of climate and vegetative cover on the earth.

9. *Inadvertent atmospheric modification:* In the previous paragraph it was pointed out that alteration of the climate was possible through manmade changes in the vegetative cover of the land. Now the possibility also exists that climatic changes can be caused by manmade air pollution. To accurately assess this theory, there is an overriding need for greatly improved and expanded methods of detecting the change in composition and determining energy budget of the atmosphere.

It is generally agreed that the total amount of carbon dioxide in the atmosphere has increased by 10 to 15 percent in this century, and that the increase is due primarily to the burning of fossil fuels. This suggests that there is a need for continual monitoring of atmospheric and oceanic carbon dioxide content, and for testing the effects of carbon dioxide on atmospheric circulation caused by changes in the heat budget of the atmosphere.

The problem of air pollution may already be growing beyond its recognized urban sources to become a widespread nuisance that possibly has effects on weather and climate over much larger areas. Little attention has been given to the effects of pollution on cloudiness, precipitation, or on the radiation balance. Urbanization is, in fact, a kind of continuing experiment in climatic modification. The meteorological effects of altering the rural landscape by reforestation, deforestation and irrigation, for example, appear to be small and localized when compared to the effects of large-scale city building, at least at their present levels.

16.9 PROBLEMS OF WEATHER MODIFICATION

In order to accomplish weather control and modification by the methods outlined in the previous section, four essential needs must be met.

1. An understanding and assessment of natural environmental weather changes is required. This means, as previously stated, that knowledge of the normal circulation and air-mass relationships of the atmosphere is necessary.

2. An understanding and assessment of the inadvertent changes in weather and climate that man's technological evolution has produced is required. This would include knowledge of the effect on climate of altering the vegetative cover as well as the influence of air pollutants on atmospheric activity.

3. Man must improve his ability to predict the behavior of the atmosphere so he can live with a minimum of danger from violent storms.

4. A variety of techniques must be devised for deliberately changing the atmospheric processes. These techniques could be used to alter the climate and weather in a manner that is favorable to mankind.

Thus, it must be recognized that a close relationship exists between weather modification and the more general problems in atmospheric sciences. An improved understanding of total atmospheric processes is necessary to accelerate progress in understanding the problems of climate and weather modification. A worldwide system of observational facilities will be useful in obtaining the necessary data for such understanding and will link weather forecasting and weather modification by a single system.

There are aspects of weather modification which transcend the simple development of techniques. Weather control and modification will have legal, economic, sociological, political, as well as scientific impact, all of which must be carefully considered before manipulating forces which can have long-range and not always obvious effects.

A long-range program of weather control and climate modification can have a direct bearing upon relations between nations. It could aid the economic and social advancement of the developing countries, many of which face problems associated with hostile climate and serious imbalance in soil and water resources. It can serve to develop common interests and thereby become an instrument for international cooperation.

Weather modification can also create international chaos. For example, melting the ice on the Arctic ocean (technologically possible) would increase the rate of evaporation of surface water and provide more precipitation in the northern middle latitudes as well as increase Arctic air temperatures. The question arises, would the Arctic Ocean refreeze the next winter? Or would a new ice age develop (see section 11.3E) covering northern North America, Europe, and Asia with an ice cap? This would wipe out all vestiges of civilization in those areas as well as lower sea level a hundred feet or more, leaving the world's seaports high and dry. With present knowledge, there is no way of determining which direction such an action would take. Any large-scale experimentation of this type by scientists before those problems are solved would be highly irresponsible.

Great uncertainty is associated with biological consequences of weather and climate modification. On the one hand, increase in rainfall over cultivated areas could partially alleviate the increasing problem of food production. However, there is an accompanying possibility that instabilities might result in the natural balances of biological communities. Such imbalances might create increases in the diseases and pests of domestic animals. In small areas of natural communities, it is possible that some wild species may be severely stressed. The timing of the atmospheric intervention relative to the reproductive cycle of the various species in the community may be more important than the magnitude of the intervention. Both field and simulated studies of these biological relationships to weather modification are needed to help avoid undesirable, unanticipated, and irreversible ecological changes.

Weather and climate modifications pose some legal questions as to the existence of "property" interests and the responsibility of weather modifiers for damage to others. Weather recognizes no property lines or international boundaries, and modification for the benefit of one party may be harmful to his neighbor. Therefore, principles for compensating losses due to weather modification will have to be determined. Principles need to be developed not only for settling claims between individuals, but also for handling cases of individuals against governments and nation against nation.

16.10 SUMMARY

Air pollution has been a by-product of man's development for centuries, but it has become particularly critical in the past few decades and has become a worldwide problem. The advancing technology, the urbanization of society, and the increasing use of fossil fuels have contributed to the production of air contaminants. Meteorological conditions have aggravated the problems in some areas until now some cities are never entirely free of air pollution.

Natural sources of air pollution include forest fires, volcanic activity, and windstorms, but these occur infrequently. Perhaps the two worst manmade air pollutants are sulfur dioxide and photochemical smog—both by-products of the combustion process.

Sulfur dioxide comes from burning fuels that contain sulfur. The sulfur dioxide combines with moisture in the atmosphere to form sulfuric acid, which is damaging to humans, plants, animals, and the materials used by man.

Photochemical smog comes primarily from the internal combustion engine and is a common air contaminant in any large city. Photochemical smog's major ingredients are sulfur dioxide, carbon monoxide, nitrogen oxides, volatile hydrocarbons, and materials such as lead, boron, and bromine from a variety of additives put in the fuel for better performance.

Air pollution has an effect on human health at a cost not yet possible to determine. Premature death of the aged and infirm have been attributed to smog. Smog also affects farm crops and animals. In some areas it is no longer possible to grow crops because of the damage caused by air pollutants. Even inert manufactured products are attacked by smog, causing the products to deteriorate.

Although the general circulation pattern of the atmosphere is known, what is not known is how this influences the movement of the airborne contaminants on a large scale. Movement of the pollutants between troposphere and stratosphere is just beginning to be explored. The implications of such movement is not yet completely understood.

The control of air pollution is currently a major worldwide concern. Since the internal combustion engine is a primary contributor to pollution, there is a great emphasis being placed on control of emissions from this source. Modification of the engine and the addition of devices to reduce harmful emissions are being explored. These changes in themselves create problems that must be solved in order to accomplish an effective solution to the smog problem.

Weather modification to suit society's needs has been a vision of man for centuries. A variety of techniques to accomplish this have been suggested and tested, but none have proved successful. The attempts at weather modification include such activities as controlling rainfall, suppression of large storms, hail suppression, and altering the climate of large areas. Some inadvertent modification may be occurring as a result of increased air pollution in the atmosphere.

To accomplish modification of weather and climate, man must first have an understanding of atmospheric dynamics. He must work toward achieving modification of atmospheric conditions without creating undesirable and irreversible changes.

International cooperation is a necessity in these endeavors, since weather and climate transcend international boundaries. The legal aspects of weather modification must be solved, for the change of weather may be beneficial for some and detrimental to others.

QUESTIONS

Air Pollution and Weather Modification

1. Briefly describe the types of air pollution you may be subject to in your community.
2. What is the effect of air pollution on health; on plant life; on inanimate things?
3. How may air pollution be controlled? Add some suggestions of your own.
4. Define the following: photochemical smog, carbon monoxide, sulfur dioxide, fly ash.
5. List the variety of ways whereby man has attempted to control or modify weather in the past.
6. Summarize briefly four ways in which attempts are being made to modify weather.

V THE OCEANS

17 Introduction to Oceanography

Importance of the oceans to man.
Exploration of the oceans.
The boundaries of the oceans.

No person who has stood on the shore and viewed the sea could help but wonder at its magnificence and scope. The sea covers 70.8 percent of the earth, which represents an area nearly fifteen times greater than the North American continent. This is an area so extensive that it has caused the earth to become known as the Water Planet, an apt description when one considers how the earth appears when viewed from space.

The oceans of the world, in addition to their extensive dimensions, are joined by wide passages of water, leading some oceanographers to speak of the *global sea*. This concept of a single ocean is indeed well suited to a world in which travel time between nations is decreasing, causing the people of these nations to come closer in their political and social relationships.

The extensive size of the oceans has long intrigued people. For centuries, many have gained a livelihood gathering food from the sea, while countless others have crossed the oceans from one land to another engaged in commerce. Despite this, the sea to date has failed to receive attention commensurate with its size. It has thus far not received the careful examinations given to the land surface and to outer space.

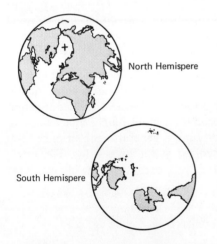

Figure 17.1 Land covers most of the Northern Hemisphere and the sea covers most of the Southern Hemisphere.

17.1 THE IMPORTANCE OF THE OCEANS

Only recently have we begun to realize the true importance of the oceans to our society. Prior to 1950, scientific investigation of the oceans was restricted to such conventional studies as composition of sea water, sedimentation studies (study of sea-floor sediments), interaction of the sea and atmosphere—including examination of currents, tides, and winds—and marine biology. Recent events have brought about new concepts for evaluating the resources of the sea, as well as a new approach which groups all previous studies within the general discipline now labeled *marine sciences*. The evaluation of these resources contained in the 350 million square kilometers of water covering the globe today seems to be challenging our imagination.

Many of the concepts essential to development of marine science have already proved useful and are widely accepted on land. Nevertheless, projections of established techniques to a relatively new and as yet scarcely penetrated environment depends upon the skill of the scientist and demands imagination and new approaches.

Discovery of untapped resources on the ocean floor and recognition of their great potential have political and social as well as economic significance to society. Therefore, a satisfactory accommodation between competing nations must be found so that efficient and harmonious development of the sea's resources may be accomplished. Besides the United States, 107 other nations border upon the sea. Use of the sea and its resources depends on the establishment of suitable offshore zones subject to coastal-nation control, while at the same time preserving the concept of "freedom of the seas."

Many problems exist concerning the breadth of territorial waters over which sovereignty is claimed by any nation. Most countries claim modest offshore zones extending out 3 to 12 nautical miles,* but a few South American nations have extended the limit out to 200 miles—a situation which has caused some jurisdictional disputes. At the present time, problems of this type are being considered at various international conferences between the involved nations.

One of the considerations in determining offshore zones is that of definitions. For example, the legal definition of the continental shelf differs substantially from the geographic definition. In a geographic sense, the term "continental shelf" refers to the shallow part of the ocean floor immediately adjacent to the continental land masses. In scientific literature, a general depth of either 100 fathoms (600 feet) or 200 meters is recognized as the outer edge. The International Law Commission at Geneva in 1958 proposed the following wording:

> The sea-bed and subsoil of the marine areas adjacent to the coast but outside the area of the territorial sea, to a depth of 200 meters or, beyond that limit, to where the depth of the superjacent waters admits of the exploitation of the natural resources of the said areas.

This appears to coincide with the geographical definition and in part it does. But the last phrase of the legal definition has been interpreted by some as extending the limits halfway to the opposite shore of the sea.

Another factor of concern is a means of identifying, jurisdictionally, that zone

---

*The United States maintains a 3-mile territorial limit and an additional 9-mile exclusive fishing zone.

of water along any coast belonging to a particular nation. Identification of this type is necessary so that any given offshore resource may be legally identifiable with one sovereign state or another. In the absence of any agreement, the resource should be subject to the regime of the high seas and thus accessible to any sovereign state. However, it is important that limits be clearly defined to permit a nation to recognize poaching of its various resources by some other agency.

From the foregoing, it is evident that many obstacles must be overcome before exploitation of the sea's wealth becomes a reality. To be able to recognize the scope of the problem, it is first necessary to identify the various resources available to us from the sea.

1. *Food:* Of fundamental importance to the earth's inhabitants—who will surpass the 4-billion mark in the mid 1970s—is the production of food. Empty land with the characteristics necessary for agricultural development has virtually disappeared. This means that existing land must meet the needs, but an increasing population will result in a land shortage. As one step in the direction of combating hunger due to the impending land shortage, the expansion of fisheries is now attracting greater attention. Fisheries have the capability of furnishing protein for the burgeoning population. Further, there is the interesting possibility that direct human consumption of plankton in addition to fish may more effectively use the sea's potential food supply.

The diversity of tastes in seafood has created some problems in that people do not agree on what is an edible fish. In Latin countries, squid cooked in its ink is considered a delicacy, and in other areas of the world, shark or eels may be desirable. To feed the hungry, a product that may be universally acceptable has been developed called *fish protein concentrate*, or FPC. The concentrate is a grayish flour which imparts no taste or odor to food when used as an additive. FPC has the advantage that it can be produced from any type of fish without waste, because the whole fish is included. One teaspoonful per day of this product is considered by nutritionists enough to provide all the minerals and vitamins necessary for normal development of an impoverished child.

Several problems face the oceanographer in the task of providing more food from the sea. In many areas of the world, fish are caught and processed much as they were centuries ago. Although a few of the more technically advanced nations use factory ships equipped with refrigeration and canning facilities, as well as electronic gear for detecting schools of fish, these methods are not yet available to the less developed countries. New methods for detecting and catching fish are required. This necessitates studying the habits of the various species used for food, as well as the habits of those species that may become a source of food. More must be learned about fish considered poisonous in one part of the world and edible in another, especially where the fish is to be used for FPC. Of particular interest should be studies on ways of quickly harvesting fish without taking too many and thereby causing their depletion or extinction. Many fishing grounds have been depleted to such a level that it will require many years before the fish are plentiful again. Some species of sea life such as

Figure 17.2 Fishing boats in San Pedro Harbor in California are equipped with radar and other electronic gear to facilitate fishing operations.

the whales are being killed at such a rapid rate that they are in serious danger of becoming extinct.

2. *Drugs:* The sea may well be the greatest store of medically important drugs available to man. Thus far, only a very small percentage of the living organisms in the sea have been investigated as sources of drugs, but these investigations have produced encouraging results. For example:

—Antibiotics have been obtained from the red sponge from Gonyaulax (one of the causes of red tide) found in Puerto Rico and from certain sargassum weeds and fungi found in the sea.

—Local anesthetics 100,000 times more powerful than cocaine may be obtained from a mussel poison, and puffer-fish poison can be converted into a pain-relieving drug for migraine headaches and terminal cancer.

—Heart stimulants may be extracted from toxins produced by scorpionfish and the hagfish, and blood coagulants from the sea-snake poison.

—A diabetes-combating substance may be contained in the trigger fish.

—Seaweed provides a whole catalogue of useful substances, ranging from fertilizer, to food and food additives, to cures for worms, to cures for radiation sickness produced by strontium 90.

These are but a few examples of the many potentially beneficial drugs that may be obtained from the flora and fauna of the sea. Most of the research thus far performed has been accomplished by universities who have been able to procure funds from governmental agencies or foundations. Little has been accomplished by the drug industry, possibly due to the enormous expense involved in this type of research before a return may be realized. There may have been some major shortcomings in educational programs related to marine medicine which could, in part, be responsible for the slowness of the drug companies to participate. Participation requires trained personnel, and to overcome this deficiency, it has been suggested that marine medicine and pharmacology be recognized as a major discipline. Further education will no doubt increase the awareness of the opportunities in this field.

3. *Petroleum and minerals:* As new technical advances become available, oil is being sought by drilling farther and farther offshore. Drilling equipment on

Figure 17.3 Platform in the Gulf of Mexico is used to drill for off-shore oil. Exxon Company USA.

platforms can now profitably operate in water depth of 100 meters or more, and this will no doubt be extended in the near future.

The continental shelf represents a rich source of oil that may be extracted from the geologically young sediments that occur there. These sediments, less than 150 million years old, contain about 30 percent of the world's oil reserves, and a major portion of this is to be found in the upper layers of these deposits. Oil is formed from diatoms and other organisms that abound in the sea, and whose remains are deposited and covered by marine sediments. Few, if any, such oil formations are thought to occur in the deep ocean basins. Cores taken from the Pacific ocean basin reveal little organic matter, and it appears that the rate of oxidation is high, preventing organic matter from accumulating.

The art of extracting minerals from the sea is as old as man. Common table salt was probably obtained from seawater in prehistoric times, and this practice is still prevalent today. Sodium chloride (table salt) comprises approximately 85 percent of the dissolved solids in seawater and may be removed by simple evaporation. Evaporation to obtain salt was practiced by the ancient Chinese, Egyptians, Greeks, and Romans, who utilized the sun's energy for this purpose. Now bromine is extracted from seawater, as well as potassium chloride and magnesium.

Use of the evaporative process is expensive and requires handling large volumes of water. Other more efficient methods are being sought, and the use of ion-exchangers or semipermeable membranes appear promising. Certain sea organisms have the ability to concentrate some substances. For example, vanadium is accumulated by sea squirts in amounts equal to 50,000 times the concentration of this element in seawater. In addition, kelp will concentrate iodine to usable quantities from seawater which contains on the average 0.06 ppm iodine. How this is accomplished is of interest to the marine scientist, for it may lead to more efficient methods of extracting minerals from the sea.

A few localized pools of hot brine have been discovered in the Red Sea. The temperature of the brine is 55°C, and the brine has a salt concentration nearly ten times that of normal seawater. Hot regions of volcanic material in the zone where the brine pools are located are thought to be responsible. The brine may be a rich source of minerals.

Also of interest and great promise are the mineral nodules that are found in abundance on the ocean floor. These are for the most part of two types, phosphorite, and manganese, and are formed by the slow precipitation of these elements from seawater. Both these materials are of economic importance to man and will, no doubt, be mined from the sea.

The manganese nodules assay as high as 80 percent manganese dioxide, and concentrations as high as 4,000 tons per square kilometer have been found. The nodules are of interest not only for the manganese but also for the copper, zinc, cobalt, molybdenum, and nickel they contain. Cost of recovery has thus far been a factor in hindering exploration for such minerals, but the day is not too far away when sea mining will be economically feasible.

One of the principal disadvantages related to extracting oil and minerals from the sea is that of pollution. Oil spills and the by-products of recovering the oil have created some locally disastrous problems. It is felt that mining operations may cause similar effects, and the necessary steps should be taken to avoid problems before they arise (see Chapter 20).

4. *Source of fresh water:* Approximately 97 percent of the earth's water is contained in the seas, 2 percent in the ice caps and the balance in the rivers and lakes, the latter being the major source of our water supply. This water supply is transitory, for it represents water returning to the sea, where the water is unfit for irrigation or human consumption because of the salt content. Many areas of the world lack even a small portion of the available fresh water and are, therefore, unable to supply the needs of an expanding population. A case in point is Australia, which has a land area equal to that of the United States but is barely able to support a population of 12 million due to a lack of readily available fresh water. A partial answer to their problem would be an inexpensive method of desalinating seawater.

Desalination of seawater is an accomplished fact and has been demonstrated to be a reliable source of excellent water in a number of areas of the world. Over 150 million gallons each day flow through plants around the world, supplying such places as Key West, Florida, and Kuwait with fresh water from the sea at a reasonable cost. To accomplish such a feat requires tremendous amounts of energy, but with the development of nuclear generating plants the necessary power would be available.

Several basic methods may be used for desalting seawater. One is by distillation. Seawater is heated and converted to steam, and the steam, which is mineral free, is passed through a condenser and changed back into fresh water. Variations of this method are used, but a fundamental problem in the method is the scaling or buildup of salts inside the system's tubing. This scale must be periodically removed by one technique or another to maintain the system at peak operating efficiency. Once removed, the scale also poses the problem of disposal.

Another system makes use of a process called *reverse osmosis* in which salt water and fresh water are separated by a membrane. Pressure is applied to the salt water, forcing the water but not the salts through the membrane to the freshwater side. This system is, as yet, not completely satisfactory for converting salt water into fresh water. However, it does function well where brackish or otherwise contaminated waters are to be purified. The development of tougher membranes in the future may eventually permit desalting sea water by this method. At the present, reverse osmosis is used to solve some water pollution problems.

Additional methods of desalting water include the *electrodialysis* process, in which salt is removed by using an electric current, and the *hydration* system, in which the salt water is frozen. When the water freezes, the salts contained therein coat the ice crystals and can be rinsed away). Melting the ice will then yield fresh water.

5. *Long-range weather forecasting:* The ocean and the atmosphere are inextricably linked—each influences the other. Tremendous amounts of heat are stored in the water and the exchange of this heat and water vapor are important factors in the origin of weather. A great deal more global information is required to fully understand the mechanisms of weather phenomena, especially data from remote areas about which very little was known until recently. An extensive network of floating weather stations is planned and is being tested by the World Meteorological Organization to collect such weather data. These stations, fully automated, will be anchored in remote ocean areas and the data telemetered ashore via satellite or other means to weather computers. Similar techniques are also being considered to improve navigational facilities for ships and aircraft in the interest of maritime safety.

6. *Waste disposal:* Improved techniques of offshore waste disposal are being considered as a means of solving water-pollution problems for lands along the seacoasts. In addition, disposal of highly dangerous nuclear wastes and chemical and biological war materials in the deeper reaches of the sea has also been the object of study to minimize the danger of contamination of the sea. A major concern is the effect waste disposal of various types, including the disposal of heat from generating plants and other sources, will have on the marine environment.

7. *Underwater travel:* Such a mode of travel is at present limited to naval vessels. However, in the future, underwater travel may be considered seriously as a means of transferring cargo free from the hazards of winds, storms, and ice. This may offer an alternative means of travel from the east coast to the west coast of North America by sailing under the polar ice.

8. *Recreation:* Attractions of the coast and offshore waters have enjoyed enormous popularity in the past, and there is great interest in stimulating further development of littoral recreation areas. In addition, more underwater activities, including below-surface resorts, may prove feasible and popular in the future.

9. *Energy from the sea:* The restless sea appears to contain an abundance of unharnessed energy. Pulled ceaselessly by the force of gravity between the earth, moon, and sun, the tides are an enormous source of potential power. A 10-second wave breaking on a coastal shore releases more than 35,000 horsepower per mile. It is estimated that moderate waves 2 to 4 meters high, striking an 85-kilometer stretch of coastline could produce as much energy as is generated by Hoover Dam; however the prospect of capturing this much energy with present technology is rather remote. With increasing interest, scientists and engineers are studying ways of harnessing tides to generate electricity for a variety of uses. The basic technology already exists, and a tidal power plant has been put in operation in France on the Rance River estuary.

17.2 EXPLORATION OF INNER SPACE

Introduction to Oceanography

Many of the topics in the foregoing section are dependent upon the solution of technological rather than scientific problems. One large aspect of the current effort in oceanography is the development of techniques and equipment that will permit man to explore the various opportunities that the sea presents. Traditionally, undersea technology has developed as a result of naval operational requirements, industrial attempts to create new business opportunities, and governmental efforts to provide a higher level of services to a variety of ocean-based users. Therefore, it is from these sources that advancement in techniques may be expected.

Because of the great water pressures that exist, even at modest depths, there is a concern for the types of materials needed to protect instruments, equipment, and personnel below the sea surface. High-strength titanium alloys,

Figure 17.4 The Nansen bottle is a commonly-used device for collecting seawater samples at various depths.

aluminum alloys, and the possible future use of high-strength glass, cast ceramics, and plastics, along with advanced construction techniques, have reduced these problems to a manageable size.

The type of equipment required for work at sea falls into two categories. First is the automated or unmanned devices used to collect data in a variety of ways. These include instruments basic to the study of oceanography and

Figure 17.5 Deep Quest is a research submarine designed to carry a crew of four men, and up to 3,000 kilograms of equipment, to a depth of 2,500 meters. Courtesy of Lockheed Missile and Space Co.

used to collect data on temperatures, sea currents, density, and salinity. The Nansen bottle for collecting water samples at various depths is an example. These instruments have been modified, improved, and automated and are valuable tools in the many studies made of the sea (fig. 17.4). Satellites are becoming valuable in oceanographic work, transmitting data from instrument buoys to shore stations.

Bottom-viewing devices enable man to see and explore the ocean floor at various depths without actually descending. Television cameras operating at depths of 70 meters can be used for underwater surveying, object location, and monitoring man-in-the-sea projects. Another device, the side-scan sonar, consists of a towed transducer that can discern bottom features or locate underwater objects such as cables or wrecks. These are the types of oceanographic "hand" tools man has developed in recent years, and many more will be available in the future.

Equipment in the second category includes a remarkable variety of manned ships known as *submersibles* that have emerged in the last decade. These have enabled man to observe and record data on the undersea world while being on the site. The list of undersea vehicles is, of course, too lengthy to be given here, and their future in ocean engineering and resource development cannot even be estimated. Submersibles may eventually be built for ocean mining,

Figure 17.6 The 75-ton, 16 meter long Aluminaut has been used to collect manganese nodules from the ocean floor and to retrieve lost underwater equipment. Courtesy of Reynolds Metals Co.

Figure 17.7 An artist's concept of the Navy's Deep Submergence Search Vehicle (DSSV). It is designed to recover small objects at depths of 6,000 meters. Courtesy of Lockheed Missile and Space Co.

petroleum and natural gas exploration, underwater salvage, fish farming and aquiculture (growth and management of sea plants), undersea archaeological exploration, and possibly even for recreation and tourism.

One of the major considerations in oceanographic work is man himself. What part will he play in underwater exploration and what are the limits of his adaptation to a deep-water environment? The oil industry has used free divers routinely to depths of 200 meters, and experimental work by the United States Navy indicates that man has the ability to survive at 300-meter depths. Other users of the sea have requirements that demand a capability for man to work and live beneath the surface for extended periods. This capability may lead to new opportunities in the production of food, either by fishing or by aquiculture, and in harvesting mineral resources.

Two major groups of problems are associated with man living and working beneath the surface of the sea:

1. Problems directly related to survival, including biomedical problems and hazards from marine organisms.
2. Problems associated with design and operation of facilities for working while underwater.

Figure 17.8 Deepstar 2000 will be used for scientific investigation of oceans, lakes and estuaries to depths of 650 meters. Courtesy of Westinghouse Electric Corp.

The biomedical problems of survival are divisible into several categories, most immediate of which are those produced directly by the wet, cold, dark, high-pressure environment. These problems include an increased resistance to breathing during exertion and at rest; central nervous system narcosis by nitrogen and probably any other inert gas; the long, slow decompression necessary for safe elimination of excessive inert gas from the tissues; the toxicity of oxygen at high pressure; the loss of body heat during prolonged submergence; and the complex interactions of all these factors. As the duration of man's underwater stay increases additional problems appear. These include nutritional requirements under the rigorous conditions, the composition and palatability of foods, psychological behavior in isolated, crowded, small spaces, and impairment of speech by unusual atmospheres (the oxygen-helium atmosphere distorted speech, giving it a "Donald Duck" quality" during the *Sealab* experiments conducted in the 1960s). Medical procedures, including the action of drugs on man-in-the-sea, also require study. The similarity of certain of these problems to those of manned space flight is obvious, and the lessons learned from one environment should be applied to the other.

The presence of other sea organisms constitutes yet another group of complications. In any marine environment, a variety of organisms are toxic if touched or eaten. Also, predatory forms such as sharks consider man fair game.

Men working under water require a wide range of support facilities. These range from diving suits and scuba (*s*elf-*c*ontained *u*nderwater *b*reathing *a*pparatus), to various underwater vehicles, such as *Deepstar* and *Aluminaut* and underwater chambers such as *Sealab*. Studies are in progress to build facilities that would accommodate up to 1,000 men for periods up to a year and at depths of 7,000 meters. Men will be expected to do useful work at these depths, and to this end the Navy and some commercial firms are experimenting with new gas mixtures to permit man to operate at these depths. The oxygen-helium mixture becomes toxic at 500 meters, and new combinations are being sought. Experiments are being conducted on a hydrogen-oxygen mixture, but the dangers from explosion are obvious. Hydrogen does have the advantage of being less narcotic than other gases and may permit reduced decompression time. As yet, no one can foresee the results of oceanographic exploration, but the opportunities, economic as well as scientific, seem boundless.

17.3 PHYSICAL CHARACTERISTICS OF THE SEA

Although the sea is considered by some as global in nature, there are some recognizable divisions which are useful in discussing the oceanographic characteristics. These divisions are the Arctic Sea, the Atlantic Ocean, the Indian Ocean, and the Pacific Ocean, all of which are more or less readily recognized because they are clearly divided by land masses. The Antarctic Ocean, which borders three of the other four oceans to the south, may be distinguished from them only by the circulation patterns and seawater characteristics. Certain smaller bodies of water, such as the Caribbean Sea and the Mediterranean Sea, are distinguishable because they are separated from the oceans by land or island chains. Some, like the Norwegian Sea or Sargasso Sea, have oceanographic features which distinguish them from surrounding bodies of water.

Characteristics of the ocean basins have been described in section 8.4, and the influence of continental drift on the formation of the ocean basins in section 8.5. The reader is referred to those sections as well as to section 8.2, which briefly discusses hypotheses on the origin of seawater in the ocean basins.

Since the continental shelf is the region that will, in all likelihood, receive the most attention by man in his endeavor to broaden his knowledge of the oceans, it should be discussed in somewhat more detail than the ocean basins. Earlier in this chapter, it was suggested that the outer edge of the continental shelf occurred at a depth of 100 fathoms or 200 meters. Beyond this relatively shallow zone, a steep continental slope leads to greater depth. Maps often show submarine contours of these values, which are then conveniently used to delineate that part of the undersea surface qualifying as a shelf. Actually, 100 fathoms equals only about 183 meters, but the location of the break in the slope is so indefinite that it cannot be precisely identified by a fixed numerical value. In fact, the criterion of 100 fathoms tends to be somewhat high, since available data show the average depth of the break in slope to lie between the 60- and 80-fathom submarine contours. On the other hand, there is evidence of continental shelves at much greater depth; the most extreme reported so far is 550 meters for the Sahul shelf off the coast of northern Australia.

The angle of slope on a physical continental shelf is incredibly small, only about 2 fathoms per mile, or 0.085°. The human eye cannot detect a slope of even double this inclination. In many instances, however, the surface of the shelf is not smooth and may be in the form of terraces, ridges, hills, depressions, and canyons. Uneven submarine topography of this type obviously makes the physical shelf difficult to identify, especially where its outer periphery is fractured and defies delineation other than by detailed charting or unrealistic generalization of actual sea-floor contours.

On the average, the continental shelf extends seaward for about 50 kilometers. But average width is not very meaningful, due to the great variation found from place to place. For example, along the west coast of South America, where mountains rise sharply from the coast, the submarine surface in turn plunges to great depths with very little trace of a ledge which may be construed as a continental shelf. At the opposite extreme, the entire Bering Strait and an area extending 1,300 kilometers north of the north coast of Siberia are less than 100 fathoms in depth. At other places, also, the width of the shelf is measured in hundreds of kilometers, including the Atlantic Ocean off the southern coast of Argentina and the South China Sea off the eastern coast of the Malay Peninsula. The Persian Gulf, some 720 kilometers long by 370 kilometers wide, is for the most part no deeper than 50 fathoms. Its seabed generally falls into the category of a continental shelf.

17.4 SUMMARY

Exploration of the sea may well rival and exceed space exploration in the next few decades. The oceans are more readily accessible, and benefits therein to man are evident and seemingly abundant. The exploration of the oceans is replete with political as well as scientific and economic ramifications. Boundaries at sea will become as important as those on land, and uniform rights at sea need to be recognized by all nations.

The sea can provide food and water, basic ingredients necessary for survival on earth. Minerals necessary in the development of society, and oil for power, are to be found in the sea, as well as drugs to heal man's ailments. And as these more basic needs are met, the sea may be enjoyed for its aesthetic values both on and under the surface.

The need to overcome many obstacles before exploration of the sea becomes commonplace is readily recognized. Suitable equipment needs to be developed, and many questions as to man's ability to adapt to such an environment must be answered. To a degree, some technological progress has been accomplished, but much more is required before man himself has the capability to move freely in this new environment.

Surprisingly little was known about the sea, especially that part hidden from view, until the last few decades. Only now is a picture of the physical characteristics of the ocean basins becoming clearer, and a better understanding of how the ocean basins were formed is now available. But many questions remain, especially about the relationship of the sea to the weather. A study is beginning, but it will be many years, if ever, before an overall understanding of these relationships is at hand.

QUESTIONS

1. Give a brief summary of three of the resources of the sea and their value to us. One of these topics may be the subject for a term paper.
2. What are some of the problems to us associated with exploration of the sea?
3. Describe the characteristics of the continental shelf.
4. What are some of the international problems related to using the resources of the sea?
5. Describe one unmanned device and one manned device used for exploration of the sea. You may need to go to other sources for this information.

18 Seawater and Its Properties

Chemical properties of seawater.
Physical properties of seawater.
The influence of the seas on climate.

Each of the 1,300 million cubic kilometers of seawater on earth contain 160 million tons of dissolved solids. If all these solids were removed from the sea they would cover the land area with a layer approximately 170 meters deep. This represents mineral resources of incalculable value which will remain for the most part suspended tantalizingly in the sea until marine science finds methods for reclaiming them economically.

How were all these minerals deposited in the sea? The hydrologic cycle would provide most of the answer (see figure 13.5). Water is evaporated from the sea by the sun's energy and carried by the atmosphere over land areas where the water falls as rain. The water returns to the sea as runoff after first having dissolved mineral matter from the earth's crust. Each year approximately 3 billion tons of dissolved material is carried by rivers to the oceans. Rain also washes gases and suspended matter from the atmosphere into the sea. Another source of mineral matter is supplied by the wind, which carries copious amounts of dust great distances out to sea before the dust slowly filters out of the atmosphere. Submarine volcanoes also supply mineral matter, as does meteorite material that constantly showers down on earth from outer space. All these materials are subject to the slow dissolving action of water and contribute to the dissolved solids in seawater.

18.1 CHEMICAL PROPERTIES OF OCEAN WATER

18.1A Salinity

The amount of salt contained in seawater is relatively large, making it simple to distinguish seawater from fresh water. Seawater contains, on the average,

Table 18.1
Mineral Content of Seawater

Element	Metric Tons per Cubic Kilometer
Chlorine	21,000,000
Sodium	12,000,000
Magnesium	1,500,000
Sulfur	1,000,000
Calcium	450,000
Potassium	430,000
Bromine	72,000
Carbon	30,000
Strontium	9,000
Boron	5,200
Fluorine	1,400
Rubidium	130
Iodine	65
Barium	30
Zinc	10
Arsenic	3
Copper	3
Uranium	3
Manganese	2
Silver	0.25
Lead	0.025
Gold	0.005

3.5% dissolved material, the amount varying from one part of the ocean to another. All dissolved mineral matter in the sea is expressed as *salinity* and is measured in parts per thousand, designated by the symbol ‰. Thus, the average salinity may be stated as 35‰, with the ocean waters ranging in value from 33 to 37‰.

Although seawater is constantly mixing, there are a number of processes which control salinity and prevent it from stabilizing at a uniform level throughout the entire ocean. For example, certain parts of the ocean are subject to high temperatures and low rainfall. The resulting high evaporative rates in these regions remove water from the sea and increases the salt concentration, thereby raising the salinity. At the same time, where high precipitation rates are experienced the seawater is diluted and below-average salinity levels prevail. Whenever large-scale masses of ocean water with different salinity levels meet, mixing will result and cause the establishment of salinity levels intermediate between that of the two masses of seawater.

Methods of lesser importance responsible for changing salinity levels include the freezing of water and the melting of ice in the sea. When pure water is removed as ice by freezing, the concentration of salt increases in the remaining seawater, thus increasing salinity. The melting of ice reverses this process by adding fresh water to the seawater. Fresh water is also added to the sea by rivers.

Although river water contains dissolved solids, it is much fresher than seawater, and thus a dilution of the seawater and lowering of salinity can be expected where rivers empty into the sea. These last three processes are responsible for local variations in salinity and do not influence salinity over broad areas.

The processes of evaporation, precipitation, and mixing are probably the most important in changing salinity levels of surface seawater over large areas. However, deep-water salinity levels are not influenced by the first two processes, only by mixing.

In the past, salinity itself was difficult to measure directly, so alternate methods were sought. The measurement of *chlorinity* (chloride content of seawater) is relatively simple, and since there appeared to be an approximate linear relationship between salinity and the chloride content of seawater, this relationship was found useful. Now salinity (S) in grams of salt per kilogram of seawater may be found by:

$$S(\permil) = 0.03 + 1.805 \, Cl \, (\permil)$$

where Cl is chlorine in grams per kilogram of seawater.

Salinity is now most frequently determined by measuring the *electrical conductivity* of seawater. Electrical conductivity measures the ability of water to conduct an electric current, the variation of which is dependent upon the amount of salt present in the water. A known relationship existing between electrical conductivity and salinity provides a faster and more precise procedure for determining salinity. Since temperature of the seawater influences the conductivity reading, a correction for temperature is required.

18.1B Elements in Seawater

Sodium and chlorine constitute approximately 85 percent of all mineral elements dissolved in seawater (table 18.2). These form sodium chloride (table salt), a product whose extraction from the sea represents a $200 million-per-year worldwide industry. Almost all the world's bromine supply (used as an antiknock ingredient in motor fuels) and approximately 80 percent of the magnesium used in the United States are extracted from seawater. Besides oxygen and hydrogen, which constitute the water, the above elements plus potassium, calcium, and sulfur are the most abundant materials in seawater. Sodium, magnesium, calcium, and potassium exist for the most part as positive ions (cations), and chlorine, sulfur, and bromine as negative ions (anions). Carbon also is present in fairly substantial amounts. Its presence results from the absorption of carbon dioxide from the atmosphere in seawater, where a series of chemical reactions yield carbonate and bicarbonate compounds.

These reactions occur essentially as follows:

$$CO_2 + H_2O \rightleftharpoons H_2CO_3 \rightleftharpoons H^+ + HCO_3^- \rightleftharpoons 2H^+ + CO_3^=$$

| carbon dioxide | water | carbonic acid | hydrogen ion | bicarbonate ion | 2 hydrogen ions | carbonate ion |

Table 18.2
Major Constituents of Seawater at Salinity of 35‰

Constituent	g/kg of Seawater	% Salt in kg of Seawater	% of Each Element of Total Solids
Water	965.650	96.565	
Chlorine	18.900	1.890	54.900
Sodium	10.500	1.050	30.600
Sulfate	2.650	0.265	7.700
Magnesium	1.270	0.127	3.800
Calcium	0.400	0.040	1.160
Potassium	0.380	0.038	1.110
Bicarbonate	0.140	0.014	0.410
Bromine	0.065	0.006	0.190
Boric Acid	0.027	0.003	0.080
Strontium	0.013	0.001	0.040
Other Elements	0.005	0.001	0.010

Some of these compounds result in large-scale sedimentary deposits on the ocean floor. For example, some limestones are deposited in this manner by the combination of calcium and carbonate ions. Carbon is also introduced into seawater as carbonate and bicarbonate compounds carried to the sea from land-surface runoff or dissolved from the ocean floor.

It is assumed that all elements occur in seawater as single or complex ions and that some may be present in such small quantities that they have not yet been detected. In addition, the chemistry of seawater appears to be much more complex than originally thought, and methods for detecting and extracting some elements are not yet available. Many marine organisms have the ability to accumulate relatively high concentrations of certain elements in their tissues (high in relation to the amounts in seawater), a fact which permits ocean-ographers to infer the presence of these elements in seawater. Oceanographers would like to know the secret of the process whereby these organisms accomplish such accumulations. If the process could be imitated or accelerated on a large scale, it might be possible in the future to extract from seawater some of the heavy metals such as gold, platinum, silver, mercury, and tin that are in short domestic supply. The cost of extracting these materials from the sea using known methods is now prohibitive. For example, one attempt to extract gold from the sea produced 0.09 milligrams of gold, worth 14/1000 of a cent, (based on $42.00 per ounce) from 15 tons of seawater at a cost far exceeding the return.

Very small amounts of certain radioactive elements such as carbon 14 also occur in the marine environment. Carbon 14, which has a half-life of 5,700 years, is useful in dating the age of recent sedimentary deposits and in this way permits the oceanographer to determine at what rate sedimentation is taking place. By the same method, it is possible to ascertain how long ago deep-water masses were exposed to the atmosphere.

Thus far, mention has been made only of inorganic matter in seawater and nothing of the possible presence of organic substances. The existence of organic compounds in seawater was recognized in the nineteenth century. However, due to a lack of the necessary techniques, little progress was made in identifying them until recently. Cellular secretions of sea animals was thought to be the source of organic compounds until the last decade, when it was suggested that organic material may originate from dead phytoplankton.

Dissolved gases such as oxygen and carbon dioxide also play a role in the chemistry of seawater. Free oxygen occurs in the sea as a result of the absorption of this gas at the interface of the sea and atmosphere, and by the natural aeration process that takes place as water flows over irregular land surfaces back to the sea. Most of the oxygen in the sea is generated by the *photosynthetic* process of the abundant plant life in the sea. Plants make use of the carbon dioxide in the sea, the water itself, and energy from sunlight in the formation of carbohydrates. Oxygen is a by-product of this metabolic process, which can be represented as follows:

$$6\,H_2O + 6\,CO_2 + \text{energy from sunlight} \rightleftharpoons C_6H_{12}O_6 + 6\,O_2$$

$$\text{water} \quad \text{carbon dioxide} \quad \quad \quad \text{carbohydrate} \quad \text{oxygen gas}$$

This is an oversimplified version of what actually occurs, as there are many intermediate stages not shown. Water is the source of oxygen in this equation. The reaction shown in the equation is reversible in the sense that animal life in the sea utilize the oxygen in metabolizing carbohydrates in their tissues to produce energy. Thus, water and carbon dioxide are by-products of this *respiration* process. The role of carbon dioxide found in seawater was discussed in a previous paragraph.

Since light, which does not penetrate to too great a depth, is an important factor in photosynthesis, most of the plant life will occur close to the sea surface. Thus, the greatest concentration of oxygen in seawater may be found there. At lower levels, oxygen is utilized by sea animals and by oxidation of mineral matter, thus reducing the concentration. The circulation patterns of the sea generally cause a periodic renewing of the oxygen level in oxygen-deficient water. However, certain areas in the sea are not replenished with oxygen because of a lack of circulation, and in effect the seawater becomes stagnant. Such areas occur in the Black Sea and in the Cariaco Trench off the Venezuelan coast. Because of a lack of oxygen, these regions are practically devoid of any form of sea life. Pollution of certain inland waters has created the same condition (see Chapter 20). A by-product of such stagnant conditions in the sea is the formation of hydrogen sulfide, notable for an odor similar to that of rotten eggs. Hydrogen sulfide is formed by the reduction of sulfates in the absence of oxygen. Only some forms of bacteria are able to survive in such an environment.

Several inert gases are also found dissolved in seawater. These include helium, nitrogen, neon, argon, krypton, and xenon, all absorbed from the atmosphere and by aeration of waters entering the sea. Of this group, only nitrogen is significant, as it is an important element in the nutrition of marine plant life. Nitrogen occurs in seawater as nitrate, nitrite, and ammonium ions, as well as nitrogen bound up in organic compounds and nitrogen as gas. Nitrate is the form in which plant life utilizes nitrogen as a nutrient.

18.2 PHYSICAL PROPERTIES OF OCEAN WATER

Physical properties of seawater such as temperature variations and density are as important in the study of oceanography as the chemical characteristics just discussed. The physical properties vary from one part of the ocean to another, and these differences are of importance because of their relationship to currents and weather. To have a complete understanding of the interrelationship of the sea and weather phenomena requires a great deal more knowledge about temperature and density variations than is currently available. Considerable effort is now being expended to accumulate such data.

18.2A Temperature

Temperature levels are not uniform throughout any given portion of the sea, despite the considerable mixing that may occur. Seawater temperatures, as a rule, decrease with depth. But they do not do so uniformly, and the temperature profile may for convenience be subdivided into three temperature zones. An *upper zone*, varying from 50 to 200 meters in depth, will have a temperature approximately that of the surface. There are, of course, expected deviations from this. For example, in a given area on a sunny afternoon with not much wind, the temperature will be higher at the immediate sea surface than at any level below.

Beneath the upper temperature zone is the *thermocline*, a layer in which a greater drop in temperature occurs (table 18.3) than in the layer above or below. The thermocline may extend 500 to 1,000 meters down as a relatively narrow "discontinuity layer" between the upper zone and the *deep zone*, where temperatures decrease more slowly again. Needless to say, the zones are occasionally difficult to define exactly within the given limits. However, in the middle and lower latitudes a "permanent" thermocline is recognized that ranges in depth from 200 to 1,000 meters. There does not appear to be a permanent thermocline in the polar seas. The thermocline may vary seasonally, being much deeper in the summer months in the Northern Hemisphere than during the winter months. The thermocline is affected by currents and waves, as well as by above-surface processes such as the wind and solar radiation. Thermoclines will in turn affect the transmission of sound through water as well as transparency.

On occasion, warmer layers have been detected beneath cool layers, but these

Table 18.3
Variation in Temperature and Salinity with Depth in Tropical Waters. From *Meteor* Expedition, South Atlantic

Depth (meters)	Temperature (°C)	Salinity (‰)
0	25.52	36.12
50	25.58	36.11
100	22.92	36.43
150	17.28	35.93
200	12.79	35.25
300	8.77	34.80
400	7.46	34.65
600	5.75	34.52
800	4.42	34.48
1000	3.95	34.60
1500	3.97	34.91
2000	3.29	34.96
2500	2.90	34.93
3000	2.775	34.92
4000	1.73	34.83
5000	0.72	34.72
5500	0.43	34.68

circumstances are rare. This would occur when a cold air mass cools the immediate surface of the sea more rapidly than the layers beneath. In the ocean trenches, water temperature at the ocean floor may be slightly higher than at levels above. Basically, temperatures on the ocean floor are near 0° C, because water at these depths originates in the polar regions. The icy water moves along the bottom and enters the ocean trench from a much shallower level. The water moves down the steep slope of the trench, and as it does, the pressure increases and compresses the water slightly. This compression results in a small temperature rise which causes the water at the bottom of the trench to have a slightly higher temperature than water entering the trench higher up.

Sea surface temperatures vary with latitude, with the *isotherms* or lines of equal temperature running approximately east and west (fig. 18.1 and 18.2). A study of the relationship of isotherms reveals that temperatures generally increase from the polar regions toward the equator, although some isotherms are diverted near coastal areas by currents. Upwelling of cool subsurface water may also deflect isotherms from their generally east-west direction. This occurs, for example, during summer months along the west coast of North America.

18.2B   Density

Density is defined as mass per unit volume and may be stated as grams per cubic centimeter or pounds per cubic foot. *Specific gravity* is the ratio of two densities, usually of the substance under discussion and pure water. This ratio yields a

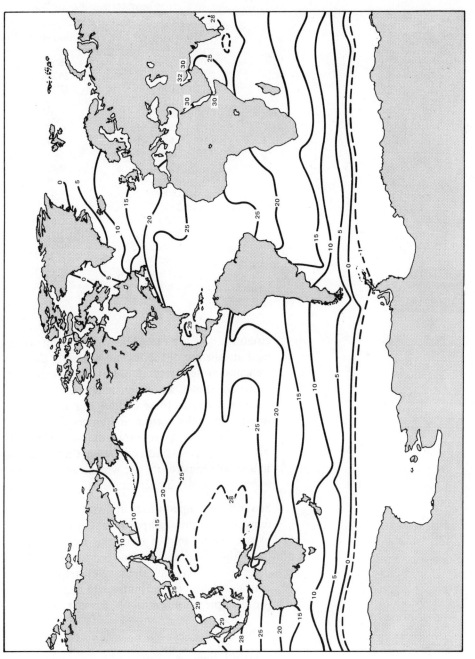

Figure 18.1 Surface temperatures of oceans in August.

405

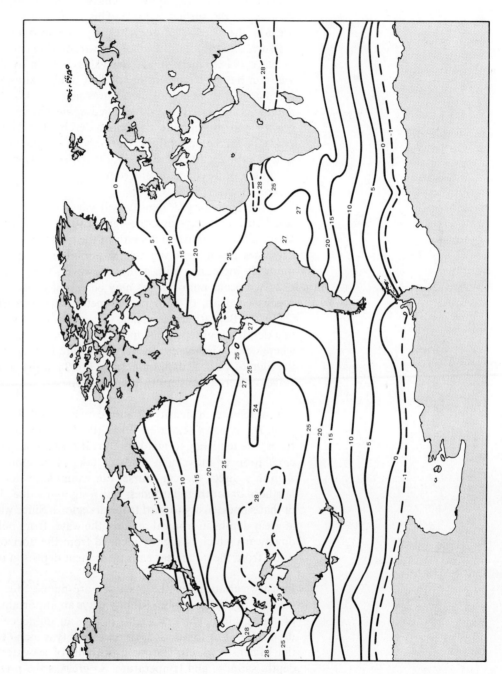

Figure 18.2 Surface temperatures of oceans in February.

dimensionless number which, if the metric system is used, is numerically equal to density. In oceanography, specific gravity is used, and the standard value for seawater at the surface is 1.025. This value will vary with latitude and depth, because the density of seawater is influenced by changes in temperature, pressure, and salinity. But since the variation in density of seawater from place to place is extremely small, the value is given to five decimal places. Thus, seawater having a temperature of 20° C and a salinity level of 35‰ will have a specific gravity at atmospheric pressure of 1.02478. For convenience, this number is converted to a value called *sigma-t* by subtracting 1 from the specific gravity and multiplying the remainder by 1,000, which in the above value yields 24.78. Sigma-t values refer only to specific gravity of seawater at atmospheric pressure (on the surface); otherwise pressure is also taken into account, yielding a value called *density in situ*.

Like the temperature, density will vary with latitude and with depth in the sea. Thus, on the average, the sigma-t value at the equator is 24 and rises to 27 at latitudes 60° north and south of the equator. At higher latitudes, the value decreases again very slightly. Variation in density also occurs vertically—normally increasing with depth. However, the increase in density with depth does not occur uniformly as may be expected. In tropical oceans, the surface zone, to a depth of perhaps 100 meters, has a nearly uniform sigma-t value (24 to 25). Beneath this is a zone, the *pycnocline*, where density increases rapidly to a depth of perhaps 2,000 meters. Underneath the pycnocline is a deep zone, where density increases at a very slow rate with depth. In the higher latitudes, the sigma-t value is approximately 27 to 28 at the surface, which is close to the value of deep-water density at any latitude.

The difference in density of surface water and deep water is a factor in the vertical mixing of seawater. When the difference is small, as in the higher latitudes, the tendency to mix is not great and the water's *stability* is high. Water circulation or movement tends to take place along density gradients rather than in a direct vertical course. For example, water at the surface in extreme northern or southern latitudes, with a sigma-t value of 27, will follow a gradient of that same density toward tropical regions. Since water of that density occurs at great depths in tropical regions, the water from polar regions follows a path downward which is slightly inclined from the horizontal. In this way, surface water from the polar seas ends up at great depths in tropical oceans.

Density variations in the open sea are mainly due to temperature changes, whereas in coastal regions salinity plays an important role. The compressibility of water, although very slight, is also an influence on density, especially in deep water. For this reason, compressibility is included for accurate determination of density *in situ*. The compressibility of seawater decreases with increasing depth, salinity, and temperature. So great is the pressure in deep oceans (500 atmospheres at 5,000 meters, 1,000 atmospheres at 10,000 meters) that if seawater were truly incompressible, the sea level would be about 30 meters (100 feet) higher than it is.

18.2C Sound

Because of a lack of visibility in the sea, sound has been used to great advantage to explore the depths and contours of the ocean floor. The speed of sound is greater in water than in air due to water's smaller compressibility, and it varies with changes in salinity, temperature, and density. The speed of sound is 1,445 meters per second at 1 atmosphere pressure, with the salinity level at 34.85‰ and temperature at 0° C. If the temperature level is 20° C, with other conditions remaining the same, the speed of sound is 1,518.5 meters per second. The speed of sound increases about 1.5 meters per second for every 1‰ increase in salinity, 18 meters per second for every 1,000 meters increase in depth, and 4.5 meters per second with each 1° C rise in temperature. All these factors must be considered when echo-sounding equipment is used in deep water for measuring depth.

The principle of the echo-sounder used to determine depths and ocean-floor contours is relatively simple. A transmitter sends a sound impulse to the bottom, the signal is reflected, and the echo is received back at the surface by a hydrophone—an instrument used to detect sound in water. An accurate timing device measures the time elapsed from the instant the signal is sent until it is received again. All the influences on the sound wave as it travels to the bottom of the sea and returns must be understood to obtain an accurate depth record. Present-day echo-sounding equipment can be used for all depths and is accurate to within $\frac{1}{2}$ of 1 percent.

18.2D Color

Seawater color ranges from deep blue (the dominant color) to green or yellowish green. The blue color is a result of the selective scattering of light by water molecules. More of the blue light is radiated by water than any other of the component colors of light. The blue color occurs in the open sea in temperate and tropical zones where there is little biological activity. The blueness indicates a degree of purity of the water and has been called the "desert color of the sea." At higher latitudes, water has a tendency to be blue green to greenish—a color which is also prominent along coastal areas over the shallow continental shelves. This coloration has been attributed to the presence of organic matter which may be yellow; this would produce a green when it combined with the natural blue of the water. Water seen at the mouths of large rivers is frequently brown as a result of the silt and clay carried by the streams to the sea. The muddy waters may sometime be detected many miles out to sea even before one is within sight of land. In the fjords of Alaska and Norway, water may be a milky color caused by the presence of the finely pulverized "rock flour" formed by the grinding action by the rocks borne in glacial ice on the bedrock.

18.2E Penetration of Light

Anyone who has experienced diving even to a modest level of a few meters is aware that light diminishes with depth in the sea. The amount of light that penetrates the surface of a body of water is progressively reduced to extinction by absorption and scattering of light, both by the water and by materials suspended and dissolved in the water. Ultimately, all light is either absorbed or reflected back to the surface, and a realm of absolute darkness is reached. The depth at which this occurs may be a hundred meters or less for a diver, who is limited by the amount of light that his eyes can detect. However, tests with photographic film exposed for a prolonged period at a depth of 1,000 meters revealed some reaction to light, but at a depth of 1,700 meters, no reaction was detected, indicating a level of total darkness.

The penetration of light into clear ocean water will vary depending upon a number of factors (table 18.4). One of these is the property of light itself, which

Table 18.4
Amount of Light Penetrating to Various Depths as a Percentage of the Light Entering Through the Surface

Depth (meters)	0	1	2	10	50	100
Clearest Ocean Water (%)	100	45	39	22	5	0.5
Turbid Coastal Water (%)	100	18	8	0	0	0

is composed of a variety of colors of differing wavelengths. Water is least transparent to red light and most transparent to blue green (table 18.5). For this reason, objects in the sea with a normally red or yellow color will appear dark or black at moderate depths because most of the light at the red end of the spectrum has been absorbed in the upper layers of water. Objects in the blue green range will appear in their normal colors at greater depths. Turbid (muddy) water along coastal areas greatly restricts the penetration of light to perhaps 10 meters or less. In this case, the yellow green wavelengths of light penetrate better than the blue green. The penetration of light in any part of the ocean may vary at different depths because of the presence of particulate matter carried by a layer of water moving through the sea, or because of a variation in the ability of light to penetrate through two adjacent bodies of water which differ in turbidity.

How has the depth of visibility been measured? An early method used was the *Secchi disk*, a white disk (named for its developer) standardized to 30 centimeters in diameter in order to obtain reproducible results (fig. 18.3). The disk was lowered into the sea until it was no longer visible and the depth measured.

Table 18.5
Percent Loss of Light (Light Absorbed) at Different Wavelengths in One Meter of Pure Water

Wavelength (angstrom units)	4000 Violet	4600 Blue	5000 Green	5400 Yellow	5800 Yellow	6400 Orange	7000 Red
Percent Loss of Light	4	1.6	2.6	4	7	21	40

Figure 18.3 A Secchi disk used to determine depth of visibility in seawater.

This method was used extensively during the nineteenth and early twentieth centuries, but with the development of the photoelectric cell, a more reliable and accurate procedure was available and the Secchi disks are no longer used. The photoelectric cell may be lowered into the sea and values of light intensity transmitted and recorded on board ship. Another use that is made of this device is to lower the photoelectric cell with a light source and measure the penetration of light through a standard distance of one meter. This will yield information on the turbidity of the seawater.

By the methods described, light intensities have been shown to diminish rapidly with depth. So rapid is the reduction that at a depth of 10 meters light intensity will only be 10 percent of the amount at the surface (table 18.6). This applies to water in the open sea. In coastal waters, light penetration is much lower due to the greater amount of suspended matter that is present.

Table 18.6
Percent Illumination at Various Depths in Average Seawater

Depth (meters)	0	10	20	50	130	200
Percent Illumination	100	9.5	3.7	0.31	0.0005	0.000,002

18.3 THE SEA AND THE CLIMATE

The Oceans

The sea and the atmosphere are inextricably joined. Ocean water circulation is dependent on its physical properties, such as temperature and salinity differences, and on the wind. The atmosphere is like a giant heat engine that is driven largely by the sea. Great quantities of heat are stored in the waters of the ocean—heat which is released to the atmosphere at different places and in varying amounts. An understanding of this phenomenon and its effects will aid greatly in an understanding of weather and possibly to its modification. Each year 350,000 cubic kilometers of water are evaporated from the sea, of which approximately 100,000 cubic kilometers falls on land in some form of precipitation. The total amount evaporated represents a layer one meter deep over the entire ocean but the mistake must not be made in thinking that water is evaporated uniformly from the entire surface of the sea. Marine air masses contribute up to 80 percent of the moisture for rain, and the balance comes from continental air masses that accumulate moisture from glacial ice, inland waters, from soil, and from plant transpiration. The total effect of heat and moisture exchange between the oceans and the atmosphere is just barely beginning to be understood.

Within the past decade, oceanographers and meteorologists have come to the realization that the ocean and the atmosphere must be considered as a total entity in the study of weather phenomena. Several studies have been undertaken to gather data by the use of buoys anchored in remote parts of the sea—areas from which information previously was unavailable (fig. 18.4). In this way, it is possible to obtain simultaneous information rather than data collected over a considerable period. In investigating the relationship of ocean to weather, data collected even several days apart in different sites in the sea is of doubtful value. Hopefully, in the near future a system will be available that would provide simultaneous data from all parts of the world through use of weather stations, ocean buoys, and weather satellites.

Such a system is slowly being developed under the auspices of the World Weather Watch which is a project of the World Meteorological Organization (WMO). The WMO is comprised of 130 national weather services and the International Council of Scientific Unions. So ambitious is the program that it will not be fully functional until at least 1980. At present, some of the hardware, such as a computer capable of handling the mountains of weather data at high speed, is not yet available.

How do the oceans affect the climate in various parts of the world? First, it is recognized that oceans modify the climatic conditions of coastal areas compared to inland areas at the same latitude. For example, Los Angeles has a mean January temperature of 13°C and mean of 21°C for July. Fifty miles to the east is Riverside, California, with a wider temperature range: a mean January temperature of 12°C and a mean July figure of 24°C. The proximity of the Pacific Ocean to Los Angeles has modified its climate compared to that of Riverside. Cities located along the coastlines may have quite different

Figure 18.4 A weather buoy will transmit weather and pollution data to a shore station at regular preset intervals. Courtesy of General Dynamics Corp.

temperature ranges depending upon whether the prevailing wind is of marine or continental origin. San Francisco, for example, which has a January mean of 11° C and a July mean of 15° C, is influenced by marine air from the Pacific Ocean which restricts the temperature range within relatively narrow limits. On the other hand, Washington, D.C., despite its proximity to the Atlantic Ocean, is influenced by air currents which have moved over the North American continent and therefore are not modified by the sea. Washington has an average January temperature of 3° C and an average July temperature of 23° C. As a general rule, marine air will moderate the climate of areas close to the sea, whereas areas in the same general latitudes influenced by continental air masses will have hotter summers and colder winters even if located in coastal areas.

The examples above are, of course, isolated sites illustrating the influence of marine or continental air masses on land climate. Are there examples of large land masses over which weather is controlled by oceanic affects? Studying the temperature patterns of the Gulf Stream and comparing the climatic conditions of the eastern and western North Atlantic permits one to observe the considerable influence of the ocean on climate. One need only compare the climate of the west coast of France with that of Maine, or the climate of England and Ireland with that of Newfoundland, or the climate of the west coast of Norway with the east coast of Greenland. Each of the locations compared are on approximately the same latitude but have markedly different climates.

The Gulf Stream flows generally in a northeasterly direction from the Gulf of Mexico toward northern Europe (a more detailed discussion of Gulf Stream activity is given in Chapter 19). Warm water brought in this fashion as far north as North Cape, Norway, results in a mean annual temperature of about $5°C$ at this site and maintains an open sea even in winter. By comparison, Greenland's east coast is blocked by pack ice even in summer. The Swedish meteorologist, J. W. Sandstrom, has stated that the Scandinavian peninsula would be a desert of ice just like Greenland were it not for the Gulf Stream.

The Gulf Stream has little direct effect upon the climate of the eastern United States and Canada. This region is influenced rather by the northwest winds that sweep down over frigid continental Canada, particularly in the winter. Winds such as these are generated by atmospheric low-pressure areas which are created where warm air (in this case warmed by the Gulf Stream) comes in contact with colder air (cooled by the cold Labrador current along the east coast). Such a low is formed off the east coast of the United States through the influence of the Gulf Stream, causing the northwesterlies to blow offshore. For this reason, the winters of northeast United States are much more rigorous than those experienced in France or Spain, although all are at comparable latitudes. A similar low-pressure area is formed off the Norwegian west coast, causing winds to blow from the southwest and bringing warm marine air over the northwestern part of Europe.

Variations in the amount of heat carried by such ocean currents as the Gulf Stream can cause changes from year to year in the climate of land areas influenced by marine air masses. However, the heating of the sea in the Gulf of Mexico and the Florida Straits, for example, will not affect the climate of northern Europe for some months hence, as it requires time for Gulf Stream water to travel this distance. As yet, the precise interrelationship between the movement of large amounts of seawater and climate is not fully understood. But it is anticipated that in the future, long-range weather predictions can be made, based, in part, on detailed information on the vicissitudes of such currents as the Gulf Stream.

18.4 SUMMARY

The fact that the sea is salty is a surprise to no one, whereas the fact that salt levels (expressed as salinity) vary from one part of the ocean to another may be less common knowledge. Evaporation rates and rainfall levels differ and will

Seawater and Its Properties

influence surface salt concentrations, or the salinity, of surface sea water. Mixing will produce changes in deep-water salinity levels.

The most common salt in seawater is sodium chloride, which comprises about 85 percent of dissolved material in the sea. This has been the most prevalent source of table salt throughout history—obtained by ponding sea water and permitting it to evaporate. Economic substances such as magnesium and bromine are currently extracted from the sea, and it is believed in the future other elements will be extracted as well.

Dissolved gases play an important role in the biological activity of the sea. Principal among these are carbon dioxide and oxygen. Carbon dioxide is important in the photosynthetic process of sea flora, whereas oxygen is vital in the metabolism of sea fauna.

Temperatures vary in the sea, generally decreasing with depth, but not at a uniform rate. Temperature and salinity levels influence the density levels of seawater, which in turn is a factor in vertical mixing of seawater. Other physical properties of seawater include its penetrability to light, which decreases with depth and varies with turbidity. Sound travels through seawater at a higher speed than in air due to the greater density of seawater. The speed of sound varies with differences in temperature, salinity, and depth.

The land climate, profoundly influenced by the oceans, is tempered by the movement of marine air masses toward the coastal regions. Some continental areas such as northwestern Europe might be uninhabitable were it not for such currents as the Gulf Stream which move huge quantities of warm water many hundreds of miles to modify air temperatures.

QUESTIONS

1. What is meant by the salinity of seawater?
2. What causes changes in salinity levels of seawater?
3. What two elements make up the major portion of the salts in seawater?
4. What is the source of dissolved oxygen in seawater?
5. Define the following: thermocline, isotherm, Sigma-t, pycnocline.
6. How does the density of seawater vary with changes in temperature; changes in salinity?
7. What influences do increases in salinity, increases in density (depth), and increases in temperature have on the velocity of sound in seawater?
8. By what means is the depth of visibility measured in seawater?
9. How do oceans modify the coastline climate as compared to inland climate?

19 The Restless Sea

How ocean currents are measured.
Circulation of ocean currents.
Factors causing the currents.
Tides—how they are generated.
How waves are formed.
Properties of a sea wave.

The interaction of the sun, moon, and earth causes dynamic, restless and ever-changing conditions on the seas. These conditions affect ocean currents. Oceanographers throughout the world have long recognized the existence of these ocean currents and are deeply interested in knowing more about them. The continuous motion of these "rivers of the sea"—generally in a clockwise direction in the Northern Hemisphere and counterclockwise in the Southern Hemisphere—continues ceaselessly (fig. 19.1). Ponce de Leon's pilot discovered the Gulf Stream in the Straits of Florida in 1513, and Benjamin Franklin drew the first chart of this current in approximately 1770 (fig. 19.2). The first comprehensive study of prevailing ocean currents and oceanic climate was made by M. F. Maury, an American naval officer, in the mid nineteenth century. His sources were the experiences of naval and merchantmen, and from these ex-experiences Maury compiled information which he analyzed and used as a basis for a pioneer study of the oceans.

19.1 MEASUREMENT OF OCEAN CURRENTS

In the past century, methods for measuring current flow have become increasingly more sophisticated, evolving from the interview and survey type of approach used by Maury to the use of automated electronic instruments

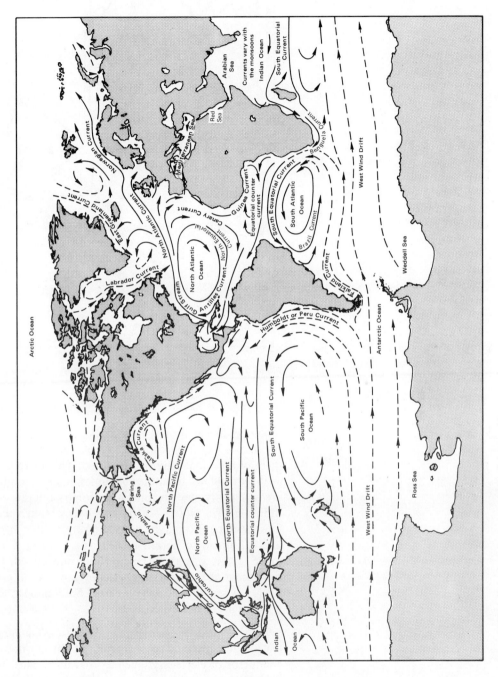

Figure 19.1 Principal ocean currents of the world.

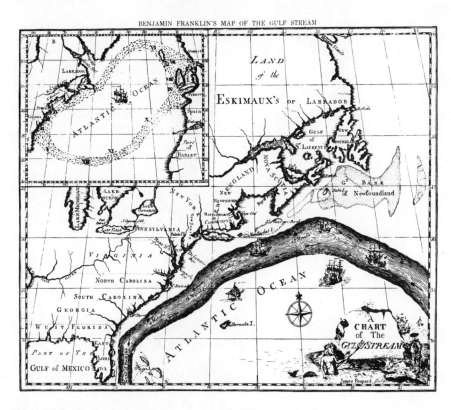

Figure 19.2 Benjamin Franklin's map of the Gulf Stream.

mounted in earth-orbiting satellites. Surface observation at sea is difficult because of a lack of readily available reference points against which the movement of a current may be measured. The movement of a ship caused by such a current will not be noticeable unless a large number of very accurate bearings are repeatedly made against a stationary reference point. Ships at anchor may utilize *current meters* and in this way obtain valid measurements of current flow. Meters of this type consist of a propeller which is turned by the current and calibrated to permit measurement of the current's speed. Measurement made in this way must be carefully interpreted, because the ship's swinging movement at the end of the anchor cable could give misleading results. Relatively few readings of this type have been made in the open sea, but the method is useful in coastal waters.

A device called the *Geomagnetic Electro-Kinetograph* (G.E.K.) has been perfected which permits recording ocean current flow from a vessel underway. A conductor of electricity (such as seawater) when moving through a magnetic field will generate electricity proportional to the speed at which the conductor moves. This principle is used in the G.E.K. for measuring oceanic currents

Drift bottles were used to determine the direction and extent of ocean currents more than one hundred years ago. The bottles were weighed to prevent them from projecting above the sea surface and by this means reduced the influence of the wind on the bottle. A card identifying the bottle's launch site was included with a request that the card be returned with discovery location and date noted. A disadvantage to this method was that only the launch and pickup sites were known with any accuracy. Little is known about the exact route taken by the bottle, although some good current data has been obtained by this method in inland seas like the North Sea. It might be noted here that plotting the movement of drifting icebergs in the North Atlantic by the iceberg patrol has contributed valuable knowledge of oceanic currents in that region of the sea.

Electronic "drift bottles" are now utilized. These can be tracked quite accurately with modern electronic devices such as LORAN (*l*ong *r*ange *n*avigation) mounted on ships. Sea currents at any depth may also be traced with a *swallow buoy*—an instrument capable of drifting at any predetermined depth. This device is fitted with a "pinger" (emitting an electronic signal) that permits accurate tracking. It was this device that revealed the counter-current flowing generally northeast to southwest beneath the Gulf Stream (fig. 19.3).

The total picture of ocean currents is by no means complete; there are still many gaps that require detailed study. The present picture of ocean currents has been the result of patiently piecing together thousands of bits of information collected over many decades. This has revealed a circulation of ocean water that changes somewhat with the seasons but, broadly speaking, maintains a fairly constant pattern. The change in circulation results primarily from seasonally changing air movements over the ocean surface.

19.2 CIRCULATION OF OCEAN CURRENTS

Looking at figure 19.1, it can be seen that the major ocean currents follow a regular pattern of flow. These currents only include movement of surface water extending at most to a depth of 1,000 meters. Needless to say, there are many details in the flow pattern that have been omitted here due to the limitations of space. It is the intended purpose here to become acquainted only with the general circulation pattern.

It appears that the pattern of ocean current flow is divided into a number of closed systems—the most obvious of which occur in the North Atlantic and North Pacific, where flow is clockwise, and in the South Atlantic, South Pacific, and Indian Ocean, where flow is generally counterclockwise. In the extreme south around the Antarctic continent is the Southern or Antarctic Ocean. With no land barriers to interrupt its flow, the general direction of circulation is west

418

The Oceans

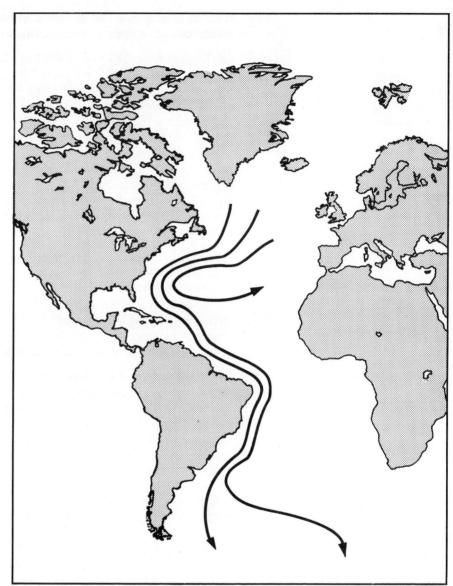

Figure 19.3 Deep ocean circulation at 2,000 meters.

to east around the continent. This is the circumpolar current attributed to the west wind which blows in this region. The west wind (i.e., blowing from the west), together with the circumpolar current flowing toward the east, made it extremely difficult for sailing ships to round Cape Horn from the Atlantic to the Pacific Ocean. In a narrow zone around the Antarctic continent, the current flow is in a westerly direction caused by the prevailing east wind that blows off the continent.

With the exception of the Antarctic Ocean, each system, or *gyre*, as the circulation systems are called, is composed of four large-scale currents. At the northern boundaries of the northern oceans the current flow is west to east, and at the southern boundaries flow is east to west. The Gulf Stream flows northwesterly on the west side of the North Atlantic gyre, as does the Kuroshio in the western Pacific. These two currents are quite deep, extending to depths of 1,000 meters, and flow at a rate of 50 to 120 kilometers per day. The flow is here somewhat swifter and confined to a narrow region compared to other oceanic currents. This is due to a westward intensification of these currents by the prevailing winds. The Canary Island and California currents, found in the eastern boundaries of the North Atlantic and North Pacific respectively, are distinctly slower (3 to 7 kilometers per day), cover a broader area, and are shallower than their western counterparts.

Similar currents occur in the Southern Hemisphere, but they move in a direction opposite to that of the currents in the north. Because of the lack of large continental masses in the Southern Hemisphere, the western components of those currents are not as conspicuous as in the Northern Hemisphere.

19.3 FORCES CAUSING SEA CIRCULATION PATTERNS

What forces are considered responsible for creating circulation patterns in the sea? Basically, solar radiation is the prime mover and the principal source of energy for moving the huge amounts of water which comprise the ocean currents. Actually this energy is responsible for considerable activity in both atmosphere and oceans, and these activities influence each other. Only the forces affecting the sea will be considered here. The forces acting upon seawater are those resulting from inherent physical properties of seawater and those that are applied to seawater by external factors.

The inherent physical properties of seawater generate certain internal forces which are in part responsible for the large-scale movements of the water. Such movement may take place when the *density* of seawater changes due to *changes in temperature or salinity*. Early theories suggested that differential heating of the sea between the polar and equatorial regions caused movement, but this is now believed to be of little consequence. However, cooling of the seawater in polar regions will result in an increase in the density of surface water, causing it to sink. In addition, the formation of ice will increase the salinity of the surface water, increasing its density further and thus increasing vertical flow. Therefore, in the polar regions *thermohaline circulation (thermo:* "*heat*"; *haline:* "*salt*"*)* may be brought about by increased density of the upper surface water. Water movement is thus originated by vertical flow downward to mid-depth to a level where water of a similar density is reached, or even to the bottom if the sea is relatively shallow. Horizontal flow then moves the water toward the lower latitudes along the density gradient.

In the equatorial regions, evaporation will increase salinity of seawater and in this respect has the same general affect as the formation of ice in the polar regions. However, the increase in density resulting from increased salinity may be approximately offset by a decrease in density due to increase in water temperature. Because of this, little or no vertical movement may be initiated.

The change in density, which gives rise to vertical flow, will also cause a change in water pressure that may in turn result in horizontal flow. It is readily apparent that water pressure increases with depth, increasing 1 atmosphere with each additional 10 meters of depth. Thus, at 30 meters the pressure is 3 atmospheres plus 1 for the air for a total of 4 atmospheres. However, exact pressure at all points of a horizontal plane at a depth of 30 meters will be determined by the weight of the column of water above each point. Variations in weight—and therefore in pressure—will depend upon the density of the water, which in turn is dependent upon the temperature and salinity of the water at that site. If the pressure at 30 meters differs from the pressure at the same depth at a point 200 kilometers away, then a *horizontal pressure gradient* exists and horizontal flow will occur.

One other inherent property of seawater that influences flow is *internal friction*. Friction can cause the transfer of energy between contiguous layers of water moving at different speeds. Thus, a slow-moving layer will be dragged along by a faster-moving layer, and conversely the faster-moving layer is slowed down in its movement as a result of its contact with the slower-moving layer.

External forces responsible for seawater movement are those applied by factors not directly related to the physical characteristics of seawater. Perhaps the most important of these is the activity of the wind, which applies a dragging force as it blows over the sea surface, initiating horizontal movement. Internal friction transfers this movement downward layer by layer, but the magnitude and direction of movement changes with depth.

The direction of water movement in the open sea does not coincide with the direction of movement of the prevailing wind. The rotation of the earth gives rise to another force—the *Coriolis force*—which is responsible for the divergence. The Coriolis force causes the upper layer of the open sea to be deflected to the right of the wind direction in the Northern Hemisphere and to the left in the Southern Hemisphere. The surface layer in the Northern Hemisphere is deflected 45° to the right of the wind direction, but this applies to the surface layer only. The layer beneath is set in motion by the surface layer (internal friction) and is in turn deflected to the right by Coriolis force. However, the deeper layers move progressively slower and are deflected to a lesser degree than the layers above into what has become known as the *Ekman spiral* (fig. 19.4). The Ekman spiral was conceived in a theoretical study on the effect of a steady wind blowing over the surface of a uniformly constituted ocean. At a depth of approximately 100 meters, according to the theory, the current is flowing at a barely perceptible rate and has been deflected in a direction opposite that of the surface movement. This is sometimes called the *depth of friction*, and it represents the limit to which movement is imparted by the wind to the seawater. If the combined water movement of all layers as represented

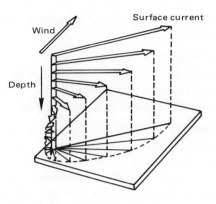

Figure 19.4 The Ekman Spiral.

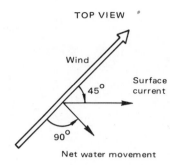

Figure 19.5 Top view of Ekman Spiral showing relationship of wind movement, surface current, and net water movement in the Northern Hemisphere.

by the Ekman spiral is considered, then the net water flow is perpendicular (90° to the right in the Northern Hemisphere) to the direction of wind movement (fig. 19.5). The Ekman spiral would appear in reverse in the Southern Hemisphere.

The maximum amount of deflection (45°) at the surface resulting from the Coriolis force may not be achieved in shallow waters because of another external force—the *friction of the sea floor*. Friction from this source has little influence on the flow of surface currents in deep oceans, but it is a factor in shallow seas and on the continental shelf. Deflection caused by Coriolis force in shallow waters may be as little as 15° to 20° as a result of bottom friction.

Another external force influencing seawater flow is the gravitational attraction of the sun and the moon, which is responsible for tidal activity. Tidal activity and tidal currents are not generally related to ocean currents, because tidal movement is alternating in nature. They are only of consequence along the continental shoreline, and hence they will be discussed in a later section. The gravitational influence of the earth has little effect on water movement in the sea, since the seawater is effected uniformly by the earth's gravitational force.

Of the forces discussed, the horizontal pressure gradients, internal friction, wind drag and the Coriolis force are of significance in large-scale seawater movement. Of these, horizontal pressure gradients and wind drag are of greatest importance. Coriolis force and internal friction becomes a factor in water movement only after movement has been generated by the former forces.

Seawater movements are generally the result of the interplay of the forces described, but in some cases it is possible to determine the dominant factor. For example, the trade winds are dominant in generating the east-to-west equatorial currents in the gyres of the Northern and Southern Hemispheres as well as the circumpolar current in the Southern Hemisphere. The Gulf Stream and Kuroshio are influenced to a greater extent by horizontal pressure gradients than by winds.

The preceding discussion has outlined the forces involved in the generation of ocean currents but has only covered the broad aspects of this type of activity. Many variations contribute to the whole of ocean current activity, of which *upwelling* is yet one more component.

Upwelling is a vertical movement of water caused by wind blowing parallel to a coastline. In this instance, movement of the water is greatly influenced by the contour of the coastline and by the shallow depth. The wind blowing parallel to the coast brings the surface water into motion. The resulting current will be deflected by the Coriolis force (depending upon the hemisphere), and where the current flows offshore, subsurface water from depths of 200 to 300 meters will come to the surface. The subsurface water compensates for the water moved by the wind, and a slow circulation from the depths is the result (fig. 19.6). Subsurface water is generally much colder than the surface water, influencing local coastal climates. Cool summers with frequent fogs are the result. Upwelling also plays a role in nutrient balance of the sea. The vertically rising waters bring important nutrients for phytoplankton to the surface, which in turn is important in maintaining an extensive food-fish population in these coastal waters. Upwelling occurs along the coast of northwest and southwest Africa, California, and Peru.

Downwelling can also occur where current flow toward shore is generated by wind blowing parallel to the coastline. This occurs in the Northern Hemisphere

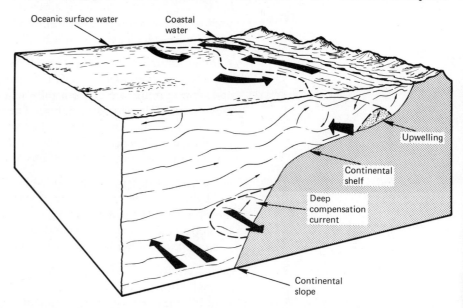

Figure 19.6 Relationship of surface and subsurface circulation in area where upwelling occurs.

where a north wind in blowing parallel to the east coast of a land mass; such is the case off the coast of Greenland.

19.4 TIDES

The tides are the alternate rise and fall of sea level—a phenomenon that can be readily observed along any shoreline. Although tides were recognized in antiquity, their cause was not known until Isaac Newton developed the universal law of gravitation. It then became apparent that the gravitational force of the moon—and to a lesser extent of the sun—attracted the waters on the earth's surface and together with the earth's rotation, created the tides.

The moon's influence is more than double that of the sun because of the moon's proximity to the earth. This has resulted in a variation in tidal magnitude according to the movement of the moon and its position relative to the sun. The largest (spring) tides occur during the full-moon phase, when the moon and sun are on opposite sides of the earth, or during the new-moon phase when the moon is directly between the earth and the sun. In these positions, the gravitational forces of the moon and sun reinforce each other. Lesser (neap) tides occur when the moon is in the first- or third-quarter phase (fig. 19.7).

The dominance of the moon in generating the tides permits us to discuss tidal action in terms of lunar movement and the rotation of the earth. The tides are measured on the basis of a lunar day of 24 hours and 50 minutes. This is the time required for the moon to arrive at the zenith the following day due to its eastward movement in orbit and the earth's rotation. Ideally, a high tide traveling across the earth would be attracted by and aligned directly under the moon (known as the *direct tide*). Friction if considered infinite between the earth's surface and ocean waters would cause the tidal bulges to shift eastward at the same rate as the earth rotates. Actually, the tidal bulge occurs at an equilibrium point between the force of the moon's gravitation and the earth's friction (fig. 19.8). Another tidal bulge occurs at the same time on the opposite side of the earth known as the *opposite tide*, thus resulting in two daily high tides approximately 12 hours and 25 minutes apart. Low tides are 90° behind high tides and occur approximately 6 hours and 12 minutes behind each high tide. This describes the semidiurnal tide which is the most familiar. In parts of the Gulf of Mexico and along the coast of Alaska and China, a diurnal tide phenomenon is experienced in that one high tide and one low tide occur each lunar day.

On the Atlantic coast of the United States, successive high tides will be at approximately the same level (fig. 19.9). Cyclical differences in the tide levels result from neap-tide and spring-tide occurrences. Tides on the Pacific coast of the United States alternate between higher high tides and lower high tides interspersed by higher low tides and lower low tides.

The explanation for the simultaneous occurrence of two high tides on opposite

The Oceans

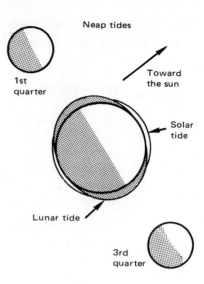

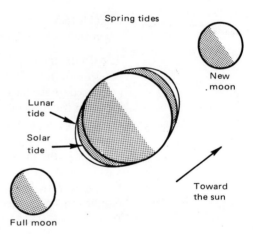

Figure 19.7 Relative positions of earth, moon, and sun during spring and neap tides.

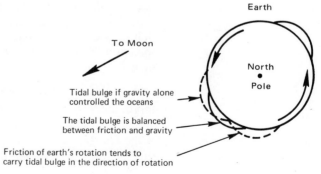

Figure 19.8 Influence of friction of the rotating earth and the gravitational attraction of the moon on the position of the ocean's tidal bulge.

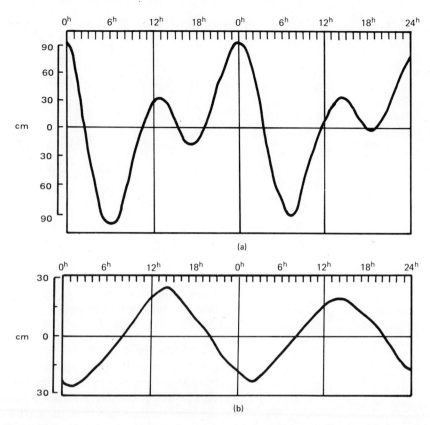

Figure 19.9 Variation in tidal configurations: (a) San Diego, California; (b) Pensacola, Florida.

sides of the earth can only be represented here in a somewhat oversimplified and imperfect fashion. It is first necessary to refer to the earth-moon motion around the barycenter (section 3.5a) and to emphasize that the earth revolves and does not rotate around this point. Therefore, all points on the earth describe equal circles with respect to the barycenter (fig. 19.10) and each circle thus formed has a radius equal to the distance from the center of the earth c to the barycenter b. The earth revolves around b, as does the moon, because of their mutual force of attraction due to gravity. However, the earth and the moon do not come together because of an opposing force (fig. 19.11)—the centrifugal force f—which is generated by their movement around the barycenter. It is the centrifugal force which causes the earth and moon to maintain their relative positions with respect to each other. The centrifugal force thus generated is equal for all points on the earth (fig. 19.11), because all these points are moving in the same radius and at the same speed. The moon's gravitational force is equal to the earth's centrifugal force at the earth's center but at no other point on the earth. The moon's force of attraction F is greater than the centrifugal force on that part of the earth's surface directly beneath the moon d and less on that portion of the earth opposite the moon o due to the difference in the distance of d and o to the moon. (Here, the inverse squares law

The Oceans

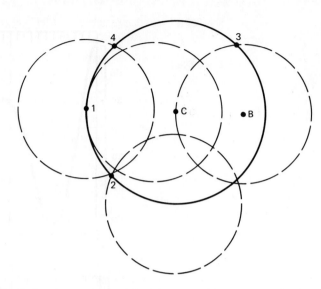

Figure 19.10 Each point on the earth's surface 1, 2, 3, 4, (or its interior) revolves around barycenter (b) with a radius equal to the distance from earth's center (c) to (b).

for gravity applies.) The difference in the magnitude of these forces t are what generate the direct and opposite tides (fig. 19.12). In reality, the vertical tide-producing force is very small, and the tidal bulge actually results from the combined effect of the tide-producing forces on all parts of the earth. The water is not lifted by the moon from the earth's surface. The earth (including the oceans) is elongated by the moon's gravitational attraction. The high rigidity of the

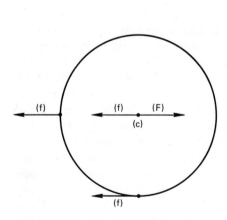

Figure 19.11 Centrifugal force generated by earth-moon revolution around the barycenter prevent the earth and the moon from being drawn together by gravity.

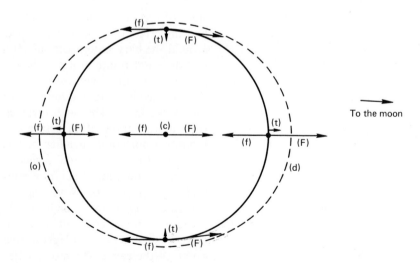

Figure 19.12 The difference in magnitude of centrifugal force and gravitational force generate direct and opposite tides.

The Restless Sea

solid earth permits it only a slight distortion, resulting in a tidal magnitude of a few centimeters. But the oceans, being fluid, move freely over the surface and pile up at sites directly under and on the opposite side of the earth from the moon. The magnitude of ocean tides may be less than a meter in most instances, but the contour of land masses and the funneling of tides into wide-mouth bays may cause sea-level changes of up to 20 meters. Tides also occur in lakes, but the amplitude of water-level change is not great. For example, tidal changes in Lake Erie amount to 8 centimeters. Even in the Mediterranean Sea, tides vary the sea level only about 30 centimeters.

While tides do not generate large-scale ocean currents, the expression, "sail with the tides" does indicate that small-scale currents do result from tidal action in harbors. Sailing vessels once depended upon tidal movement to make their entrance to or exit from a harbor. The rising tide or *flood tide* created tidal currents moving toward shore and into harbors, into rivers, and through some channels. The *ebb tide* is manifested by the retreat of water from the land and by a reversal of flow through a channel.

An example of one such channel flow occurs in the East River, which connects the lower harbor of New York City with Long Island Sound (fig. 19.13).

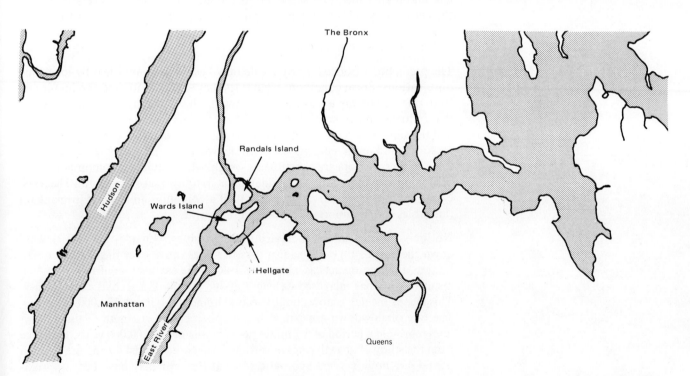

Figure 19.13 New York Harbor and Long Island Sound are connected by the East River. A narrow strait in the East River, Hell Gate, restricts movement of water resulting from differences in tide levels in the harbor and the sound.

The tide in the lower harbor crests several hours before high tide occurs in the Sound, and the difference in water level causes a strong current to flow north from the harbor to the Sound. Six hours later, the flow is reversed. A narrow strait in the East River restricts the movement of water, causing considerable turbulence. The alternate north-south flow of water through this strait at 8 to 9 kilometers per hour has earned it the name Hell Gate.

Occasionally conditions exist where a tide enters a large river or estuary in a wave. When the wave drags the river bottom, a wall of water like the surf rushes upstream. This *tidal bore*, as it is called, moves upstream at speeds of 5 to 15 kilometers per hour, and the wave may be 5 to 8 meters high. The bore at the entrance of the Amazon River sometimes reaches a height of 5 to 6 meters. Some rivers in Europe, such as the Seine, have been dredged, and the bores have virtually disappeared.

The Coriolis force also has an effect on the course of tides in some seas. Tidal currents are caused to rotate, because as the tidal wave moves, the Coriolis force deflects the wave to the right (to the left in the Southern Hemisphere). In a channel, this action will result in a higher high tide on the right-hand side of the channel (viewed in the direction of tidal-wave movement) than on the left, and higher low tide on the left as the tidal movement is reversed. This is the reason why tides of greater amplitude occur on the French coast than the English coast of the English Channel. These predictable tidal currents were and still are useful in navigation.

The prediction of tides for any locale may best be accomplished by long-time observations. Local conditions yield tides that may vary in time and amplitude from tides occurring a short distance away. This is the result of many factors that influence movement of the sea. These factors include not only the moon and the sun, but also the contours of the shoreline, depth of the sea, the earth's rotation, and even changes in seawater density and barometric pressure. Each of these factors are constituents which, when fed into a computer or tide-predicting machine, will yield tidal data for any place on earth. The U.S. Coast and Geodetic Survey operates such a computer and uses it to make up tide tables for all important harbors in the world.

Not only do tides influence navigational activities, but tides are also slowing down the earth's time mechanism. As the tides move over the shallow ocean bottom, the friction acts as a brake and slows the earth's rate of rotation. As a result the day has increased in length about 1/1000 of a second each century. This means the day is now about 1 second longer than it was 100,000 years ago. The rate of slowdown appears to be an extremely short length of time, but it does represent a period of 6 minutes per year during the 100,000 years. Evidence from the study of growth rings of certain fossil coral formed during the Silurian period 400 million years ago indicates that the year may have been approximately 400 days long. It is assumed that the earth's annual period of revolution has remained unchanged. The greater number of days during the same annual period indicates that the days were once shorter and have since increased in length.

The moon's distance is also affected by this type of activity and is gradually receding from the earth. Half a billion years ago, the moon was half its present distance. Now it is estimated to be receding approximately one centimeter per year. Ultimately, the moon will reach a point where the sun's gravitational influence upon the moon will be greater than that of the earth's and the moon will become an independent object in orbit around the sun.

What is the magnitude of the force responsible for slowing the earth's rotation and causing the moon to recede? It is difficult to determine the total, but K. F. Bowden, a British oceanographer, estimates that only 13 percent of the tidal energy coming into the English Channel from the Atlantic reaches the shore. The other 87 percent or an estimated 210 million horsepower is lost through friction on the shallow floor of the Channel. In reality, such energy is not lost but accomplishes work—in this instance resulting in a slow, constant change in the earth-moon relationship.

19.5 WAVE ACTIVITY

Restless winds blowing over thousands of miles of water create complex wave patterns that may ultimately escalate from small ripples into major storm waves and become a real hazard to man's activities. The same storm waves piling up water along a coast in a powerful surf can cause widespread damage to manmade structures as well as cause extensive erosion of the shoreline. Oceanographers are attempting to learn more about wave formation, the anatomy of waves, and about the power of wave activity.

19.5A Wave Formation

The sea is in constant motion: Even under conditions of absolute calm some *swells* may be detected. Swells are long, smooth waves observed at sea unrelated to any apparent storm activity. The swells are usually regular and uniform in shape and move in a lazy, rolling manner. Swells are not the result of local wind activity but rather are caused by winds which prevail elsewhere. Waves caused by winds do not subside when a storm is over, nor do the waves disappear when they move beyond the range of the storm. Wave action continues to move across the sea as a swell, diminishing in size with distance. Eventually the swells may break upon some distant shore, causing a high surf even in the absence of any local wind.

Most wave activity upon the sea is extraordinarily complex. The turbulence of the wind can change the smooth sea surface by raising ripples. Air impinging on the windward sides of these ripples will increase the wave activity and generate waves of ever-increasing height. The waves initially spread in all directions, but it appears that only those waves moving in the same direction as the wind are strengthened by it. Pressure is directed against the windward side of the crest and a pressure deficit prevails to leeward (fig. 19.14). This effect pushes the wave on the windward side and pulls the wave from the lee,

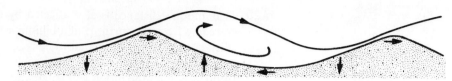

Figure 19.14 Wind pressure on the windward side of wave is augmented by a pressure deficit on the leeward side.

reinforcing the waves running with the wind. In order for the waves to grow in size, it is necessary that the energy supplied by the wind exceed the energy needed to overcome the internal friction of the water. The waves will continue to grow until they reach the size of wave capable of being generated by the prevailing speed of the wind. When wind speed is reduced, wave activity will die down accordingly, and the waves, moving beyond the influence of the wind, become swells.

Wave height is dependent upon the strength of the wind, the length of time it has been blowing, and the distance of the sea surface over which the wind has been blowing. Under these circumstances, the larger, faster-moving waves tend to overtake smaller waves, and their energies are combined to form still larger waves. In this manner, storm waves are formed that exceed 6 to 7 meters in height. Although waves of 12 to 15 meters do occur and are often spoken of, their occurrence is unusual and is the result only of very heavy storm winds. In the North Atlantic Ocean and the Antarctic Ocean, considered the most turbulent of the seas, waves in excess of 6 to 7 meters (up to 12 to 15 meters) occur on an average of 75 days per year. Waves of 20 to 30 meters have been reported but are extremely rare. The maximum height ever recorded was 34 meters (112 feet), observed in 1933 from a Navy tanker in the Pacific Ocean (fig. 19.15). Generally speaking, the equatorial Pacific and Atlantic oceans, as well as the Indian Ocean, have waves in excess of 4 meters only 36 days per year. This would indicate that large waves at sea tend to be the rare exception rather than the rule.

19.5B Anatomy of a Sea Wave

Characteristics of sea waves may differ from waves of light and sound. For example, light and sound waves are generally pictured as being symmetrical.

Figure 19.15 The maximum wave height ever recorded was 34 meters, measured from a naval vessel in 1933.

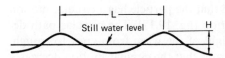

Figure 19.16 A symmetrical light or sound wave (sine wave) compared with a sea wave. A sea wave's crests are steeper and narrower than its troughs.

In the case of sea waves, the crests are steeper and narrower than the troughs (fig. 19.16) and the calm water level is closer to the bottom of the trough than to the top of the crest. Sea waves borne of the wind may be described in the same terms as are used for light and sound, but they are much more complex in structure than the classical description of light and sound waves found in physics texts. In speaking of sea waves, it is convenient to describe them in terms of simple wave motion, which in the case of sea waves is not one crest-to-crest segment but a running wave or wave train with a whole series of successive crests.

By way of quick review of wave characteristics, the *wavelength* (λ) represents the distance between two successive similar points in the wave as for example, from crest to crest. The *wave height* (H) is the distance from crest to trough. The length of time required for one wave length to pass to a given point is the *period* (P). The *wave speed* (C) or phase speed is the speed with which the wave moves. A relationship exists between these characteristics which may best be shown by an example. A wave moving past a fixed object such as a pole or dock piling will exhibit an alternate up and down motion, indicating the successive passing of crest and trough. The length of time between successive high points or crests is the period, and for this example it can be taken as 5 seconds. Now, if the rate of propagation, or speed, is 2 meters per second, then the first crest will have moved 2 meters per second for 5 seconds, or 10 meters. The 10 meters represents the distance that the first crest has apparently moved beyond the piling when the next crest passes it. Since this distance from crest to crest is one wavelength, it is possible to establish the following relationship:

$$\text{wavelength} = \text{wave speed} \times \text{period}$$

or

$$\lambda = C \times P$$

In any case where two of the three quantities are known, it is possible to calculate the third. Thus, if wavelength and period are known, wave speed may be determined. Wave speed (C) in meters per second may also be calculated if the wavelength in meters is known from the following relationship:

$$C = \sqrt{1.56\lambda}$$

There does not seem to be a relationship between wave height and any of the other characteristics. However, the ratio of wave height and length is sometimes measured as the steepness of the wave. This ratio is scarcely ever greater than 1:10.

An object floating on the surface appears to be little influenced by the movement of the wave train. Such an object will move in the direction of wave motion when at the crest and reverse its direction in the trough. The same is true for marine plants that can be seen immediately beneath the surface. These plants move with the wave when it crests and reverse again as the crest passes. Swimmers experience the same up-and-down, to-and-fro motion, which demonstrates that little water movement takes place in the direction of the wave train. It is the wave form that moves.

Viewing the above action in terms of the movement of a particle, we find that the particle moves in the direction of the wave train at the crest and down as the crest passes, or counter to the direction of the wave train in the trough and up as the next crest approaches. We find that the particle has a circular motion in a vertical plane during one wave period (fig. 19.17). The circular path described has a diameter equal to the height of the wave from crest to trough. Further study of this phenomenon indicates that the circular motion decreases with depth and vanishes at a depth equal to ½ wavelength.

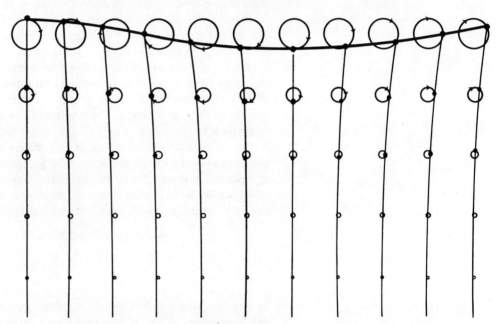

Figure 19.17 A water particle moves in the direction of the wave train at the crest and counter to the wave train in the trough. This action gives the particle a circular motion in a verticle plane during one wave period.

Close study of wave action indicates that the water particle motion in a wave is not truly circular, for it has been shown that the particle does not return exactly to the starting point. After each cycle, the particle will have moved slightly in the direction of the wave train. The reverse motion in the trough is not quite equal to the forward movement at the crest, resulting in a slight net forward motion. The amount of this displacement for one cycle (period) is approximately equal to the product of ten times the steepness and the wave height or:

$$\text{Displacement} = 10 \frac{H}{\lambda} H$$

Example:

If the wavelength is 50 meters and wave height is one meter, the displacement is as follows:

$$\text{Displacement} = 10 \frac{1 \text{ meter}}{50 \text{ meters}} 1 \text{ meter}$$

$$= 0.20 \text{ meters or 20 centimeters}$$

Following the example through, it is possible to determine wave speed as 8.8 meters per second and the period of the 50-meter wave as 5.7 seconds. Thus the displacement of the water particle in the direction of wave movement will be about 3.5 centimeters per second. This mechanism of water-particle movement is probably an important factor in the development of ocean currents generated by wind activity.

Wave activity of the type described occurs in deep water, which is for this purpose generally considered water deeper than ½ wavelength. When water is shallower, the wave will *sound* or touch bottom, interfering with the circular motion of the wave. The dominant motion then is to and fro, causing the orbital path of the water particle to become elliptical (fig. 19.18). As such a wave approaches the beach, the velocity and wavelength decreases, the wave height increases, and the period remains the same. The velocity at the bottom of the wave becomes less than the average velocity of the wave as a whole because of bottom friction. This causes the top of the wave to break and be transformed into a turbulent mass of water advancing on the beach (fig. 19.19). Usually the wave will break before its height is ⅐ of its wavelength.

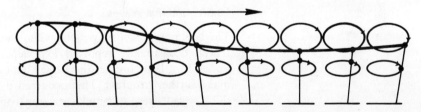

Figure 19.18 The path of a water particle becomes elliptical in shallow water.

Waves approaching a coastline have their speed affected by the depth. The waves are slowed when the depth shallows, and a long-crested wave approaching the shore at an oblique angle is gradually *refracted* or bent until the wave is almost parallel with the beach (fig. 19.20). The refraction of sea waves occurs in the same manner as the refraction of light. The wave front of a ray of light is analogous to the crest of a wave. When the light-wave front enters a denser medium (e.g., from air to water) the wave front is slowed as it touches the interface, thus bending the light ray. The same occurs when a sea wave approaches the shore, causing the direction of wave movement to become almost perpendicular to the shoreline.

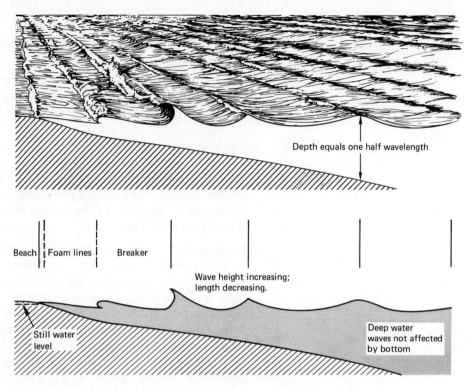

Figure 19.19 Friction reduces velocity of the bottom portion of a wave, causing the top to break in shallow water.

19.5C Internal Waves

The captains of sailing vessels, when these were the principal mode of ocean travel, occasionally reported experiencing a phenomenon known as *dead water*. In a light breeze, the vessels would stop dead in the water, unable to make further progress. This occurred mostly in or near arctic seas, where a thin layer of fresh or brackish water from melting ice formed on top of more dense saline water. The density discontinuity between the two layers of water is in some instances quite sharp, and internal waves in the denser layer of water have their surface at the interface. Slow-moving ships may generate internal waves when the upper layer of water is no deeper than the ship's draft. The exact mechanism for this is not completely understood but in a light breeze, a sailing vessel would remain stationary, as would a slow-moving powered ship, because the energy used to move the ship forward is dissipated in generating internal waves. At a slight increase in speed, the generation of internal waves ceases and the loss of speed due to internal waves stops. Huge internal waves are also generated by other causes, although there is some degree of uncertainty as to the mechanism. Storms, certain underwater currents, and tidal action are thought to be related to internal waves, which may be found at greater depths than those influencing surface craft.

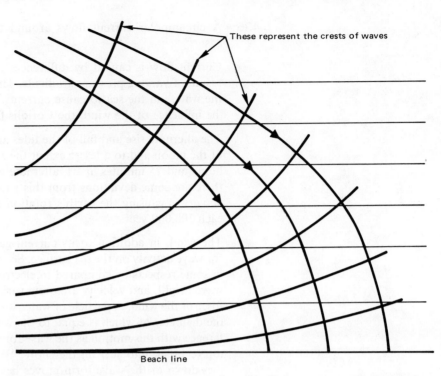

Figure 19.20 A series of wave crests are refracted in shallow water, turning waves almost parallel to the beach.

Internal waves differ from surface waves in several ways. First, the propagation of internal waves is much slower for the same wavelength than surface waves. This is due to the forces within the internal wave being weaker and its period longer, thus causing the wave's speed to be smaller. Another difference has to do with the direction of rotation of a particle within an individual wave. The rotation of a particle in an internal wave may be moving in a circular or elliptical orbit as does a particle in a surface wave. But a particle above the interface is moving in the opposite direction from a particle below the interface.

19.6 SUMMARY

Many factors influence the oceans of the earth, causing a variety of activities in the sea—some of which have been studied for centuries.

The existence of ocean currents have been recognized for hundreds of years because of their effect on shipping. A number of devices of varying degrees of sophistication are now used to determine with greater exactitude than in the past the many facets of ocean current flow. Major currents exist in all oceans, circulating clockwise north of the equator and counterclockwise to the south.

A circumpolar current flows around the Antarctic continent from west to east.

The Oceans

Current flow is caused by differences in density and temperature, and, as a result, of varying pressure gradients. These are internal forces which act upon the waters in the sea to cause currents. External forces affecting currents are the influence of the wind, the Coriolis force, and friction of the sea floor.

The alternate rise and fall of the tides are the result of the gravitational forces of the moon and to a lesser extent the sun. Two daily tides, approximately 12 hours and 25 minutes apart, alternate with two low tides each day, although there are some deviations from this pattern. Tidal friction resulting from this activity is causing the earth's rotation to slow down by one second per day each 100,000 years.

The wind, in addition to its current-generating function, is also responsible for wave activity on the sea surface. Sea waves are quite complex, but they may in some respects be compared to electromagnetic wave activity. Wave length, wave height, and velocity are quite variable in sea waves, depending upon the force of the wind. A particle of water within the wave moves in a circular orbit, the diameter of which is equal to the wave height from crest to trough. Interference with this motion as the wave form enters shallow water causes the wave top to break into surf on the shore. Waves do not disappear when the wind dies down or the wave form moves beyond the range of wind action. Waves continue to move across the sea as swells.

In some instances, seawater of different densities may be stratified, and internal waves are generated below the interface of two such layers. Such internal waves occasionally caused difficulties for sailing vessels by holding them stationary in the water if only a light breeze prevailed.

QUESTIONS

1. Briefly describe several means by which the movement of ocean currents are measured.
2. What is the pattern of ocean current flow in the Northern Hemisphere; the Southern Hemisphere?
3. How does horizontal density gradient, internal friction, wind drag, and Coriolis force contribute to large scale seawater movement?
4. How is upwelling initiated and of what importance is it?
5. Define: opposite tide, direct tide, spring tide, neap tide.
6. What forces are responsible for the direct and opposite tide, and how do these forces generate the tides?
7. Describe the mechanics of wave motion.
8. What causes the formation of the surf?

20 Water Pollution

The importance of water to human beings.
Water pollution—a world-wide problem.
Types of water pollution.
Solving the problem of water pollution.

Human beings have become a potent influence in nature—an influence which can no longer be ignored because of increasing population density and the resulting increase in pollution. George Perkins March, who wrote *Man and Nature* over one hundred years ago, pointed this out when he stated, "Man is everywhere a disturbing agent. Wherever he plants his foot the harmonies of nature are turned to discord." He also pointed out that the web of life was made up of every single living organism, and he warned that destruction of any part of that web might disrupt the whole biological community. Water is perhaps the most vital ingredient in this community.

20.1 WATER AND MAN

Water is central to human history. Civilizations have persisted or disappeared depending on whether there was enough, too much, or too little of this precious fluid. Our history and mythology are marked by catastrophic floods, droughts, famines, and periods of plenty. But until a century ago, our water-rich planet was rather sparsely populated and only the cataclysms seemed significant.

Water is also central to our view today, but the emphasis has changed. Society has become more conscious of the steady growth in the use of water by industry and population—an increasing demand on the planet's fixed supply of water. The catastrophic changes in the supply of water described by early historians were less dramatic and affected fewer people than the 1965 and 1966 droughts in the northeastern United States or the seasonal flooding in the West and Midwest. We are beginning to understand that our prosperity and prospects for survival vary with the amount and distribution of fresh water, and that each year there is no more water than before but millions more people to use it. It is beginning to be realized through harsh experience how easily we can pollute or obstruct or diminish the available water resources. As chemicals, synthetics, sprays, and gadgets proliferate in the interests of pest control, cleanliness, or merely to satisfy the passing whims of growing numbers of humans, we are altering the environment so drastically and at such a rapid rate that the experience of the father is often too late to help the son. "Nature never makes haste," Thoreau observed. "Her systems evolve at an even pace." Man's progress, on the other hand, has accelerated at an incredible rate.

In the early years of the United States, population was sparse and resources, especially water, were plentiful. There was little need for concern. But as population densities grew and the country became more and more industrialized, pollution of the waters became an increasingly greater problem. The public philosophy that grew out of the highly individualistic habits of the developing society allowed men to act without concern or responsibility with respect toward their environment and each other. Water pollution is one of the more undesirable by-products of this philosophy—and of the industrial and technological progress that has occurred particularly since the start of the twentieth century.

20.2 POLLUTED WATERS

Water pollution is anything that makes the water unfit or undesirable for use. Long stretches of most American rivers, particularly those in heavily populated areas, are shamefully dirty. The Merrimack, on which the earliest textile mills were built, has turned filthy brown and bubbles with nauseating gases. Along the lower reaches of the Mississippi, millions of fish died each winter for several consecutive years due to pollution. On the Hudson River near Troy, New York, scavenger eels attacked engineers taking samples of the water. The Missouri sometimes flows red with blood from slaughter houses, and carries greaseballs almost as big as footballs. The Mahoning, an important waterway in Pennsylvania and Ohio, is rust colored from the "pickle liquor" of steel mills, and it is strewn with debris along a good portion of its length. Most beaches on the Great Lakes, the largest freshwater bodies in the world, are closed to swimmers, and Lake Erie is considered to be a dying lake.

The pollution problem is not restricted to the United States but is a problem in all industrialized countries and in many of the underdeveloped countries. The Baltic Sea is rapidly going the way of Lake Erie, being poisoned by pesticides, mercury, and sewage. In Japan and in Sweden, deformities in babies

Water Pollution

have been blamed on mercury poisoning in fish. The fish in the Red Sea have been killed by chlorinated hydrocarbons leaking from barrels dumped by industry. Even the earth's oceans are not immune to pollution. Various forms of sea life have been found to contain such pollutants as DDT and mercury in their tissues, indicating the presence of these materials in seawater. In addition oil spills are a frequent offshore problem in many parts of the world.

The marine environment is particularly subject to pollution because most avenues of waste-water disposal terminate in the oceans. In the past, pollution of the oceans has been of little concern because the oceans have always been considered so large. However, most pollution occurs at the continental margins where human activities are centered, and the concentrated wastes remain in this region for varying lengths of time before they are dispersed into the vast open sea. Moreover, the potential for pollution is increasing as more of man's activity is concerned with the oceans. It was once thought that the rivers could not be seriously polluted, but the truth is now obvious. It is also becoming evident that large bodies of water such as the Great Lakes can be drastically altered and reduced in value as a natural resource. We have paid a great price to learn these lessons and should not make similar mistakes as the oceans become populated and are exploited.

20.3 FORMS OF POLLUTION

Pollution of water may take a variety of forms—some readily perceived, others unseen and somewhat insidious. Effective action in resolving the problem of water pollution first requires that the cause and source of the problem be recognized. Space here does not permit more than a brief discussion of an assortment of pollution sources and some of the efforts being made to alleviate the problems.

20.3A Sewage

Sewage disposal into waterways is probably one of the more critical pollution problems. The traditional method of dumping raw sewage into the nearest river or lake for dilution is no longer acceptable, and even disposing of treated sewage creates some unwanted problems.

Sewage is approximately 99.9 percent water, which can be purified for reuse, and 0.1 percent solids (fecal matter, soaps, detergents, fats, vegetable matter). It is this small percentage of solids which, if not treated, will use up oxygen in the stream and thus kill aquatic plants and fish.

Another source of danger is the many infectious organisms, usually transmitted in human wastes, which contaminate the streams and cause a serious health threat. For example, a sample of water tested from the Connecticut River near Hartford revealed the presence of twenty-six different forms of infectious bacteria of the types found in human wastes. Such pollution has made many beaches unsuitable for swimming and certainly unsafe for drinking or other domestic use.

Treatment of sewage for all forms of waste material is the only answer to the problem of sewage disposal. But sewage is being treated adequately for only about 60 percent of the urban population of the United States, whereas for the rural population little or no treatment is available. Urban sewer systems are frequently combined with storm-drain systems, and when these overflow, treatment plants must let the overflow, including raw sewage, pass unprocessed into the streams.

A number of techniques for eliminating sewage pollution are being investigated, but none may be suitable for all circumstances. The use of large storage tanks for holding overflow is being tested where combined sewer and storm-drain systems are used. During periods of flooding, the raw sewage would be held in huge tanks until the flood is past, at which time the sewage would be pumped to the treatment plant for processing. But this may not be adequate, especially where large systems are necessary. The issue may possibly be resolved only at tremendous expense by constructing separate sewer and storm-drain systems.

The primary problem with sewage disposal is that certain nutrients, notably phosphates and nitrates, are passed into waterways from treated and untreated sewage alike. Lake Erie is a prime example of what can occur when excessive amounts of these materials are discharged into the water. The lake is presently undergoing a process of *eutrophication*, or aging, from the over-enrichment of the water by phosphorous and nitrates which fertilize tremendous growths of algae. The algae uses enormous supplies of oxygen in its growth, and this has caused about 25 percent of the lake to become devoid of oxygen. The reduction of oxygen seriously upsets the ecological balance of the lake, since no aquatic organism can survive without oxygen in the water. Algae litters the shoreline in large unsightly masses and adds an objectionable taste to the drinking water drawn from the lake by many communities. During the summer, a 2000-square-kilometer area in the western basin is covered by an algae mass 60 centimeters thick.

A major step in saving Lake Erie and returning it to its previous healthy condition is the elimination of phosphates and nitrates from the sewage effluent. About seventy tons of phosphates are delivered into the lake daily, and about three-quarters of this comes from sewage. Approximately two-thirds of the phosphates in sewage stems from detergents, and research is now in progress to determine methods for removing phosphates from sewage. One promising technique is to precipitate phosphates with aluminum sulfate during the sewage treatment process and removing the precipitate with the sludge from the settling tanks. Tests are also being conducted to determine the feasibility of using the treated sewage effluent for irrigation of forest and crop lands. Plants of all forms could ultilize phosphates and nitrates in the effluent and permit clean water to return to the ground-water supply instead of being lost via the river to the sea.

Tests have been conducted at Pennsylvania State University for some years, and these tests have shown that trace elements in the effluent caused no ill effect to the soil.

Ground-water resources are also in danger of becoming polluted by sewage. The rapid growth of the population and the flight to the suburbs has overtaxed existing sewer facilities and the ability to construct and expand them. Septic tanks are therefore built in new housing projects and the discharge of detergents and human wastes from these tanks is seriously reducing the value of the ground-water supply.

20.3B Industrial Pollution

Industrial pollution is a principal source of controllable waterborne wastes brought about by the multitude of products manufactured for man's use. In terms of volume, wastes from manufacturing plants are about three times the volume of the nation's sewage. In addition, the volume of industrial wastes is rising at about 4.5 percent per year, or three times that of the increase in population. The primary-metals industries (steel, aluminum, copper) contribute the largest share of waste water in that they utilize about 30 percent of the water required for manufacturing in the United States. Chemical-industry wastes have the highest *biochemical oxygen demand* (BOD), which is a measure of the oxygen required to break down masses of contaminants put into the water. Most of the settleable and suspended solids are contributed by the food-processing and paper industries.

Many of the industrial wastes differ markedly in chemical composition and toxicity from wastes normally found in domestic sewage. Thus, the BOD or solids contents often are inadequate indicators of the quality of industrial effluent. Industrial wastes frequently contain strong acids or alkalies or persistent organics which resist treatment procedures normally applied to domestic sewage. When the cumulative quantity of these materials becomes great enough to exceed the natural neutralizing capacity of the stream, damage occurs. The severity of the damage depends upon the acidity of the affected waters and the effect upon fish, fish-food organisms, and the structures and equipment exposed to the acidified water. This means that the cost of pretreatment of water intended for municipal or industrial use is increased. It is estimated that over 4,300 miles of major streams in the United States are polluted significantly by acid drainage coming from mining and manufacturing activities. In addition to neutralizing acids and alkalies, some industrial effluents require that specific organic compounds be stabilized or that trace elements be removed as part of the treatment process.

20.3C Oil Pollution

One of the major industry-related problems is that of oil pollution. Oil in its many forms and in vast quantities is one of the necessities of a modern industrial society. Under control and serving its intended purpose, oil is an efficient, versatile product. Out of control, it can be a very devastating substance in the environment. Spilled into water, it spreads havoc for miles around. The destructive characteristics of oil out of control and the inadequacy of current measures for dealing with it have never been better illustrated than when the *Torrey Canyon*, with 119,000 tons of crude oil in her tanks, ran aground and broke up off the southern coast of England in March, 1967, or when submarine oil-well leakage occurred in the Santa Barbara Channel off the coast of California in 1969. The desperate efforts of the British and the French to cope with the first tragedy showed the shortcomings of existing control measures. This was matched by the less than completely successful efforts to clean up after the Santa Barbara channel spill in 1969.

Oil spills large and small, as well as the careless or accidental release of other hazardous materials into the environment, have long been of concern to those interested in the control of pollution. Once an area has been contaminated by oil, the whole character of the environment is changed. Afloat, even a relatively small quantity of oil goes where the water goes. By its nature, oil on water is a *seeker*. Once it has encountered something to cling to—whether it be a beach, a rock, a piling, or a bather's hair—it does not readily let go.

Cleaning up an oil-contaminated area is time consuming, difficult, and costly. To the costs of cleanup must be added the costs of the oil invasion itself—the destruction of fish and other wildlife, damage to property, contamination of public water supplies and beaches, and any number of other material and esthetic losses. Depending on the quantities and kinds of oil involved, these losses may extend for months or years, with correspondingly heavy costs of restoring the area to its prior condition.

The risks of pollution by oil and other hazardous substances are as numerous and varied as the uses of the many materials involved and the means of transporting them. These risks involve terminals, wayside chemical and other industrial plants, loading docks, refineries, barges, pipelines, tank cars, truck filling stations, in short, everywhere that oil is used, stored, or moved. All are subject to mechanical failures or to human carelessness and mistakes. There are countless opportunities for oil to get out of control.

The acute effects of an oil spill upon the marine environment by such a disaster as the grounding of the *Torrey Canyon* are economic to man but lethal to marine and bird life. Oil slicks on water seem to have an irresistible attraction for water birds. Once a bird alights on the oil mass, its feathers become matted and oil soaked. The almost inevitable result is death by drowning through loss of bouyancy, by toxicosis from ingested oil, or from exposure caused by loss of body heat insulation; or, because they are unable to fly, the birds may slowly starve to death or be eaten by predators.

Successful treatment of rescued birds is extremely difficult. A month after oil was spilled by the *Torrey Canyon*, the Nature Conservacy in London reported that 7,000 birds (guillemuts, puffins, shags, and herring gulls) had been rescued and treated, but only a few hundred survived. Near annihilation of several colonies of waterbirds was predicted in both England and France.

When surface-feeding fish swim into floating oil, their bodies and gills become coated. If death does not result from such contact, their flesh absorbs the odor-producing fractions of the oil, rendering the fish unfit for human consumption.

As the oil mass moves toward shore, toxic oil fractions can bring death to both larval and adult forms of invertebrate marine life that inhabit the shallow near-shore areas. Marine life valuable to man as a food resource may be totally destroyed, or at least it is likely to acquire disagreeable tastes and odors from the oil. Beds of seaweed, valuable as food or industrial material, can be totally destroyed by the oil.

Finally, as the oil hits the shoreline and collects in harbors and port areas, it blankets everything in its path. The usefulness of beaches for recreation suddenly ends. Navigational and fire hazards are created in harbors, ports, and marinas. Shorefront properties are despoiled, and the air reeks from the fumes.

When chemical compounds are used in the shallow or littoral areas to emulsify, precipitate, or sink oily materials, or otherwise cleanse the surface of the water, the effect on aquatic community may be more deadly than the floating oil itself.

Oil that has been recovered from the sea and from the shore areas presents yet another problem. If it is disposed of near drainages, the next rain could return the pollutant into water supplies and fishing grounds. If it is buried where it could affect ground water, future domestic water supplies could be polluted.

It is difficult to measure with precision the economic costs associated with an oil spill. Such projections involve assumptions concerning the extent and location of the spills, and the effectiveness of containment and cleanup. Some data does exist indicating that the potential magnitude of these costs may run into many millions of dollars annually.

Beyond these aspects lie potential costs whose magnitude cannot yet be readily estimated. A major oil spill or the accumulation of related polluting substances may destroy the potential store of food or other resources in the offshore waters.

20.3D Agricultural Pollution

One of the serious agriculturally related sources of water pollution is sedimentation—the transport of soil by water. Silting and discoloration of streams by sediments is a major pollution problem. Sedimentation also promotes other forms of pollution by acting as a vehicle to transmit to waterways such pollutants as pesticides, nutrients, and bacteria, as well as decomposable organic matter. The solution to this problem lies in the whole pattern of land-use activities required to fix soil in place, and soil conservation measures are practiced toward this end in many areas.

The movement of pesticides into streams and lakes is a well-recognized hazard which has accelerated since World War II. About 500,000 tons of various pesticide materials are currently being used annually in the United States, although the use of DDT has been restricted and all major uses are expected to be phased out by 1973.

Much of the pesticide material, as well as a good portion of the 25 million tons of fertilizer used annually, find their way into the streams and lakes. The pesticides have a deleterious effect on birds and marine life. For example, DDT has been linked to reproductive failures in bald eagles, pelicans, peregrine falcons, dungeness crabs, and other forms of wildlife. The fertilizers, on the other hand, contribute to the nutrient enrichment of waters, thereby stimulating algae growth and a subsequent reduction in oxygen in the water.

Pesticide control prospects seem to rest largely with stringent regulation of the types of materials allowed and of application practices. With accelerated research, it may be possible to develop degradable pesticides that break down readily in water or that do not produce harmful effects in wildlife and man.

One other formidable farm-related source of water pollution is the manure from the ever-increasing numbers of livestock that are being produced. On the farm, this waste can generally be disposed of by distributing it on the land, where the manure acts as a valuable soil additive. But disposal of the manure is a problem in the feedlots, where large numbers of animals are raised in a small area, sometimes near urban centers. Each animal produces about a ton of manure per month, and it has been determined that a feedlot raising 10,000 head of cattle produces the same sewage disposal problem as a city with a population of 165,000. It has been estimated that there were about 11 million cattle in feedlots in the United States in 1970, and the number is expected to increase.

Although installation costs of feedlots vary widely, the few feedlots with recommended treatment lagoons incurred costs ranging from one to five dollars per head of cattle. Thus, an operator setting up a feedlot with a 20,000-head capacity could spend a sizeable amount in construction of waste-treatment facilities. Although considerable research is needed in this area, it is known that the potential costs of controlling water pollution from feedlot operations is high.

20.3E Thermal Pollution

Thermal pollution refers generally to the degrading of streams by the addition of heat. The changes in water temperatures which result may affect aquatic life—either directly, through the harmful effect of the warmer water upon organisms which cannot tolerate such temperature increases, or indirectly, by lowering the dissolved-oxygen concentrations. Dissolved-oxygen levels in water decrease as temperature increases. Oxygen starvation and heat are suspected in the killing of millions of alewives in Lake Michigan in 1969. This is, however, not the only means by which thermal pollution affects marine life. In one experiment with brook trout, it was found that the trout became extremely lethargic in water at 20° C and were unable to catch their food. High water temperature may cause spawning salmon to lose their urge to spawn or hamper downstream migration of the young. It is not necessary for thermal pollution to raise water temperatures to a lethal level to destroy marine life. It is possible to eliminate a species by raising temperatures just enough to favor competition and disease.

There are several major sources of water-temperature increase. Temperature change can be brought about by reservoir impoundment, by depletion of a stream through diversion, by warming of irrigation return waters in fields, or by industrial process and thermal electric power generation. Thermal pollution resulting from power generation and industrial processes are the significant sources of thermal pollution. Water for cooling represents 70 percent of all water used for industrial purposes.

Thermal electric power generation is a far greater source of thermal pollution (in effect it is heat that is wasted to the cooling water) than is industrial processing. Electric-power production in this country has doubled every ten years during this century, and most of the increase has been achieved through use of thermal generating methods. The number of new plants has increased the total production of waste heat at a rate far in excess of the rate which growing generating efficiency has reduced unit heat loss. In addition, the discharge of heated cooling water at discharge locations has increased as power plants have grown in size. In the early 1970s, it is estimated that power plants discharged approximately 50 trillion gallons of hot water each year into streams, lakes, and oceans, and this is expected to double by 1980. Accordingly, thermal pollution has become an increasingly serious water-pollution problem. A major portion of the expected increase in heat discharge is being attributed to the increased construction of nuclear-fueled power plants with their attendant higher unit heat loss. These new power plants represent about half of the new generating capacity in the next decade.

20.3F Radiation Pollution

The rise in the rate of construction of nuclear power plants has been paralleled by the fear of radioactive contamination of the cooling water used in nuclear power plants and its discharge back to the source of the cooling water. Only one instance of serious contamination of a stream has thus far occurred, and this was not related to a nuclear power facility. The pollution resulted from the discharge of radioactive wastes from an ore mill into the Animas River, a tributary of the Colorado River. The many precautions taken in the construction and operation of nuclear power plants have thus far prevented radioactive materials from polluting the bodies of water into which cooling water is discharged. But accidents will occur, and some pollution of waterways by radioactive material can be anticipated. Awareness of the potential hazard should lead those concerned to take steps to minimize the danger.

The foregoing are some of the major contributors to water pollution, a list which is of necessity incomplete. It indicates most clearly that the triumphs of science and technology have not been the harbingers of unalloyed blessings. Indestructible materials may, over the short term, improve the quality of life, but they can result over the long run in the destruction of other vital resources, principal among which is water.

20.4 SOLVING THE PROBLEM

The chief cause of water pollution is the introduction of waste materials from a variety of sources. The reduction or elimination of these wastes requires an awareness of their existence, the means by which they are formed, and their general effect upon the quality of life, human and otherwise. Such an awareness could lead to the realization that water conservation is a key factor in reducing pollution and that a new set of values with regard to natural resources may be required. The new set of values would set quality above quantity and require that all segments of society take a responsible part in reestablishing a desirable environment.

Since a rapidly advancing technology is responsible for much of the pollution now in evidence, it would appear that this same technology could solve the problem. This is true to a large extent, but there are some conflicting opinions about the harmful effects of pollution that may delay positive and immediate action. For example, sewage is considered a major source of water pollution,

Figure 20.1 An oil skimmer designed to take in oil-water mixture at a rate of 6,000 liters per minute and concentrate the oil for pumping to a collection tank and ultimate disposal. Courtesy of Reynolds Metals Company.

Figure 20.2 A lightweight, corrugated aluminum boom, which may be stored on a small spool aboard a vessel or at dockside, is useful in containing small-scale oil spills. Courtesy of Reynolds Metals Company.

and approximately one billion gallons of sewage per day are discharged into the Pacific Ocean along the Southern California coast alone. Such a volume must have some impact upon the marine environment.

A survey made in the vicinity of the Los Angeles sewer outfall revealed that many of the economically important species such as lobster and abalone had mostly disappeared. This was attributed to a reduction in the marine algae caused by organic sediment from the sewage. The algae served as food for many of the fish that disappeared when the source of food was reduced. On the other hand, another study in the same area indicated that inorganic nutrients in the sewage stimulated the growth of microorganisms to four times that normally found in offshore water. This, in turn, attracted fish into these areas where the food they required was plentiful.

A difference of opinion also exists on the effect of thermal pollution on marine life. Several deleterious effects of thermal pollution on marine life were discussed above. However, in a pilot study being conducted in Great Britain and the United States, water that was warmer than normal stimulated growth of fish and shellfish. It appears that raising the water temperature above normal

Figure 20.3 Testing effluent from refinery which employs bacteria to consume waste material in water used to process petroleum products. Exxon Company, USA.

(it varies for each species) increases the metabolic rate and the rate of development, but that raising the temperature too high too fast is lethal. The problem appears to revolve around how much thermal pollution is permissible and what the environmental significance of the damage is. It may well be that the condition which forces the growth of fish may be desirable for producing food,

Water Pollution

but that it could be detrimental to the long-term maintenance of the species. Marine life in a natural environment would not survive for long if the water temperature were drastically changed.

Despite these apparent conflicts, it is best to assume the proposition that any water pollution, regardless of the amount, is harmful to the delicate balance that exists between the multitude of organisms in nature. Water pollution represents an alarming condition, not merely because polluted water insults both the eye and nose and prevents healthful recreation, but because in a densely populated, urbanized, and industrialized land, the use of rivers, lakes, and the sea solely for the transport and disposal of wastes is no longer tolerable. There are too many important uses for these waterways—particularly, for domestic water supply, irrigation, industrial uses, and recreation. Clean water is essential to our existence and to all other forms of life.

To remedy the problem of pollution will require comprehensive projects for pollution control, some of which have already been initiated. These projects seek to find answers to such questions as: How much water is available and what is it being used for? What are the sources and nature of the pollution? What is the quality of the water and how are present uses affecting it? What

Figure 20.4 A waste-water recovery system utilizing a small Acro-Pac module for reclaiming polluted water. Courtesy Aqua-Chem, Inc.

will be the water demands and resultant waste loadings in twenty to fifty years? What is required to make the used water reusable? How much will the necessary pollution-control measures cost, and how soon can they be implemented? As an initial step toward implementing some control over water pollution, an executive order by the president of the United States requiring federal permits for about 40,000 industries discharging waste into U.S. waterways was put into effect. By July 1, 1971, all industries were ordered to obtain a permit which required them to comply with state and federal water-quality standards. This action was taken under the 1899 Refuse Act, which covers industrial pollution of navigable waters. Not included in this step were municipal sewage systems, agricultural, and other nonindustrial waste dischargers. But new proposed legislation will bring these water users under permit requirements.

The reduction and ultimate prevention of pollution will be costly, but this expense must be weighed against the costs—both tangible and intangible—of continued pollution of the nation's waterways. On the whole, economics and good sense dictate the need to prevent water pollution rather than accept the costs of it occurring.

20.5 SUMMARY

Water is one of the most vital ingredients necessary for sustaining life on planet earth. Man needs fresh water, but a large percentage of the available water has been polluted by him in a variety of ways.

The disposal of sewage represents a major source of pollution, and the problem increases with the increase in population density. Untreated sewage is detrimental to man and other forms of animal life, while treated sewage has been found damaging to marine life. The treatment of sewage has not kept pace with the burgeoning urban areas, leading in some instances to contamination of ground-water supplies by seepage from increasing numbers of septic tanks.

Industrial pollution, stemming from the effluent of a multitude of manufacturing processes, is increasing faster than the increase in population. Many of these pollutants are highly toxic and difficult to dispose of. They will require expensive treatment units to return water to a usable state.

Oil pollution, a worldwide problem, has had a disastrous influence on the environment in a number of places. Not only is it unsightly, but it is extremely harmful to a variety of wildlife.

Agricultural pollution includes the transport of soil, as well as a variety of chemicals including fertilizers, herbicides, and pesticides, carried to streams by runoff. In the case of fertilizers, eutrofication of lakes is promoted. The herbicides and pesticides, notably DDT, are deposited in the tissue of a number of animals to their detriment. Animal manure is a problem as a pollutant in that much of this material finds its way into streams. This is especially true where animals are raised in feedlots.

Water Pollution

Thermal pollution is essentially a by-product of the industrial process where large amounts of water are utilized for cooling purposes. Excess heat is eliminated by transferring it to streams by means of circulating water. The heat will warm the streams, in some cases to a sufficiently high level to damage or kill the fish population.

Radiation pollution in water has not yet been a severe problem, but many are concerned over the effect of an accidental release of radioactive materials into a stream.

To solve the problem of water pollution will require an informed and concerned citizenry. It will require the cooperation of all those responsible for creating the pollution problem and a government to coordinate the activities necessary to return water standards to an acceptable level.

QUESTIONS

1. Summarize the effect sewage has had on a body of water like Lake Erie.
2. What type of sewage treatment facility does your community have? Is it adequate for future needs? Visit the treatment plant as a class or individually and find out about the problems of its operation. How pure is the effluent? Where does it go?
3. Define: eutrofication, biochemical oxygen demand (BOD), acid equivalent, septic tank.
4. What are the sources of oil pollution? Which of these exists in your locality?
5. Of what influence has water pollution been on recreation in your community?
6. What steps are being taken by local government to combat water pollution in your community?

Appendix I

Units of Measurement: Metric and English Equivalents

The Metric system of measurement is universally used except by the United States and a few of the small emerging nations of the world. The United States is gradually converting unofficially from the English to the Metric system. Official conversion may be accomplished within a decade. Those using this book will see the conversion become a reality during their lifetime. Even now many agencies of the federal government, private industries doing business with foreign governments or businesses, and those involved in scientific endeavors are using the Metric system.

The most commonly used units of length, mass, and volume in the Metric system are included here along with conversion to the more familiar English equivalents.

Length

1 meter = 100 centimeters = 1000 millimeters = 1.094 yards = 39.37 inches
1000 meters = 1 kilometer = 0.6214 miles
1 centimeter = 0.01 meter = 10 millimeters = 0.3937 inches
1 mile = 1.6093 kilometers
1 inch = 2.540 centimeters

Mass

1000 milligrams = 1 gram
1000 grams = 1 kilogram = 2.2046 pounds
1000 kilograms = 1 metric ton = 2,204.6 pounds
1 pound = 453.6 grams
1 ounce = 28.3495 grams

Volume

1 cubic centimeter = 1 milliliter
1000 milliliters = 1 liter

Appendix I

1 liter = 1.057 quarts
1 quart = 0.946 liters = 946 millileters
1 gallon = 3.785 liters

Appendix 2

Temperature Scales

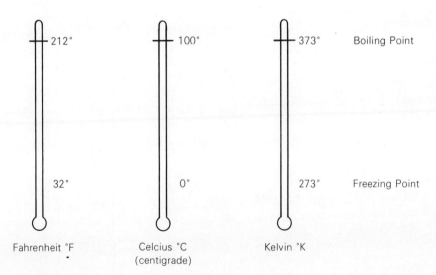

°F = 1.8°C + 32

°C = $\frac{(°F - 32)}{1.8}$

°K = °C + 273

Absolute zero = −273.18°C = −459.72°F = 0°K

Appendix 3

Nomograms of Height, Length, and Temperature

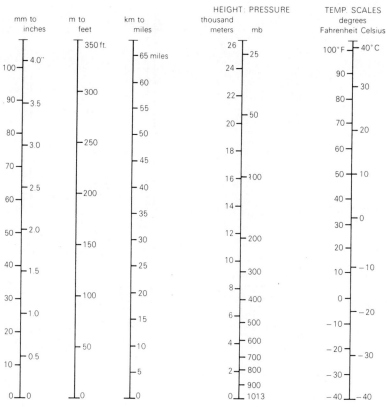

The left-hand scale of metric units can be read against those for inches, feet, or miles.

Appendix 4

The Beaufort Scale of Wind (Nautical)

Beaufort No.	Name of Wind	Wind Speed knots	km/hr	Description of Sea Surface	Sea Disturbance Number	Average Wave Height ft	m
0	Calm	<1	<1	Sea like a mirror.	0	0	0
1	Light air	1–3	1–5	Ripples with appearance of scales are formed, without foam crests.	0	0	0
2	Light breeze	4–6	6–11	Small wavelets still short but more pronounced; crests have a glassy appearance but do not break.	1	0–1	0–0.3
3	Gentle breeze	7–10	12–19	Large wavelets; crests begin to break; foam of glassy appearance. Perhaps scattered white horses.	2	1–2	0.3–0.6
4	Moderate breeze	11–16	20–28	Small waves becoming longer; fairly frequent white horses.	3	2–4	0.6–1.2
5	Fresh breeze	17–21	29–38	Moderate waves taking a more pronounced long form; many white horses are formed; chance of some spray.	4	4–8	1.2–2.4
6	Strong breeze	22–27	39–49	Large waves begin to form; the white foam crests are more extensive everywhere. Probably some spray.	5	8–13	2.4–4
7	Moderate gale	28–33	50–61	Sea heaps up and white foam from breaking waves begins to be blown streaks along the direction of the wind. Spindrift begins to be seen.	6	13–20	4–6
8	Fresh gale	34–40	62–74	Moderately high waves of greater length; edges of crests break into	6	13–20	4–6

Appendix IV

The Beaufort Scale of Wind (Nautical) —*continued*

Beaufort No.	Name of Wind	Wind Speed knots	km/hr	Description of Sea Surface	Sea Disturbance Number	Average Wave Height ft	m
				spindrift. The foam is blown in well-marked streaks along the direction of the wind.			
9	Strong gale	41–47	75–88	High waves. Dense streaks of foam along the direction of the wind. Sea begins to roll. Spray affects visibility.	6	13–20	4–6
10	Whole gale	48–55	89–102	Very high waves with long overhanging crests. The resulting foam in great patches is blown in dense white streaks along the direction of the wind. On the whole the surface of the sea takes on a white appearance. The rolling of the sea becomes heavy. Visibility is affected.	7	20–30	6–9
11	Storm	56–65	103–117	Exceptionally high waves. Small- and medium-sized ships might be for a long time lost to view behind the waves. The sea is covered with long white patches of foam. Everywhere the edges of the wave crests are blown into foam. Visibility is affected.	8	30–45	9–14
12–17	Hurricane	above 65	above 117	The air is filled with foam and spray. Sea is completely filled with driving spray. Visibility very seriously affected.	9	over 45	over 14

Source: After R. C. H. Russell and D. H. Macmillan (1954), *Waves and Tides*, London, Hutchinson's Sci. and Tech. Publ., p. 54, Table 7; and N. Bowditch (1958), U.S. Navy Oceanographic Office Publ. No. 9.

Appendix 5

MINERAL IDENTIFICATION KEY

METALLIC LUSTER	**Hard—Not Scratched by Knife**	Yellow brown to brown to black. H: 5.5–6.0, G: 3.5–4.0. Streak, yellow brown to brown. May also appear non-metallic and dark colored.	Limonite $Fe_2O_3 \cdot nH_2O$
		Red to brown, steel gray to black. H: 5–6, G: 5.3. Streak, reddish brown. May also appear non-metallic and dark colored.	Hematite Fe_2O_3
		Brown to black. H: 6, G: 5.2. Streak, black. Magnetic.	Magnetite $FeO \cdot Fe_2O_3$
		Brassy yellow. H: 6–6.5, G: 5. Streak, black. Occurs as cubes with striations.	Pyrite FeS_2
	Soft—Scratched by Knife	Brass yellow, may tarnish to purple. H: 3.5–4, G: 4.2. Streak, green black.	Chalcopyrite $CuFeS_2$
		Silvery gray, cubic structure. H: 2.5, G: 7.5. Streak, gray black.	Galena PbS
		Gray black, smudges fingers. H: 1, G: 2.2. Streak, black. May also appear as non-metallic dark colored.	Graphite C

Appendix V

MINERAL IDENTIFICATION KEY

			Description	Mineral
NON-METALLIC DARK COLORED	*Hard—Not Scratched by Knife*	*Prominent Cleavage*	Black to dark green. H: 5–6, G: 3.2–3.6, C: 2 planes at nearly 90°.	Augite Ca, Mg, Fe, Al Silicate
			Black to dark green. H: 5–6, G: 3–3.3, C: 2 planes at 60°.	Hornblende Ca, Na, Mg, Fe, Al silicate
		Poor to No Cleavage	Red to red brown. H: 6.5–7.5, G: 3.5–4.3. Fractures may resemble cleavage. Streak white.	Garnet Fe, Mg, Ca, Al silicate
			Green to yellow, granular masses, generally glassy luster. H: 6.5–7, G: 3.3–3.4. Streak, white.	Olivine $(MgFe)_2SiO_4$
			Red to brown. H: 7, G: 2.65. Conchoidal fracture.	Jasper SiO_2
			Vary colored. H: 5.5–6.5, G: 1.9–2.2. Conchoidal fracture.	Opal $SiO_2 \cdot nH_2O$
	Soft—Scratched by Knife	*Prominent Cleavage*	Black to dark brown. H: 2.5–3, G: 2.7–3.2. Excellent cleavage one plane in thin sheets.	Biotite K, Mg, Fe, Al silicate
			Yellowish brown to brown. H: 3.5–4, G: 3.9–4.2. Good cleavage, 6 planes. Resinous luster. Streak, white to yellow.	Sphalerite ZnS
			Variety of colors to black. H: 4, G: 3.2. Good cleavage, cubic crystals.	Fluorite CaF_2
			Green to black green. H: 2–2.5, G: 2.6–2.9. Streak, white. Cleavage in one plane.	Chlorite Mg, Fe, Al silicate
		Poor to No Cleavage	Green, brown, purple, blue. H: 5, G: 3.1. Streak, white.	Apatite Ca, F, phosphate
			Yellow brown to black. H: 5–5.5, G: 3.5–4. Streak, yellow brown. Earthy luster.	Limonite $Fe_2O_3 \cdot nH_2O$
			Red to brown. H: 5–6, G: 5.2. Streak, red brown. Earthy luster.	Hematite Fe_2O_3

Appendix V

MINERAL IDENTIFICATION KEY

NON-METALLIC LIGHT COLORED	**Hard—Not Scratched by Knife** — *Prominent Cleavage*	Light pink or flesh colored. H: 6–6.5, G: 2.6–2.75. Cleavage 2 planes at nearly 90°.	Orthoclase $K(AlSi_3O_8)$
		White to gray. H: 6–6.5, G: 2.6–2.75. Cleavage 2 planes at nearly 90°.	Plagioclase $NaCa(AlSi_3O_8)$
	Poor to No Cleavage	White (milky), rose, smoky. H: 7, G: 2.65. Vitreous luster, conchoidal fracture.	Quartz SiO_2
		Various colors. H: 7, G: 2.65. Waxy luster, conchoidal fracture.	Chalcedony SiO_2 (agate)
	Soft—Not Scratched by Knife — *Prominent Cleavage*	White, colorless. H: 2, G: 2.2. Streak, white. Cubic cleavage. Salty taste.	Halite $NaCl$
		Colorless to white. H: 2, G: 2.32. May be platy, fibrous or massive.	Gypsum $CaSO_4 \cdot 2H_2O$
		Colorless, white to yellow. H: 3, G: 2.72. Cleavage, 3 planes not at right angles. Effervesces in acid.	Calcite $CaCO_3$
		Colorless, white to pink. H: 3.5–4, G: 2.85. Cleavage 3 planes not at right angles. Powder effervesces in acid.	Dolomite $CaMg(CO_3)_2$
		Colorless to yellowish brown. H: 2–2.5, G: 2.76–3. Cleavage in one plane in thin sheets.	Muscovite K,Al silicate
		White to green. G: 2.6. Fibrous.	Asbestos Mg,Al silicate
		White to green. H: 1, G: 2.7. Pearly luster. Feels slippery.	Talc Mg silicate
		White to yellow and green. H: 4, G: 3.2. Good cleavage, cubic crystals.	Fluorite CaF_2
	Poor to No Cleavage	White, earthy luster. H: 2–2.5, G: 2.6.	Kaolinite Al silicate
		White to green. H: 1, G: 2.7. Pearly luster. Feels slippery.	Talc Mg silicate
		Yellow to greenish. H: 1.5–2.5, G: 2. Streak, yellow, resinous luster.	Sulfur S

Glossary

Å Abbreviation for angstrom, a unit of length, 10^{-8} cm.

Ablation As applied to glacier ice, the process by which ice below the snow line is wasted by evaporation and melting.

Abrasion Erosion of rock material by friction of solid particles moved by water, ice, wind, or gravity.

Abyssal Referring to the bottom waters of the ocean.

Abyssal plain A very flat portion of the ocean floor underlaid by sediments. The slope of this feature is less than 1:1000.

Activity, solar Increased temperature, strongly enhanced magnetic fields, etc., in localized areas of the sun.

Adiabatic process A process in which no heat passes into or out of a substance.

Aftershock An earthquake that follows a larger earthquake and originates at or near the focus of the larger earthquake. Generally, major shallow earthquakes are followed by many aftershocks. These decrease in number as time goes on, but may continue for many days or even months.

Albedo A measure of the part of the incoming solar radiation that is reflected from the earth and the atmosphere.

Alluvial fan The land counterpart of a delta. An assemblage of sediments marking the place where a stream moves from a steep gradient to a flatter gradient and suddenly loses its transporting power. Typical of arid and semiarid climates, but not confined to them.

Alpha particle A helium atom lacking electrons and therefore having a double positive charge.

Alpine glacier A glacier confined to a stream valley. Usually fed from a cirque. Also called *valley glacier* or *mountain glacier*.

Amphibole group Ferromagnesian silicates with a double chain of silicon-oxygen tetrahedra. Common example: hornblende. Contrast with *pyroxene group*.

Anaerobic A condition where oxygen is absent in a body of water. The Black Sea is an example.

Andesite A fine-grained igneous rock with no quartz or orthoclase, composed of about 72 percent plagioclase feldspars and the balance ferromagnesian silicates. Important as lavas, possibly derived by fractional crystallization from basaltic magma. Widely characteristic of mountain-making processes around the borders of the Pacific Ocean. Confined to continental sectors.

Anemometer An instrument for measuring the force or the speed of wind.

Aneroid barometer A barometer on which the change of pressure is indicated by the motion of the elastic top of a box from which the air has been partly exhausted. In contrast to the mercury barometer, the aneroid contains no liquid.

Angstrom A unit of length, equal to one hundred-millionth of a centimeter, 10^{-8} cm. Abbreviation, Å.

Angular momentum A vector quantity, the product of mass times radius of orbit times velocity. The energy of motion of the solar system.

Anticline A configuration of folded, stratified rocks in which the rocks dip in two directions away from a crest, as the principal rafters of a common gable roof dip away from the ridgepole. The reverse of a *syncline*. The "ridgepole" or crest is called the axis.

Anticyclone System of winds blowing around a center of high pressure.

Aphotic zone That part of the ocean where not enough light is present for photosynthesis by plants.

Aquaculture Farming of the ocean, whereby organisms such as fish, algae, and shellfish are grown under controlled conditions. At present this technique is only used in nearshore areas.

Aquifer A permeable material through which ground water moves.

Arete A narrow, saw-toothed ridge formed by cirques developing from opposite sides into the ridge.

Arroyo Flat-floored, vertically-walled channel of an intermittent stream typical of semiarid climates. Often applied to such

Glossary

features of southwestern United States. Synonymous with *wadi* and *wash*.

Artesian water Water that is under pressure when tapped by a well and is able to rise above the level at which it is first encountered. It may or may not flow out at ground level.

Asbestos A general term applied to certain fibrous minerals that display similar physical characteristics although they differ in composition. Some asbestos have fibers long enough to be spun into fabrics with great resistance to heat, such as those used for automobile brake linings. Types with shorter fibers are compressed into insulating boards, shingles, etc. The most common asbestos mineral (95 percent of U.S. production) is chrysotile, a variety of serpentine, a metamorphic mineral.

Asymmetric fold A fold in which one limb dips more steeply than the other.

Atmosphere The gaseous envelope surrounding any heavenly body, especially the earth.

Atoll A ring of low coral islands arranged around a central lagoon.

Atom A combination of protons, neutrons, and electrons. Ninety-two kinds are found in nature; 102 kinds are now known.

Atomic energy Energy associated with the nucleus of an atom. It is released when the nucleus is split, or is derived from mass that is lost when two nucleii are fused together.

Atomic number The number of positive charges on the nucleus of an atom; the number of protons in the nucleus.

Atomic size The radius of an atom (average distance from the center to the outermost electron of the neutral atom). Commonly expressed in angstroms.

Aureole A zone in which contact metamorphism has taken place.

Authigenic deposits Deposits that have formed in place before the sediment is buried. They usually have precipitated directly from the seawater.

Autotrophic bacteria Bacteria that produce their own food from inorganic compounds.

Axial plane A plane through a rock fold that includes the axis and divides the fold as symmetrically as possible.

Axis The ridge, or place of sharpest folding, of an anticline or syncline.

Barchan A cresent-shaped dune with wings or horns pointing downwind. Has a gentle windward slope and steep lee slope inside the horns. About 100 feet in height and 1,000 feet wide from horn to horn. Moves with the wind at about 25 to 50 feet per year across a flat, hard surface where a limited supply of sand is available.

Baroclinic A state in which the surfaces of equal density are inclined relative to the isobaric surfaces.

Barograph A self-recording barometer, usually of the aneroid type.

Barometer An instrument for measuring atmospheric pressure. The pressure is determined by the weight of the air column. See also *aneroid barometer*.

Barrier island A low, sandy island near the shore and parallel to it, on a gently sloping offshore bottom.

Barrier reef A reef that is separated from a landmass by a lagoon of varying width and depth, opening to the sea through passes in the reef.

Basalt A fine-grained igneous rock dominated by dark-colored minerals, consisting of more than 50 percent plagioclase feldspars and the balance ferromagnesian silicates. Basalts and andesites represent about 98 percent of all extrusive rocks.

Base level For a *stream,* a level below which it cannot erode. There may be temporary base levels along a stream's course, such as those established by lakes, or resistant layers or rock. Ultimate base level for a stream is sea level. For a *region,* a plane extending inland from sea level sloping gently upward from the sea. Erosion of the land progresses toward this plane, but seldom, if ever, reaches it.

Batholith A discordant pluton that increases in size downward, has no determinable floor, and shows an area of surface exposure exceeding 40 square miles. Compare with *stock*.

Bathyal That part of the ocean between approximately 200 to 2,000 meters in depth.

Bathymetry Measurement of the ocean depth.

Bay barrier A sandy beach, built up across the mouth of a bay, so that the bay is no longer connected to the main body of water.

Bed load Material in movement along a stream bottom, or, if wind is the moving agency, along the surface. Contrast with material carried in suspension or solution.

Bedding A collective term used to signify the existence of beds or layers in sedimentary rocks. Sometimes synonymous with *bedding plane*.

Benthic or benthonic The area of the ocean bottom inhabited by marine organisms.

Bergschrund The gap or crevasse between glacier ice and the headwall of a cirque.

Berm Flat portion of a beach formed by wave action.

Berms In the terminology of coastlines, berms are stormbuilt beach features that resemble small terraces; on their seaward edges are low ridges built up by storm waves.

Binary star Twin star; two stars, presumably formed together and gravitationally bound to each other.

Binding energy The amount of energy that must be supplied to break an atomic nucleus into its component fundamental particles. It is equivalent to the mass that disappears when fundamental particles combine to form a nucleus.

Bioluminescence The production of light by living organisms as a result of a chemical reaction.

Biomass The amount of living organisms in grams per unit area or unit volume.

Biosphere A collective term for the area of habitat of the organisms of the earth.

Biotope An area where the principal habitat conditions and the living forms adapted to the conditions are uniform.

Black body Idealized light source; the energy distribution inherent in its radiation spectrum is described by a specific relation between brightness, temperature, and wavelength (Planck's law).

Glossary

Blowout A basin, scooped out of soft, unconsolidated deposits by the process of deflation. Ranges from a few feet to several miles in diameter.

Body wave Push-pull or shake earthquake wave that travels through the body of a medium, as distinguished from waves that travel along a free surface.

Bottomset bed Layer of fine sediment deposited in a body of standing water beyond the advancing edge of a growing delta. The delta eventually builds up on top of the bottomset beds.

Boulder size A volume greater than that of a sphere with a diameter of 256 millimeters, or 10 inches.

Boulder train A series of glacier erratics from the same bedrock source, usually with some property that permits easy identification. Arranged across the country in the shape of a fan with the apex at the source and widening in the direction of glacier movement.

Bowen's reaction series A series of minerals for which any early-formed phase tends to react with the melt that remains, to yield a new mineral further along in the series. Thus early-formed crystals of olivine react with remaining liquids to form augite crystals; these in turn may further react with the liquid then remaining to form hornblende.

Braided stream A complex tangle of converging and diverging stream channels separated by sandbars or islands. Characteristic of flood plains where the amount of debris is large in relation to the discharge.

Breccia A clastic sedimentary rock made up of angular fragments of such size that an appreciable percentage of the volume of the rock consists of particles of granule size or larger.

Caldera A roughly circular, steep-sided volcanic basin with a diameter at least three or four times its depth. Commonly at the summit of a volcano. Contrast with *crater*.

Calving As applied to glacier ice, the process by which a glacier that terminates in a body of water breaks away in large blocks. Such blocks form the icebergs of polar seas.

Capacity The amount of material that a transporting agency such as a stream, a glacier, or the wind can carry under a particular set of conditions.

Cavitation A process of erosion in a stream channel caused by sudden collapse of vapor bubbles against the channel wall.

Cementation The process by which a binding agent is precipitated in the spaces between the individual particles of an unconsolidated deposit. The most common cementing agents are calcite, dolomite, and quartz. Others include iron oxide, opal, chalcedony, anhydrite, and pyrite.

Central vent An opening in the earth's crust, roughly circular, from which magmatic products are extruded. A volcano is an accumulation of material around a central vent.

Chalcedony A general name applied to fibrous cryptocrystalline silica, and sometimes specifically to the brown, translucent variety with a waxy luster. Deposited from aqueous solutions and frequently found lining or filling cavities in rocks. *Agate* is a variety with alternating layers of chalcedony and opal.

Chemical energy Energy released or absorbed when atoms form compounds. Generally becomes available when atoms have lost or gained electrons, and often appears in the form of heat.

Chemical weathering The weathering of rock material by chemical processes that transform the original material into new chemical combinations. Thus chemical weathering of orthoclase produces clay, some silica, and a soluble salt of potassium.

Chert Granular cryptocrystalline silica similar to flint but usually light in color. Occurs as a compact massive rock, or as nodules.

Chute or Chute Cutoff As applied to stream flow, the term "chute" refers to a new route taken by a stream when its main flow is diverted to the inside of a bend. A chute occurs as a trough formed by the deposition of material on the inside of the bend where water velocities are reduced.

Cinder Rough, slaglike fragment from four to thirty-two millimeters across, formed from magma blown into the air during an eruption.

Cinder cone Built exclusively or in large part of pyroclastic ejecta dominated by cinders. Parasitic to a major volcano, it seldom exceeds 500 meters in height. Slopes up to 30° to 40°. Example: Paricutin.

Cirrus A form of high cloud having diverging filaments like locks of hair or wool.

Cirque A steep-walled hollow in a mountainside at high elevation, formed by ice-plucking and frost action, and shaped like a half-bowl or half-amphitheater. Serves as principal gathering ground for the ice of a valley glacier.

Cleavage, mineral A property, possessed by many minerals, of breaking in certain preferred directions along smooth plane surfaces. The planes of cleavage are governed by the atomic pattern, and represent directions in which atomic bonds are relatively weak.

Climate The weather conditions which, in combination, characterize a region or a place.

Closed system One that has negligible interaction with its surroundings.

Cloud, interstellar Localized volume in which gas and dust are denser than average in the universe.

Cluster, open and globular Group of stars, presumably formed at about the same time from one large cloud of uncondensed material; open and globular clusters differ in spatial arrangement, age, and other characteristics. Cluster of galaxies is a grouping of galaxies presumably bound together by gravitational interaction.

Coal A sedimentary rock composed of combustile matter derived from the partial decomposition and alteration of cellulose and lignin of plant materials.

Cobble size A volume greater than that of a sphere with a diameter of 64 millimeters (2.5 inches), and less than that of a sphere with a diameter of 256 millimeters (10 inches).

Col A pass through a mountain ridge. Created by the

Glossary

enlargement of two cirques on opposite sides of the ridge until their headwalls meet and are broken down.

Color Term referring to and resulting from the wavelength of light. Because of the relation between the surface temperature of, for example, a star, and the wavelengths of light of its surface, the color is indicative of the temperature of the light source.

Columnar jointing A pattern of jointing that blocks out columns of rock. Characteristic of tabular basalt flows or sills.

Community An integrated group of organisms inhabiting a common area. These organisms may be dependent on each other or possibly upon the environment. The community may be defined by its habitat or by the composition of the organisms.

Compaction Reduction in pore space between individual grains as a result of pressure of overlying sediments or pressures resulting from earth movement.

Competence The maximum size of particle that a transporting agency, such as a stream, a glacier, or the wind, can move.

Composite volcanic cone Composed of interbedded lava flows and pyroclastic material. Characterized by slopes of close to 30° at the summit, reducing progressively to 5° near the base. Example: Mayon.

Conchoidal fracture A mineral's habit of breaking in which the fracture produces curved surfaces like the interior of a shell (conch). Typical of glass and quartz.

Concordant pluton An intrusive igneous body with contacts parallel to the layering or foliation surfaces of the rocks into which it was intruded.

Concretion An accumulation of mineral matter that forms around a center or axis of deposition after a sedimentary deposit has been laid down. Cementation consolidates the deposit as a whole, but the concretion is a body within the host rock that represents a local concentration of cementing material. The enclosing rock is less firmly cemented than the concretion. Commonly spheroidal or disk-shaped, and composed of such cementing agents as calcite, colomite, iron oxide, or silica.

Conglomerate A detrital sedimentary rock made up of more or less rounded fragments of such size that an appreciable percentage of the volume of the rock consists of particles of granule size or larger.

Connate water Water that was trapped in a sedimentary deposit at the time the deposit was laid down.

Constellation Area on the celestial sphere, used to identify the approximate position of celestial objects. The term originally referred to sets of bright stars which were thought to outline mythological beings or animals in the sky.

Contact metamorphism Metamorphism at or very near the contact between magma and rock during intrusion.

Continental crust Portion of the earth's crust composed of two layers: first layer, sialic rock, 10 to 15 miles thick; second layer, simatic rock, 10 to 15 miles thick.

Continental drift A theory that an original single continent, sometimes referred to as *Pangaea*, split into several pieces that "drifted" laterally to form the present-day continents.

Continental glacier An ice sheet that obscures mountains and plains of a large section of a continent. Existing continental glaciers are on Greenland and Antarctica.

Continental margin That portion of the ocean adjacent to the continent and separating it from the deep sea. The continental margin includes the continental shelf, continental slope, and continental rise.

Continental rise An area of gentle slope (usually less than a half-degree or 1:100) at the base of the continental slope.

Continental shelf Shallow, gradually sloping zone from the sea margin to a depth where there is a marked or rather steep descent into the depths of the ocean down the continental slope. The seaward boundary of the shelf is about 200 meters (100 fathoms) in depth, but may be either more or less than this.

Continental slope Portion of the ocean floor extending from about 200 meters (100 fathoms), at the seaward edge of the continental shelves, to the ocean deeps. Continental slopes are steepest in their upper portion, and commonly extend more than 400 meters (2,000 fathoms) downward.

Convection A mechanism by which material moves because its density is different from that of surrounding material. The density differences are frequently brought about by heating.

Convection cell A pair of *convection currents* adjacent to each other.

Convergence Flowing together. In fluid motion the term has the strict meaning of physical shrinking. Thus there is convergence if the motion is such that a volume of the fluid shrinks. Similarly, in two dimensions, there is convergence if the area within a closed chain of particles decreases with time. The converse of convergence is divergence; the latter is postive, and convergence is negative.

Core The innermost zone of the earth. Surrounded by the mantle.

Cosmology Study of the universe at large, its structure, and its development.

Crater A roughly circular, steep-sided volcanic basin with a diameter less than three times its depth. Commonly at the summit of a volcano. Contrast with *caldera*.

Creep As applied to soils and surficial material, slow downward movement of a plastic type. As applied to elastic solids, slow permanent yielding to stresses that are less than the yield point if applied for a short time only.

Crevasse A deep crevice or fissure in glacier ice, or a breach in a natural levee.

Crust The outermost zone of the earth, composed of solid rock between 30 to 45 km. thick. Rests on the mantle, and may be covered by sediments.

Cryptocrystalline A state of matter in which there is actually an orderly arrangement of atoms characteristic of crystals, but in which the units are so small (that is, the material is so fine-grained) that the crystalline nature cannot be determined with the aid of an ordinary microscope.

Crystal A solid with orderly atomic arrangement. May or may not develop external faces that give it crystal form.

Crystal form The geometrical form taken by a mineral, giving an external expression to the orderly internal arrangement of atoms.

Crystalline structure The orderly arrangement of atoms in a crystal. Also called crystal structure.

Crystallization The process through which crystals separate from a fluid, viscous, or dispersed state.

Cumulus Heap-shaped cloud or set of rounded masses of cloud heaped on each other.

Cyclone System of winds blowing around a center of low pressure.

Debris slide A small, rapid movement of largely unconsolidated material that slides or rolls downward to produce an irregular topography.

Decibar A measure of pressure equal to 1/10 normal atmospheric pressure and approximately equal to the pressure change of one meter depth in sea water.

Deep-sea trenches See *island arc deeps*.

Deflation The erosive process by which the wind carries off unconsolidated material.

Degeneracy Particular state of matter, at great densities, in which component parts such as electrons or neutrons are fixed in their lowest energy states.

Delta A plain underlaid by an assemblage of sediment that accumulates where a stream velocity is reduced as it flows into a body of standing water. Originally so named because many deltas are roughly triangular in shape, like the Greek letter delta (Δ), with the apex pointing upstream.

Dendritic pattern An arrangement of stream courses that, on a map or viewed from the air, resembles the branching habit of certain trees, such as oaks or maples.

Density current A current due to differences in the density of seawater from place to place, caused by changes in temperature and variations in salinity or the amount of material held in suspension.

Density, electron (ion) Number of electrons (ions) per unit volume. Density (of matter): mass per unit volume.

Desalination A variety of processes whereby the salts are removed from seawater, resulting in water that can be used for human consumption.

Desiccation Loss of water from pore spaces of sediments through compaction, or through evaporation caused by exposure to air.

Detrital deposits Sedimentary deposits resulting from erosion and weathering of rocks.

Detrital sedimentary rocks Rocks formed from accumulations of minerals, and rocks derived either from erosion of previously existing rock, or from the weathered products of these rocks.

Diamond A mineral composed of the element carbon. Used as a gem and industrially in cutting tools. The hardest substance known.

Diatomaceous ooze A siliceous deep-sea ooze made up of the cell walls of one-celled marine algae known as diatoms.

Dike A tabular discordant pluton.

Dip The acute angle that a rock surface makes with a horizontal plane. The direction of the dip is always perpendicular to the strike. See *magnetic declination*.

Disc, galactic The flattened portion of our star system (galaxy).

Disc, solar Projection of the sun onto the celestial sphere.

Discharge With reference to stream flow, the quantity of water that passes a given point in unit time. Usually measured in cubic feet per second, abbreviated *cfs*.

Discordant pluton An intrusive igneous body with boundaries that cut across surfaces of layering or foliation in the rocks into which it has been intruded.

Distributary channel or stream A river branch that flows away from a main stream and does not rejoin it. Characteristic of deltas and alluvial fans.

Diurnal Referring to tides, one low and one high tide within one lunar day (about 24 hours, 50 minutes).

Divergence The flow of water in different directions away from a particular area or zone; often associated with areas of upwelling.

Divide Line separating two drainage basins.

Doldrums Region of calms and light baffling winds near the equator.

Double star Two stars that appear very close to each other on the celestial sphere; most such stars are unrelated, their closeness being a consequence of the projection effect.

Dome An anticlinal fold without a clearly developed linearity of crest, so the beds involved dip in all directions from a central area, like an inverted but usually distorted cup. The reverse of a basin.

Drift Any material laid down directly by ice, or deposited in lakes, oceans, or streams as a result of glacial activity. Unstratified glacial drift is called *till* and forms *moraines*. Stratified glacial drift forms *outwash plains*, *eskers*, *kames*, and *varves*.

Drumlin A smooth, streamlined hill composed of till. Its long axis is oriented in the direction of ice movement. The blunt nose points upstream and a gentler slope tails off downstream with reference to the ice movement. In height, drumlins range from 10 meters to 25 meters, with the average somewhat less than 30 meters. Most drumlins are a quarter to a half mile in length. The length is commonly several times the width. Diagnostic characteristics are the shape and the composition of unstratified glacial drift, in contrast to kames, which are of random shapes and stratified glacial drift.

Dune A mound or ridge of sand piled by wind.

Dust Interstellar mixture of very small solid grains, such as ice crystals, metal particles, graphite.

Dwarf Designation for stars that are significantly smaller than some comparison group.

Earthquake Waves in the earth generated when rocks break after

Glossary

being distorted beyond their strength.

Eccentricity (of an ellipse) Describes degree of an ellipse's departure from circularity.

Ecliptic The apparent path of the sun in the heavens; the plane of the planets' orbit.

Elasticity A property of materials that defines the extent to which they resist small deformation from which they recover completely when the deforming force is removed. Elasticity = stress/strain.

Element A unique combination of protons, neutrons, and electrons that cannot be broken down by ordinary chemical methods. The fundamental properties of an element are determined by its number of protons. Each element is assigned a number that corresponds to its number of protons.

End moraine A ridge or belt of till marking the farthest advance of a glacier. Sometimes called *terminal moraine*.

Energy The capacity for producing motion. Energy holds matter together. It can become mass, or can be derived from mass. It takes such forms as kinetic, potential, heat, chemical, electrical, and atomic energy, and can be changed from one of these forms to another.

Entrenched meander A meander cut into underlying bedrock when regional uplift allows the originally meandering stream to resume downward cutting.

Envelope Gas cloud surrounding a star, a star system, a galaxy, etc.

Erosion A term that describes the physical and chemical breakdown of a rock and the movement of these broken or dissolved particles from one place to another.

Erosional flood plain A flood plain that has been created by the lateral erosion and the gradual retreat of the valley walls.

Erratic In the terminology of glaciation, an erratic is a stone or boulder carried by ice to a place where it rests on or near bedrock of different composition.

Esker A widening ridge of stratified glacial drift, steepsided, 3 to 30 meters high, and from a fraction of a mile to more than a hundred miles in length.

Eustatic change of sea level A change in sea level produced entirely by an increase or a decrease in the amount of water in the oceans, hence worldwide.

Estuary A semienclosed coastal body of water having a free connection with the open sea and within which sea water is diluted by fresh water derived from land drainage.

Euphotic zone That part of the water which receives sufficient sunlight for plants to be able to photosynthesize.

Evaporation The process by which a liquid becomes a vapor at a temperature below its boiling point.

Evaporite A rock composed of minerals that have been precipitated from solutions concentrated by the evaporation of solvents. Examples: rock salt, gypsum, anhydrite.

Exfoliation The process by which plates of rock are stripped from a larger rock mass by physical forces.

Exosphere The outermost layer of the atmosphere.

Extragalactic object One whose position is outside the Milky Way star system.

Extrusive rock A rock that has solidified from a mass of magma that poured or was blown out upon the earth's surface.

Fathom A common unit measure of depth equal to 6 feet (1.83 meters).

Fault A surface of rock rupture along which there has been differential movement.

Fault-block mountain A mountain bounded by one or more faults.

Field (electric, gravitational, magnetic, etc.) The presence of forces.

Field, radiation The presence of photons.

Fiord A glacially deepened valley that is now flooded by the sea to form a long, narrow, steep-walled inlet.

Firn Granular ice formed by the recrystallization of snow. Intermediate between snow and glacier ice. Sometimes called *névé*.

Fissility A property of splitting along closely spaced parallel planes more or less parallel to the bedding. Its presence distinguishes shale from mudstone.

Fission Breakup of large atomic nuclei into smaller units.

Flare, solar Multifaceted phenomenon in the solar atmosphere.

Flare, stellar Postulated phenomenon in stellar atmospheres corresponding to solar flare.

Flood plain Area bordering a stream, over which water spreads in time of flood.

Flux, energy (radiation) Amount of energy (radiation) passing through a unit surface area per unit time.

Focus The source of a given set of earthquake waves.

Focus (of a conic section) Reference point of a conic section.

Focus (of a telescope) Point of convergence of light after reflection by a mirror, or after passage through a lens.

Fold A bend, flexure, or wrinkle in rock produced when the rock was in a plastic state.

Foliation A layering in some rocks caused by parallel alignment of minerals. A textural feature of some metamorphic rocks. Produced rock cleavage.

Food chain A complex system that involves many different organisms, each of which is the food for an organism higher up in the chain or sequence.

Footwall One of the blocks of rock involved in fault movement. The one that would be under the feet of a person standing in a tunnel along or across the fault. Opposite the hanging wall.

Foreset beds Inclined layers of sediment deposited on the advancing edge of a growing delta or along the ice slope of an advancing sand dune.

Foreshock A relatively small earthquake that precedes a larger earthquake by a few days or weeks and originates at or near the focus of the larger earthquake.

Fossil Evidence of past life, such as the bones of a dinosaur, the shell of an ancient clam, the footprint of a long extinct

Glossary

animal, or the impression of a leaf in a rock.

Fossil fuels Organic remains (once living matter) used to produce heat or power by combustion. Includes petroleum, natural gas, and coal.

Fracture As a mineral characteristic, the way in which a mineral breaks when it does not have cleavage. May be conchoidal (shell-shaped) fibrous, hackly, or uneven.

Fracture zone A large linear and irregular area of the sea floor, characterized by ridges and seamounts. These features are commonly associated with the median ridge common to most ocean basins.

Fringing reef A reef attached directly to a landmass.

Front A sloping boundary between two air masses of different temperature. Originally, the term polar front was used to indicate the forward side of an advancing mass of polar air.

Galaxy Very large group of stars tied together by mutual gravitational interaction, usually with a very complex structure. All stars visible to the naked eye belong to a small portion of our galaxy.

Gas A state of matter that has neither independent shape nor volume, can be compressed readily, and tends to expand indefinitely. In geology, the word "gas" is sometimes used to refer to natural gas, the gaseous hydrocarbons that occur in rocks, dominated by methane. Compare with use of the word *oil* to refer to petroleum.

Geode A roughly spherical, hollow or partially hollow accumulation of mineral matter from a few inches to more than a foot in diameter. An outer layer of chalcedony is lined with crystals that project inward toward the hollow center. The crystals, often perfectly formed, are usually quartz, although calcite and dolomite are also found and, more rarely, other minerals. Geodes are most commonly found in limestone, and more rarely in shale.

Geologic time-scale A chronologic sequence of units of earth time.

Geology An organized body of knowledge about the earth. It includes both *physical geology* and *historical geology* (q.v.).

Geomagnetic poles The dipole best approximating the earth's observed field is one inclined $11\frac{1}{2}°$ from the axis of rotation. The points at which the ends of this imaginary magnetic axis intersect the earth's surface are known as the geomagnetic poles. They should not be confused with the magnetic, dip poles, or the virtual geomagnetic poles.

Geomorphology The study of the shape of the earth's surface and the processes that control and modify these features.

Geophysics The physics of the earth.

Geosyncline Literally, an "earth syncline." The term now refers, however, to a basin in which thousands of feet of sediments have accumulated, with accompanying progressive sinking of the basin floor explained only in part by the load of sediments. Common usage of the term includes both the accumulated sediments themselves and the geometrical form of the basin in which they are deposited. All folded mountain ranges were built from geosynclines, but not all geosynclines have become mountain ranges.

Geyser A special type of thermal spring which intermittently ejects its water with considerable force.

Giant Designation for stars that are significantly larger than some comparison group.

Glacier ice A unique form of ice developed by the compression and recrystallization of snow, and consisting of interlocking crystals.

Globular cluster Specific type of star cluster, characterized by highly symmetrical structure.

Globule Small, dense area in an interstellar cloud involved in star formation.

Gondwanaland Hypothetical continent thought to have broken up in the Mesozoic. The resulting fragments are postulated to form present-day South America, Africa, Australia, India, and Antarctica. See *Laurasia*.

Gradation Leveling of the land. This is constantly being brought about by the forces of gravity and such agents of erosion as water at the surface and underground, and wind, glacier ice, and waves.

Graphic structure An intimate intergrowth of potassic feldspar and quartz with the long axes of quartz crystals lining up parallel to a feldspar axis. The quartz part is dark and the feldspar is light in color, so the pattern suggests Egyptian hieroglyphs. Commonly found in pegmatites.

Gravity The force of attraction that causes objects on earth to fall toward the center of the earth. The universal law of gravitation as first given by Newton states that every particle in the universe attracts every other particle with a force that is proportional to the product of their masses, and inversely proportional to the square of the distances between the particles.

Gravity fault A fault in which the hanging wall appears to have moved downward relative to the footwall. Also called *normal fault*.

Groundmass The finely crystalline or glassy portion of a porphyry.

Ground moraine Till deposited from a glacier as a veneer over the landscape and forming a gently rolling surface.

Groundwater Underground water within the zone of saturation.

Groundwater table The upper surface of the zone of saturation for underground water. It is an irregular surface with a slope or shape determined by the quantity of ground water and the permeability of the earth materials. In general, it is highest beneath hills and lowest beneath valleys. Also referred to as *water table*.

Guyot A flat-topped *seamount* rising from the floor of the ocean like a volcano but planed off on top and covered by appreciable depth of water. Synonymous with *tablemount*.

Halo Spherical space surrounding a galaxy, populated by globular cluster or cosmic rays.

Halocline A zone, usually 50 to 100 meters below the surface and extending to a depth of perhaps 1,000 meters, where the salinity changes rapidly. The salinity change is greater in the halocline than in the water above or below it.

Hanging valley A valley that has a greater elevation than the valley to which it is tributary, at the point of their junction. Often (but not always) created by a deepening of the main valley by a glacier. The hanging valley may or may not be glaciated.

Hanging wall One of the blocks involved in fault movement. The one that would be hanging overhead for a person standing in a tunnel along or across the fault. Opposite the footwall.

Hardness A mineral's resistance to scratching on a smooth surface. The Mohs scale of relative hardness consists of ten minerals. Each of these will scratch all those below it in the scale and will be scratched by all those above it: (1) talc, (2) gypsum, (3) calcite, (4) fluorite, (5) apatite, (6) orthoclase, (7) quartz, (8) topaz, (9) corundum, (10) diamond.

Heterotrophic bacteria Bacteria that use organic material, produced by other organisms, for their food.

Historical geology The branch of geology that deals with the history of the earth, including a record of life on the earth as well as physical changes in the earth itself.

Horn A spire of bedrock left where cirques have eaten into a mountain from more than two sides around a central area. Example: Matterhorn of the Swiss Alps.

Hot spring A spring that brings hot water to the surface. A *thermal spring*. Water temperature usually is 15°F or more above mean air temperature.

Hydrologic cycle The general pattern of movement of water from the sea by evaporation to the atmosphere, by precipitation onto the land, and by movement under the influence of gravity back to the sea again.

Hydrothermal solution A hot, watery solution that usually emanates from a magma in the late stages of cooling. Frequently contains and deposits in economically workable concentrations minor elements that, because of incommensurate ionic radii or electronic charges, have not been able to fit into the atomic structures of the common minerals of igneous rocks.

Hygroscope An instrument that indicates changes of moisture.

Ice sheet A broad, moundlike mass of glacier ice of considerable extent with a tendency to spread radially under its own weight. Localized ice sheets are sometimes called *icecaps*.

Infiltration The soaking into the ground of water on the surface.

Interface The boundaries or surfaces between two different materials; the major interfaces in the ocean are water-atmosphere, water-biosphere, and water-sediment.

Intermittent stream A stream that carries water only part of the time.

Intrusive rock A rock solidified from a mass of magma that invaded the earth's crust but did not reach the surface.

Island arc deeps Arcuate trenches bordering some of the continents. Some reach depths of 10,000 meters or more below the surface of the sea. Also called deep-sea trenches or trenches.

Isostasy The ideal condition of balance that would be attained by earth materials of differing densities if gravity were the only force governing their heights relative to each other.

Joint A break in a rock mass where there has been no relative movement of rock on opposite sides of the break.

Juvenile water Water brought to the surface or added to underground supplies from magma.

Kame A steep-sided hill of stratified glacial drift. Distinguished from a drumlin by lack of unique shape and by stratification.

Kame terrace Stratified glacial drift deposited between a wasting glacier and an adjacent valley wall. When the ice melts, this material stands as a terrace along the valley wall.

Karst topography Irregular topography characterized by sinkholes, streamless valleys, and streams that disappear into the underground, all developed by the action of surface and underground water in soluble rock such as limestone.

Kettle A depression in the ground surface formed by the melting of a block of ice buried or partially buried by glacial drift, either outwash or till.

Kilometer A metric measure of distance equal to 1,000 meters, 0.62 statute miles, or 0.54 nautical miles.

Knot A unit of velocity equal to one nautical mile (6,080 feet) per hour. It is approximately equal to 50 cm/sec or 1.69 ft/sec.

Laccolith A concordant pluton that has domed up the strata into which it was intruded.

Lagoon A shallow pond or lake separated from the sea by a shallow bar or bank.

Landslide A general term for relatively rapid mass movement, such as slump, rock slide, debris slide, mudflow, and earthflow.

Lateral moraine A ridge of till along the edge of a valley glacier. Composed largely of material that fell to the glacier from valley walls.

Laurasia Hypothetical continent whose resulting fragments are postulated to constitute North America, Europe, and part of Asia. See *Gondwanaland*.

Lava Magma that has poured out onto the surface of the earth, or rock that has solidified from such magma.

Leeward The direction toward which the wind is blowing.

Levee (natural) Bank of sand and silt built by a river during floods, where the suspended load is deposited in greatest quantity close to the river. The process of developing natural levees tends to raise river banks above the level of the surrounding flood plains. A break in a natural levee is sometimes called a crevasse.

Light Form of energy vested in photons. Some light is visible to the human eye, other "light" is not (ultraviolet, infrared, X-ray, etc.).

Limb One of two parts of an anticline or syncline on either side of the axis.

Lithification The process by which unconsolidated rock-forming materials are converted into a consolidated or coherent state.

Littoral The benthic zone

between high tides out to a depth of about 200 meters.

Load The amount of material that a transporting agency, such as a stream, a glacier, or the wind, is actually carrying at a given time.

Loess An unconsolidated, unstratified aggregation of small, angular mineral fragments, usually buff in color. Generally believed to be wind-deposited. Characteristically able to stand on very steep to vertical slopes.

Longitudinal dune A long ridge of sand oriented in the general direction of wind movement. A small one is less than 3 meters in height and 75 meters in length. Very large ones are called *seif dunes*.

Longshore currents Currents in the nearshore region that run essentially parallel to the coast.

Long waves Earthquake surface waves.

Magma A naturally occurring silicate melt, which may contain suspended silicate crystals or dissolved gases, or both. These conditions may be met in general by a mixture containing as much as 65 percent crystals, but no more than 11 percent dissolved gases.

Mantle The intermediate zone of the earth. Surrounded by the crust, it rests on the core at a depth of about 2,900 km.

Marsh gas Methane, CH_4, the simplest paraffin hydrocarbon. The dominant component of natural gas.

Massive pluton Any pluton that is not tabular in shape.

Matter Anything that occupies space. Usually defined by describing its states and properties: solid, liquid, or gaseous; possesses mass, inertia, color, density, melting point, hardness, crystal form, mechanical strength, or chemical properties. Composed of atoms.

Meander *n.*, A turn or sharp bend in a stream's course. *v.i.*, To turn, or bend sharply. Applied to stream courses in geological usage.

Mechanical weathering The process by which rock is broken down into smaller and smaller fragments as the result of energy developed by physical forces. Also known as *disintegration*.

Medial moraine A ridge of till formed by the junction of two lateral moraines when two valley glaciers join to form a single ice stream.

Meniscus The convex top of mercury (or other liquids) in a glass tube.

Mesosphere An atmospheric layer between the stratosphere and the thermosphere.

Metamorphic rock "Changed-form rock." Any rock that has been changed in texture or composition by heat, pressure, or chemically active fluids after its original formation.

Metamorphic zone An area subjected to metamorphism and characterized by a certain metamorphic mineral that formed during the process.

Metamorphism A process whereby rocks undergo physical or chemical changes, or both, to achieve equilibrium with conditions other than those under which they were originally formed. Weathering is arbitrarily excluded from the meaning of the term. The agents of metamorphism are heat, pressure, and chemically active fluids.

Metasomatism A process whereby rocks are altered when volatiles exchange ions with them.

Meteorology Study of the state and processes of the atmosphere.

Mineral A naturally occurring solid element or compounds, exclusive of biologically formed carbon components. It has a definite composition, or range of composition, and an orderly internal arrangement of atoms known as crystalline structure, which gives it unique physical and chemical properties, including a tendency to assume certain geometrical forms known as *crystals*.

Mohorovičić discontinuity The sharp change in seismic velocity occurring at about 11 kilometers depth in the ocean and 35 kilometers depth under land, that defines the top of the earth's mantle. This discontinuity, commonly called the *Moho*, may represent either a chemical or a phase change in the layering of the earth.

Monadnock A hill left as a residual of erosion, standing above the level of a peneplain.

Monsoon A term for seasonal winds, usually applied to the changing wind patterns in the Indian Ocean.

Moraine A general term applied to certain landforms composed of till.

Mountain Any part of a landmass that projects conspicuously above its surroundings.

Mountain glacier Synonymous with *alpine glacier*.

Mountain range A series of more or less parallel ridges, all of which were formed within a single geosyncline or on its borders.

Mudcracks Cracks caused by the shrinkage of a drying deposit of silt or clay under surface conditions.

Mudflow Flow of a well-mixed mass of rock, earth, and water that behaves like a fluid and flows down slopes with a consistency similar to that of newly mixed concrete.

Neap tide Low tides which occur about every two weeks when the moon is in its quarter positions. See *spring tides*.

Nebula Gas cloud, visible to the naked eye, or detectable in visible light by means of telescopes.

Névé Granular ice formed by the recrystallization of snow. Intermediate between snow and glacier ice. Sometimes called *firn*.

Nimbus Cloud from which rain falls.

Nodule An irregular, knobby-surfaced body of mineral that differs in composition from the rock in which it is formed. Silica in the form of chert or flint is the major component of nodules. They are commonly found in limestone and dolomite.

Normal fault A fault in which the hanging wall appears to have moved downward relative to the footwall. Opposite of a thrust fault. Also called *gravity fault*.

Oblateness Flattening of a celestial object, such as a planet or a star, due to its rotation.

Occlusion A complex front which forms when a cold front overtakes a warm front; also, a cyclone having such a front.

Ocean basin That portion of the ocean seaward of the continental margin which includes the deep-sea floor.

Glossary

Oceanic crust Portion of the earth's crust composed of one layer of simatic rock 30 to 45 kilometers thick.

Oil In geology, refers to petroleum.

Oölites Spheroidal grains of sand size, usually composed of calcium carbonate, $CaCO_3$, and thought to have originated by inorganic precipitation. Some limestones are made up largely of oölites.

Open cluster Star cluster located in a spiral arm of a galaxy.

Orbit Path of a celestial object.

Ore A metaliferous mineral deposit.

Outwash Material carried from a glacier by meltwater. Laid down in stratified deposits.

Outwash plain Flat or gently sloping surface underlaid by outwash.

Overturned fold A fold in which at least one limb is overturned—that is, has rotated more than 90°.

P Symbol for earthquake primary waves.

Paired terraces Terraces that face each other across a stream at the same elevation.

Paleomagnetism The study of variations in the earth's magnetic field, as recorded in ancient rocks.

Pangaea A hypothetical continent from which all others are postulated to have originated through a process of fragmentation and drifting.

Parabolic dune A dune with a long, scoop-shaped form that, when perfectly developed, exhibits a parabolic shape in plan, with the horns pointing upwind. Contrast *barchan*, in which the horns point downwind. Characteristically covered with sparse vegetation, and often found in coastal belts.

Peat Partially reduced plant or wood material containing approximately 60 percent carbon and 30 percent oxygen. An intermediate material in the process of coal formation.

Pediment Broad, smooth erosional surface developed at the expense of a highland mass in an arid climate. Underlaid by beveled rock, which is covered by a veneer of gravel and rock debris. The final stage of a cycle of erosion in a dry climate.

Pelagic A division of the marine environment, including the entire mass of water. The pelagic environment can be divided into a neritic (water that overlies the continental shelf) province and an oceanic (the water of the deep sea) province.

Perched water table The top of a zone of saturation that bottoms on an impermeable horizon above the level of the general water table in the area. Generally near the surface, it frequently supplies a hillside spring.

Permeability For a rock or an earth material, the ability to transmit fluids. Permeability for underground water is sometimes expressed numerically as the number of gallons per day that will flow through a cross section of one square foot, at 60°F, under a hydraulic gradient of 100 percent. Permeability is equal to velocity of flow divided by hydraulic gradient.

Perturbation of an orbit Deviation of a celestial object from its path.

Phenocryst A crystal significantly larger than the crystals of surrounding minerals.

Physical geology The branch of geology that deals with the nature and properties of material composing the earth, distribution of materials throughout the globe, the processes by which they are formed, altered, transported, and distorted, and the nature and development of landscape.

Phytoplankton Those organisms carried by the ocean that are plants. See *plankton*.

Piedmont glacier A glacier formed by the coalescence of valley glaciers and spreading over plains at the foot of the mountains from which the valley glaciers came.

Pirate stream One of two streams in adjacent valleys that has been able to deepen its valley more rapidly than the other, has extended its valley headward until it has breached the divide between them, and has captured the upper portion of the neighboring stream.

Plankton Floating or weakly swimming organisms that are carried by the ocean currents. The plankton range in size from microscopic plants to large jellyfish.

Plankton bloom A large concentration of plankton within an area, due to a rapid growth of the organisms. Large numbers of plankton can color the water, causing in some instances a red tide.

Playa The flat-floored center of an undrained desert basin.

Playa lake A temporary lake formed in a playa.

Plunge The acute angle that the axis of a folded rock mass makes with a horizontal plane.

Pluton A body of igneous rock that is formed beneath the surface of the earth by consolidation from magma. Sometimes extended to include bodies formed beneath the surface of the earth by the metasomatic replacement of older rock.

Porosity The percentage of open space of interstices in a rock or other earth material. Compare with *permeability*.

Porphyry An igneous rock containing conspicuous phenocrysts in a fine-grained or glassy groundmass.

Pothole A hole ground in the solid rock of a stream channel by sands, gravels, and boulders caught in an eddy of turbulent flow and swirled for a long time over one spot.

Precipitation The discharge of water, in the form of rain, snow, hail, sleet, fog, or dew, on a land or water surface. Also, the process of separating mineral constituents from a solution by evaporation (halite, anhydrite) or from magma to form igneous rocks.

Pressure The force per unit area upon an object.

Primary shorelines Shorelines where the coastal region has been mainly formed by terrestrial agents, such as rivers, glaciers, deltas, volcanoes, folding, and faulting.

Primary wave Earthquake body waves that travel fastest and advance by a push-pull mechanism. Also known as longitudinal, compressional, or P-waves.

Prominence Relatively cool,

dense gas suspended in magnetic fields in the solar corona.

Proper motion Change in position of an object on the celestial sphere; that is, the component of the object's motion at right angle to the line connecting the object with the observer (the line of sight); measured in angular units per unit of time.

Protoplanets The early planets which preceded and developed (according to the condensation theory of the origin of the planets) into the present planets.

Psychrometer An instrument for measuring the moisture content of the air by use of a dry- and a wet-bulb thermometer. When water evaporates from the wet bulb, the thermometer cools.

Pynocline A zone where the water density rapidly increases. The increase is greater than that in the water above or below it. The density change, or pycnocline, is due to changes in temperature and salinity.

Pyroxene group Igneous-rock-forming silicate minerals containing calcium, sodium, magnesium, iron, or aluminum. Contrast with *amphibole group*.

Quartz A silicate mineral, SiO_2, composed exclusively of silicon-oxygen tetrahedra with all oxygens joined together in a three-dimensional network. Crystal form is a six-sided prism tapering at the end, with the prism faces striated transversely. An important rock-forming mineral.

Radial drainage An arrangement of stream courses in which the streams radiate outward in all directions from a central zone.

Radiation Visible or invisible light.

Radial velocity Component of the velocity of an object along the line connecting the body with the observer (line of sight).

Radioactive elements Those elements that are capable of changing into other elements by the emission of charged particles from their nuclei.

Rare gases Those gases (such as krypton, xenon, and argon) that are present in the earth's atmosphere in very small quantities.

Rays, cosmic High-energy particles such as electrons, or protons, or high-energy photons.

Rays, gamma High-energy photons typically of wavelength 0.1 Å.

Rays, X High-energy photons typical of 10 Å wavelength.

Recessional moraine A ridge or belt of till marking a period of moraine formation, probably in a period of temporary stability or a slight re-advance, during the general wastage of a glacier, and recession of its front.

Recumbent fold A fold in which the axial plane is more or less horizontal.

Regional metamorphism Metamorphism occurring over tens or scores of kilometers.

Rejuvenation A change in conditions of erosion that causes a stream to begin more active erosion and a new cycle.

Relativistic Very energetic. Relativistic particles move with speeds close to 300,000 km/sec (speed of light).

Relativity, theory of general Theory describing the behavior of physical quantities in very strong gravitational fields.

Relativity, theory of special Theory describing the space and time behavior of physical quantities; includes the exchange of mass and energy forms.

Reverse fault A fault in which the hanging wall appears to have moved upward relative to the footwall. Also called *thrust fault*. Contrast with *normal* or *gravity fault*.

Revolution Periodic motion of one body about another, such as the earth's motion about the sun.

Rift zone A system of fractures in the earth's crust. Often associated with extrusion of lava.

Rip current A narrow seaward-flowing current that results from breaking of waves and subsequent accumulation of water in the nearshore zone.

Rock An aggregate of minerals of different kinds in varying proportions.

Rock cycle A concept of the sequences through which earth materials may pass when subjected to geological processes.

Rock flour Finely divided rock material pulverized by a glacier and carried by streams fed by melting ice.

Rock glacier A tongue of rock waste found in the valleys of certain mountainous regions. Characteristically lobate and marked by a series of arcuate, rounded ridges that give it the aspect of having flowed as a viscous mass.

Rotation Spin.

Rotation, differential State of motion in which the parts of a system spin at different speeds. Contrast rigid body rotation.

Rotation, rigid body Spinning of a body such that all parts of the body remain in the same positions relative to each other.

Runoff Water that flows off the land.

Salinity The total amount of dissolved material in sea water. It is measured in parts per thousand by weight in one kilogram of sea water.

Saltation Mechanism by which a particle moves by jumping from one point to another.

Salt dome A mass of NaCl generally of roughly cylindrical shape and with a diameter of about a mile near the top. These masses have been pushed through surrounding sediments into their present positions, sometimes as far as 7,000 meters. Reservoir rocks above and alongside salt domes sometimes trap oil and gas.

Satellite, of a planet A moon bound to a planet by gravity.

Satellite, of earth The moon and man-made bodies in orbit about the earth.

Savanna A type of warm climate with a wet and dry season.

Scuba Self-contained Underwater Breathing Apparatus.

Sea arch The roof of a cave by the sea through a headland.

Seamount An isolated, steep-sloped peak rising from the deep ocean floor but submerged beneath the ocean surface. Most have sharp peaks, but some have flat tops and are called *guyots* or *tablemounts*. Seamounts are probably volcanic in origin.

Secondary shorelines Secondary shorelines are those where the coastal region has been mainly formed by marine or biological agents, like coral reefs, barrier beaches, and marshes.

Secondary wave An earthquake body wave slower than the primary wave. Wave that advances by causing particles in its path to move at right angles to the direction of the wave's advance, a shake motion. Also called *shear wave, shakewave*, or *S-wave*.

Sedimentary rock Rock formed from accumulations of sediment, which may consist of rock fragments of various sizes, the remains or products of animals or plants, the product of chemical action or of evaporation, or mixtures of these. *Stratification* is the single most characteristic feature of sedimentary rocks, which cover about 75 percent of the land area of the world.

Sedimentation The process by which mineral and organic matter is laid down.

Seismic seawave A large wave in the ocean generated at the time of an earthquake. Popularly, but incorrectly, known as a *tidal wave*. Sometimes called a *tsunami*.

Seismology The scientific study of earthquakes and other earth vibrations.

Shear Change of shape without change of volume.

Shield volcano A volcano built up almost entirely of lava, with slopes seldom as great as 10° at the summit and 2° at the base. Examples: the five volcanoes on the island of Hawaii.

Shelf break The sharp break in slope which marks the edge of the continental shelf and beginning of the continental slope.

Shoreline The place where land and water meet.

Shower, meteor Entry of a group of meteoroids with the earth's atmosphere.

Sinkhole Depression in the surface of the ground caused by the collapse of the roof over a solution cavern. Also called *sink*.

Slip-face The steep face on the lee side of a dune.

Slump The downward and outward movement of rock or unconsolidated material as a unit or as a series of units. Also called *slope failure*.

Slumping The sliding or moving of sediments down a submarine slope.

Snowfield A stretch of perennial snow existing in an area where winter snowfall exceeds the amount of snow that melts away during the summer.

Soil The superficial material that forms at the earth's surface as a result of organic and inorganic processes. Soil varies with climate, plant and animal life, time, slope of the land, and parent material.

Solifluction Mass movement of soil affected by alternate freezing and thawing. Characteristic of saturated soils in high latitudes.

Solubility The degree to which a substance mixes with another substance.

Spheroidal weathering The spalling off of concentric shells from rock masses of various sizes as a result of pressures built up during chemical weathering.

Spit A sandy bar built by currents into a bay from a promontory.

Spontaneous liquefaction When water-saturated sediments are subjected to a sudden shock, shear, or increase in pore water pressure, the internal grain-to-grain contacts within the sediments may change. If this happens, the entire sediment mass may move or flow, similar to an avalanche.

Spring A place where the water table crops out at the surface of the ground and where water flows out more or less continuously.

Spring tides High tides that occur about every two weeks, when the moon is full or new. See *neap tides*.

Stack A small island that stands as an isolated, steep-sided rock mass just off the end of a promontory. Has been isolated from the land by erosion and by weathering concentrated just behind the end of a headland.

Stalactite Icicle-shaped accumulation of dripstone hanging from a cave roof.

Stationary wave A type of wave where the wave form does not move forward but the surface moves up and down. At certain fixed points, called nodes, the water surface will remain stationary.

Stock A discordant pluton that increases in size downward, has no determinable floor, and shows an area of surface exposure less than 40 square miles. Compare with *batholith*.

Storm surges Abnormally high water levels due to strong winds blowing on the water surface.

Stratification The structure produced by the deposition of sediments in layers or beds.

Stratosphere The second lowest of the main layers of the atmosphere; it is characterized by more or less isothermal conditions and a highly stable stratification.

Stratus Continuous horizontal layer of cloud.

Streak The color of the fine powder of a mineral. May be different from the color of a hand speciment. Usually determined by rubbing the mineral on a piece of unglazed porcelain (hardness about 7), known as a streak plate, which is, of course, useless for minerals of greater hardness.

Stream terrace A surface representing remnants of a stream's channel or flood plain when the stream was flowing at a higher level. Subsequent downward cutting by the stream leaves remnants of the old channel or flood plain standing as a terrace above the present level of the stream.

Striation A scratch or small channel gouged by glacial action. Bedrock, pebbles, and boulders may show striations produced when rocks trapped by the ice were ground against bedrock or other rocks. Striations along a bedrock surface are oriented in the direction of ice flow across the surface.

Strike The direction of the line formed by intersection of a rock surface with a horizontal plane. The strike is always perpendicular to the direction of the dip.

Strike-slip fault A fault in which movement is almost in the direction of the fault's strike.

Superposition, law of If a series of sedimentary rocks has not been overturned, the topmost layer is always the youngest and the lowermost is always the oldest.

Surf Breaking waves in a coastal area.

Surface wave Wave that travels along the free surface of a medium. Earthquake surface waves are sometimes represented by the symbol *L*.

Symmetrical fold A fold in which the axial plane is essentially vertical. The limbs dip at similar angles.

Syncline A configuration of folded stratified rocks in which the rocks dip downward from opposite directions to come together in a trough. The reverse of an *anticline*.

Synoptic measurements Numerous measurements taken simultaneously over a large area.

Talus A slope established by an accumulation of rock fragments at the foot of a cliff or ridge. The rock fragments that form the talus may be rock waste, sliderock, or pieces broken by frost action. Actually, however, the term "talus" is widely used to mean the rock debris itself.

Tarn A lake formed in the bottom of a cirque after glacier ice has disappeared.

Temporary base level A base level that is not permanent, such as that formed by a lake.

Terminal moraine A ridge or belt of till marking the farthest advance of a glacier. Sometimes called *end moraine*.

Terminal velocity The constant rate of fall eventually attained by a grain when the acceleration caused by the influence of gravity is balanced by the resistance of the fluid through which the grain falls.

Terrace A nearly level surface, relatively narrow, bordering a stream of body of water, and terminating in a steep bank. Commonly the term is modified to indicate origin, as in *stream* terrace and *wave-cut* terrace.

Texture The general physical appearance of a rock, as shown by the size, shape, and arrangement of the particles that make up the rock.

Thermocline A zone where the water temperature decreases more rapidly than the water above or below it. This zone usually starts from 10 to 500 meters below the surface and can extend to more than 1,500 meters in depth.

Thrust fault A fault in which the hanging wall appears to have moved upward relative to the footwall. Also called *reverse fault*. Opposite of *gravity* or *normal fault*.

Tidal bore A large wave of tidal origin that will travel up some rivers and estuaries.

Tidal wave Popular but incorrect designation for *tsunami*.

Tide Alternate rising and falling of the surface of the ocean, other bodies of water, or the earth itself, in response to forces resulting from motion of the earth, moon, and sun relative to each other.

Till Unstratified and unsorted glacial drift deposited directly by glacier ice.

Tombolo A sand bar connecting an island to the mainland, or joining two islands.

Topographic deserts Deserts deficient in rainfall either because they are located far from the oceans toward the center of continents, or because they are cut off from rainbearing winds by high mountains.

Topset bed Layer of sediment constituting the surface of a delta. Usually nearly horizontal, and covers the edges of inclined foreset beds.

Tornado A violent, destructive storm of very small horizontal extent.

Transverse dune A dune formed in areas of scanty vegetation and in which sand has moved in a ridge at right angles to the wind. It exhibits the gentle windward slope and the steep leeward slope characteristic of other dunes.

Tropopause The upper limit of the troposphere.

Troposphere The lowest main layer of the atmosphere; it is characterized by a steep lapse rate, a low degree of hydrostatic stability, and, as a result, overturnings occur frequently.

Truncated spur The beveled end of a divide between two tributary valleys where they join a main valley that has been glaciated. The glacier of the main valley has worn off the end of the divide.

Tsunami (plural, tsunami) A large wave in the ocean generated at the time of an earthquake. Popularly, but incorrectly, known as a *tidal wave*. Sometimes called *seismic seawave*.

Tundra Mossy and/or marshy, treeless plain on the fringe of the arctic.

Turbidity current A turbid, relatively dense current composed of water and sediment that flows downslope through less dense seawater. The sediment eventually settles out, forming a turbidite.

Typhoon Violent hurricane in the China seas.

Ultimate base level Sea level, the lowest possible base level for a stream.

Unpaired terrace A terrace formed when an eroding stream, swinging back and forth across a valley, encounters resistant rock beneath the unconsolidated alluvium and is deflected, leaving behind a single terrace with no corresponding terrace on the other side of the stream.

Upwelling The movement of water from depth to the surface.

Valley glacier A glacier confined to a stream valley. Usually fed from a cirque. Sometimes called *alpine glacier* or *mountain glacier*.

Valley train Gently sloping plain underlaid by glacial outwash and confined by valley walls.

Variable star A star that undergoes changes in one or more of its properties (brightness, surface temperature, radius, etc.)

Varve A pair of thin sedimentary beds, one coarse and one fine. This couplet of beds has been interpreted as representing a cycle of one year, or an interval of thaw followed by an interval of freezing in lakes fringing a glacier.

Ventifact A pebble, cobble, or boulder that has had its shape or surface modified by wind-driven sand.

Velocity of a stream Rate of motion of a stream measured in terms of the distance its water travels in a unit of time, usually in feet per second.

Viscosity An internal property of rocks that offers resistance to flow. The ratio of deforming force to rate at which changes in shape are produced.

Volcanic block Angular mass of

Glossary

hardened lava. Contrast with *volcanic bomb*.

Volcanic bomb A rounded mass of newly congealed magma blown out in an eruption. Contrast with *volcanic block*.

Volcanic dust Pyroclastic detritus consisting of particles of dust size.

Volcanic neck The solidified material filling a vent or pipe of a dead volcano.

Volcano A landform developed by the accumulation of magnetic products near a central vent.

Water gap The gap cut through a resistant ridge by a superimposed or antecedent stream.

Water table The upper surface of the zone of saturation for underground water. It is an irregular surface with a slope or shape determined by the quantity of ground water and the permeability of the earth materials. In general, it is highest beneath hills and lowest beneath valleys.

Wave attenuation The decrease in the wave form or height with respect to distance from its origin.

Wave crest The highest part of a wave.

Wavelength The horizontal distance between two wave crests (or similar points on the wave form) measured parallel to the direction of travel of the wave.

Wave period The time required for successive wave crests to pass by a fixed point.

Wave refraction The change in direction of waves which occurs when one portion of the wave reaches shallow water and is slowed down while the other portion is in deep water and moving relatively fast.

Wave trough The lowest part of a wave between two successive crests.

Wave velocity The velocity at which the wave form proceeds. It is equal to the wavelength divided by the wave period.

Wind gap The general term for an abandoned water gap.

Wind, solar Material of the outermost atmosphere of the sun which is continuously streaming away from the sun. Called *stellar wind* when referring to stars other than our sun.

Xenolith A strange rock broken from the wall surrounding a magma chamber and frozen in the intrusion as it solidified.

Yardang A sharp-edged ridge between two troughs or furrows excavated by wind action.

Yazoo-type river A tributary that is unable to enter its main stream because of natural levees along the main stream. The Yazoo-type river flows along the back-swamp zone parallel to the main stream.

Zone of aeration A zone immediately below the surface of the ground, in which the openings are partially filled with air, and partially with water trapped by molecular attraction. Subdivided into (a) belt of soil moisture, (b) intermediate belt, and (c) capillary fringe.

Zone of saturation Underground region within which all openings are filled with water. The top of the zone of saturation is called the water table. The water contained within the zone of saturation is called groundwater.

Zooplankton Those organisms carried by the ocean that are of the animal kingdom. See *plankton*.

Index

Abbe Moro, 177
Abrading action, 231
Abrasion, 211–213
Absolute magnitude, 76
Absolute time, 250
Abyssal plain, 137
Acadian orogeny, 260
Adams, John, 63
Adria, 240
Advection, 280
Aeolian erosion, 210
Aeolus, 177
Aerolite, 69
Aerosol, 274
Aignay-de-Duc, 229
Air masses, 307–310
   arctic, 308
   polar, 308
   tropical, 308
Air pollution, 357–367
Airy, G. B., 133
Albedo, 49
Alluvial fan, 237
Alpha Centauri, 76
Alpha Draconis, 38
Aluminaut, 393
Amethyst, 146
Amphibians, 262
Amphibole, 156
Anaxagorus, 145, 192
Anaximenes, 192
Andromeda galaxy, 94, 101, 102, 107
Andromeda nebula, 100
Anemometer, 289
Aneroid barometer, 287

Angular unconformity, 253
Annular eclipse, 41
Antarctic, 216
Antarctic ocean, 394
Antares, 85
Anticlines, 186
Anticyclones (highs), 303
Antioch, 239
Antler orogeny, 261
Aphelion, 15, 331
Apogee, 39
Apollo group, 67
Appalachian mountains, 176, 187, 190–191, 261, 262, 263, 265
Apparent magnitude, 75
Aquiculture, 392
Aquifer, 241
Aragonite, 146
Arctic ocean, 394
Arcturus, 85
Arete, 222
Aristarchus, 12, 110
Aristotle, 11, 12, 128, 145, 177, 192, 229
Artesian spring, 242
Artesian well, 242
Ash, volcanic, 179
Asteroids, 17, 66–67
Asthenosphere, 139, 141
Astronomical unit, 15, 17
Astronomy, defined, 9–11
   Babylonian, 10, 11
   Chinese, 10, 11, 110
   Egyptian, 10
   Greek, 11, 25, 110
   objectives, 9

Atlantic ocean, 394
Atmosphere, 273–277
 pressure, 142, 286–287
 temperature, 277–281, 284
Atome Primitif, 111
Aureole, 164
Autumnal equinox, 332

Baade, Walter, 107, 111
Bacon, Francis, 137
Baltic sea, 438
Barnard's star, 80
Barometer, aneroid, 287
Barometer, mercurial, 287
Barringer crater, 70
Bars, 246
Barycenter, 39, 425
Base level, 202, 235
Basin, 187
Basin, drainage, 229
Batholith, 185
Bauxite, 157
Beach, 246
Bergeron, Tor, 369
Bessel, F. W., 73
Betelgeuse, 26, 77, 85, 86
Binary stars, 79
Binary star hypothesis, 24
Biochemical oxygen demand, 441
Biotite, 156
Black Hills, 187
Black hole, 87
Black sea, 401
Blocks, volcanic, 179
Bode's law, 17
Bombs, volcanic, 179
Bondi, Hermann, 113
Borates, 154
Bowden, K. F., 429
Bowen reaction series, 160
Brachiopods, 261
Brahe, Tycho, 14, 86, 89
Breccia, 162, 184
Bridal Veil Falls, 221
Brontosaurus, 264
Bruno, G., 110
Buffon, G., 24
Bush, Vannevar, 368

Calcite, 146
Caldera, 181
California current, 419
Cambrian period, 258, 261
Canary Island current, 419
Canis Major, 76

Capacity, 232
Capella, 85
Carbonates, 154
Carbonation, 168
Cariaco trench, 401
Caribbean Sea, 394
Cascade mountains, 266
Cassini's gap, 62
Cast, 254
Catastrophists, 129
Cave, 242–243
Cenozoic era, 256, 265–267
Centers of action, 303
Centrifugal force, 302
Cepheid variables, 75, 86, 100, 101, 102
Ceres, 66
Chamberlain, T. C., 24
Chaparral, 339
Charles Bal, 180
Chinook, 307
Chlorinity, 399
Chromosphere, 29–30
Cinders, volcanic, 179
Cirque, 222
Cleavage, 150
Cleopatra's Needle, 169
Climate, 333–348
Climax vegetation, 339
Close encounter hypothesis, 24
Clouds, 292–294, 295
Coesite, 70
Col, 222
Cold front, 307, 311, 312–314
Color, 150
Color index, 77
Columbia plateau, 183, 266
Column, 242
Columnar jointing, 188
Coma, 68
Comet cloud hypothesis, 67
Comets, 62–69
Competence, 232
Conduction, 278
Cone of depression, 244
Conglomerates, 162
Connate water, 180
Constellations, 89–90
Contact metamorphism, 164
Continental divide, 229
Continental drift, 137–142
Continental glaciers, 216
Continental mass, 132
Continental shelf, 132, 136
Continental slope, 132, 136
Convection, 278, 280

Index

Core, 25, 134
Coriolis force, 301, 420
Copernicus, 9, 12, 13–14
Corona, 30–31, 41
Coronagraph, 30
Cosmology, 110
Crater, 177
Crater lake, 181
Creep, 203
Cretaceous period, 263, 264, 265
Crinoids, 262
Crust, 133
Crystal systems, 146–148
Culebra Cut, 207
Cumulus, 293
Cumulus stage, 325
Current meter, 416
Cusanus, N., 110
Cutbank, 233
Cyclones (lows), 303
Cygnus A, 107

Daily retardation, 40
Dead water, 434
deBury, Richard, 125
Decomposition, 168
Deepstar, 393
Deep zone, 402
Deflation, 211–213
Deimos, 58
Delta, 237
Delta Cephei, 86
Dendritic pattern, 230
Density, 79, 152, 403–406, 419
Depth of friction, 420
Desalination, 387
Desert pavement, 211
Des Voeux, Harold, 360
de Vancouleurs, G., 105
Devil's Postpile, 188
Devonian period, 260, 261, 262
Dew point, 282, 286
Digges, Thomas, 110
Dikes, 184
Dinosaurs, 262, 264
Direct tide, 423
Disconformity, 253
Discontinuity, Moho, 134
Disintegration, 165
Distributaries, 237
Divide, drainage, 229
Domes, 187
Donn, W. L., 219
Donora, Penn., 360
Doppler shift, 102

Downwelling, 422
Drainage basin, 229
  divide, 229
  system, 229
Draper, Henry, 81
Drift, 223
Drift bottles, 417
Drumlins, 225
Dry adiabatic lapse rate, 281
Ducktown, Tenn., 365
Durham castle, 210
Dust bowl, 215
Dynamic metamorphism, 164
Dyrenforth, Robert, 368

Earth, 33–37
  diameter, 33
  magnetic field, 34
  magnetic tail, 34
  magnetopause, 34
  magnetosphere, 34
  transition zone, 34
  Van Allen radiation belt, 35
Earthquakes, 192–197
Ebb tide, 427
Eclipse, 41–43
  annular, 41, 42
  lunar, 43
  partial, 41
  total, 41
Ecliptic, 17
Einstein, Albert, 16, 33, 111
Ekman spiral, 420
Electrical conductivity, 399
Electrodialysis, 387
Elliptical galaxy, 97
Elliptical orbit, 14–15
English Channel, 428, 429
Entrainment, 325
Eocene epoch, 265
Epeirogeny, 192
Epicenter, 195
Epsilon Aurigae, 79
Epsilon Eridani, 117, 118
Eratosthenes, 34
Erosion, 202–247
  gravity, 202–210
  ice, 215–228
  water, 228–247
  wind, 210–215
Erosion cycle, 202
Eruptive variable, 86
Esker, 226
Espy, James P., 369

Index

Eugeosyncline, 260
Euphrates river, 238
Eutrophication, 440
Evolutionary universe, 111–113
Ewing, Maurice, 219
Exfoliation, 167
Exhalation, dry, 128
   moist, 128

Fault, gravity, 190
   normal, 190
   reverse, 190
   scarp, 190
   thrust, 190
Faulting, 189–190
Faults, 186, 189
Fiery cloud, 182
Findeisen, Walter, 369
Fireball, 69
Fish protein concentrate (FPC), 383
Fissure, 177
Flammarion, C., 64
Flood plain, 235
Flood tide, 427
Fluvial, 228
Fly ash, 361
Focus, 195
Foehn winds, 307
Folding, 186–188
Folds, 186, 187
Foot wall, 189
Fossils, 254–256
Foucault, J. B. L., 37
Fracto, 293
Fracture, 150
Franklin, Benjamin, 414, 416
Friction, 303
Front, weather, 307
Frost heaving, 166, 204
Frost wedging, 165, 220

Gaillard Cut, 207
Galactic (open) cluster, 83
Galaxy, 93–107
   distance, 100–103
   distribution, 105–106
   elliptical, 97
   evolution, 97–100
   irregular, 97
   mass, 96
   motion, 94–95, 103–104
   normal, 107
   peculiar, 107
   population, 94
   radio, 106, 107

Galaxy (*cont'd*):
   shape, 93
   size, 93
   spiral, 96
   structure, 95
   type, 96–97
Galena, 146
Galileo, 9, 15–16, 26, 61, 62, 86, 92
Galle, Johann, 63
Gamow, George, 112
Geocentric, 11, 12
Geology, defined, 125
Geomagnetic electro-kinetograph, 416
Geostrophic winds, 302
Geosyncline, 174, 191, 258
Giants, stellar, 83, 88
Glacial stage
   Illinoian, 217
   Kansan, 217
   Nebraskan, 217
   Wisconsin, 217
Glaciation, 220–222
Glaciers, 215–228
Glacio-fluvial deposits, 225, 226
Globular cluster, 83, 94, 95
Gold, Thomas, 113
Gondwanaland, 137
Goodriche, John, 86
Graben, 263, 265
Grand Canyon, 192
Graphite, 150
Graptolites, 261
Gravity, 302
Gravity erosion, 202–210
Gravity fault, 190
Great Dunes National Monument, 213
Great Lakes, 438, 439
Great Red Spot, 60
Great Salt Lake, 212
Greenhouse effect, 51
Greenland, 216
Ground moraine, 225
Ground water, 180, 241–245
Ground water zone, 241
Gulf Stream, 412, 419, 421
Gulliver, 20
Guyots, 136
Gyre, 419

Hail suppression, 372
Hale, George, 29
Halides, 154
Halite, 146
Hall, Asaph, 58
Halley's comet, 65, 67, 69

Index

Hardness, 149
Harmonic law, 15
Hanging valley, 221
Hanging wall, 189
Helium, 26, 32
Heliocentric theory, 13–17
Hell Gate, 428
Henry, Patrick, 239
Herculaneum, 182
Hermes, 67
Herschel, Sir William, 62, 73, 80, 93, 94
Hertzsprung, Einor, 82
Hertzsprung-Russell diagram, 82, 83, 84, 85, 87, 88
Heterosphere, 275
Hexagonal system, 148
Highs (anticyclone), 303
Hills, 173
Himalaya mountains, 174
Hipparchus, 75
Homocline, 187
Hooke, Robert, 289
Horn, 222
Horse latitudes, 301
Horsts, 265
Hoyle, Fred, 113, 114
H-R diagram, 82–85
Hsuan yeh, 110
Hubbard glacier, 216
Hubble, Edwin, 99, 100, 101, 102, 105
Humason, M., 102
Humidity, relative, 285–286
Hurricane, 320
Hutton, James, 130
Huygens, Christian, 62
Hydralic action, 231
Hydration, 168, 387
Hydrogen, 32
Hydrologic cycle, 282
Hydrothermal metamorphism, 164
Hygrometer, 286
Hygroscopic, 274

Igneous rock, 159–161
Illinoian stage, 217
Index fossils, 256
Indian ocean, 394
Insolation, 277, 331
Interferometer, 77
Internal friction, 420
Internal waves, 434–435
Ionosphere, 30, 275
Irregular galaxy, 97
Island arcs, 136
Isobars, 302

Isometric system, 147
Isostacy, 133
Isotherms, 403

Jeans, Sir James, 116
Jodrell Bank, 107
Joints, 188
Jorullo, 178
Juno, 67
Jupiter, 11, 13, 16, 18, 25, 48, 58–61, 67, 115
Jurassic period, 263, 264
Juvenile water, 180

Kansan stage, 217
Kant, Immanuel, 24, 93, 100
Karst topography, 242
Kepler, Johannes, 9, 14–15, 16, 86
Kettle, 226
Kilauea, 179
Krakatoa, 180
Kuiper, G. P., 24, 117
Kuroshio current, 419, 421

Laccolith, 184
Lake Erie, 438, 440
Lambert, Johann H., 93
Landslides, 205
Langley, 289
Langmuir, Irving, 370
LaPlace, Pierre, 24
Lapse rate, 280
Laramide orogeny, 263, 265
Large Magellanic cloud, 94
Lassen Peak, 266
Lateral moraine, 226
Laurasia, 137
Lava, 159
Leap year, 37
Leavitt, Henrietta, 75
LeMaitre, George, 111, 112
Leverrier, U., 29, 63
Life zone, 115
Lightning, 326
Lightning suppression, 372
Light year, 18, 74
Limb, 26
Limb darkening, 27
Limestone, 161
Lippershey, Hans, 15
Lithophagus, 192
Lithosphere, 139
Load, 232
Local group, 105
London, 360
Long (L) wave, 195

Loran, 417
Lowell, Percival, 64
Lows (cyclones), 303
Luminosity, 32, 83
Lunar crater, 43
Luster, 148
Lystrosaurus, 137

Magma, 159
Magnetite, 145
Magnetosphere, 277
Magnitude, 75–76
Mahoning river, 438
Main sequence, 83
Mammals, 262, 267
Mantle, 25, 134
Mares, 43
Mariner II, 50
  IV, 53, 54
  V, 50
  VI, 53, 54
  VII, 53, 54
  IX, 56
Marine sciences, 382
Mars, 11, 13, 18, 48, 52–58, 66, 115
Marsh, George P., 437
Massive plutons, 184
Mass movement, 202
Mass wasting, 202
Matterhorn, 222
Mature stage (river), 233
Maury, M. F., 414
Meander, 233
Medial moraine, 226
Mediterranean sea, 394
Magellanic clouds, 97
Mecalli scale, 196
Mercurial barometer, 287
Mercury, 11, 13, 17, 18, 29, 48, 49–50, 115
Merrimack river, 438
Mesopause, 275
Mesosphere, 275
Mesozoic era, 256, 262–265
Messier, *13*, 83, 85
  16, 84
  44, 85
Metamorphic rock, 163–165
Metamorphism, 164
Meteorites, 25, 69–70
Meteoroids, 69–70
Meteors, 69–70
Meteor showers, 69
Meuse Valley, Belgium, 360
Michell, J., 193
Michelson, A. A., 77

Milky Way, 10, 16, 92, 93, 94, 95, 96, 97, 100, 101, 102, 103, 105, 107
Millibar, 287
Mineral defined, 145
  classification, 153
Miogeosyncline, 260
Mississippian period, 261, 262
Mississippi river, 438
Missouri river, 438
Miocene epoch, 265, 266
Moho discontinuity, 134
Mohorovicic, A., 134
Mohs, F., 149
Moist adiabatic lapse rate, 281
Mold, 254
Monocline, 187
Monoclinic system, 148
Moon, 11, 13, 38–45, 76, 116
Moraine, 225–226
Morrison formation, 263
Moulton, F. R., 24
Mountain breeze, 307
Mountain building, 172–199
Mountain glaciers, 216
Mountains, 173
Monte Somma, 182
Mount Mazama, 181, 266
Mount Pelee, 182
Mount Shasta, 266
Mount Wilson, 29, 77
Mudflow, 204
Multivator, 20
Muscovite, 156

Nansen bottle, 391
Native elements, 153
Natural levee, 236
Neap tide, 423
Nebraskan stage, 217
Nebular hypothesis, 24
Neck (volcanic), 184
Neptune, 17, 18, 48, 63–64
Neptunists, 129
Neutron star, 87
Nereid, 64
Nevadan disturbance, 263
Nevadan orogeny, 263
Newton, Isaac, 9, 16, 24, 423
Nile delta, 237
Nimbus, 293
Nitrates, 154
Nonconformity, 253
Normal fault, 190
Northeast trade wind, 301
Norwegian sea, 394

Nova, 86, 88
Nuclear fission, 32
Nuclear fusion, 32
Nuee ardente, 182

Obsidian, 160
Occluded front, 316–319
Ocean basins, 132, 135
Ocean ridge, 136
Ocean trench, 136
Old age stage (river), 235
Oligocene epoch, 265, 267
Olivine, 156
Opal, 146
Open (galactic) cluster, 83
Opposite tide, 423
Ordivician period, 260, 262
Orion, 95
Orionids, 69
Orthoclase, 156
Orthorhombic system, 147
Outwash plain, 226
Oxbow, 235
Oxidation, 168
Oxides, 154
Ozone, 275

Pacific ocean, 394
Paleocene epoch, 267
Paleozoic era, 242, 256, 258–262
Palisades disturbance, 263
Pallas, 67
Parallax, 14, 73
Paricutin, 177
Parsec, 74
Partial eclipse, 41
Peculiar galaxy, 107
Peneplain, 265
Pennsylvanian period, 261, 262
Penumbra, 28, 41
Perigee, 39
Perihelion, 15, 331
Permian period, 261, 262
Permineralization, 254
Perrault, Pierre, 229
Phobus, 58
Phosphates, 155
Photochemical smog, 360, 361
Photodissociation, 51, 273
Photosphere, 26–29
Piazzo, G., 17, 66
Pickering, E. C., 81
Pickering, W. H., 64
Piedmont glacier, 216
Pillar, 242

Pioneer, *10*, 61
Pitted outwash plain, 226
Plagioclase, 156
Planet X, 64
Plasma, 31
Plateau basalts, 183
Pleistocene epoch, 217, 265, 266, 267
Pliny the Elder, 128, 145, 172, 182
Pliny the Younger, 182
Pliocene epoch, 265, 266, 267
Plucking, 220
Plug (volcanic), 184
Pluto, 17, 18, 48, 64–65
Plutonists, 129
Polaris, 17, 38
Polar-front cell, 299
Pollution, 439–446
 agricultural, 443–444
 industrial, 441
 oil, 441–443
 radiation, 445–446
 sewage, 439–441
 thermal, 444–445
Pompeii, 182
Ponce de Leon, 414
Population I, II, 95, 97
Pratt, J. H., 133
Precambrian era, 257–258
Precession, 37–38
Precipitation, 287–288
Pressure gradient, 302, 420, 421
Primary (P) wave, 194
Prominences, 30, 41
Proper motion, 80
Proton–proton chain, 32
Protoplanets, 24
Proxima Centauri, 18
Psychrometers, 284
Pulido, Dionisio, 178
Pulsar, 87
Pulsating variable, 86
Pumice, 181
Pycnocline, 406
Pyranometer, 289
Pyrite, 146, 168
Pyroclastic rock, 179
Pythagorus, 11

Qattara depression, 211
Quarrying, 220
Quartz, 144, 146, 156
Quasar, 107, 114
Queen Elizabeth I, 5

Radial velocity, 80

Radiant, 69
Radiation, 278
Radiosonde, 289
Rain gauge, 288
Rainmaking, 368
Rapid flowage, 204
Recessional moraine, 226
Red giant, 85
Red Sea, 386, 439
Red shift law, 102
Regional metamorphism, 164
Relative humidity, 285
Relative time, 250
Reptile, 262
Respiration, 401
Retrograde motion, 11
Reverse fault, 190
Reverse osmosis, 387
Revolution, 37
Richter, C. F., 197
Richter scale, 197
Ring of fire, 175, 184
Rivers, 233–236
Rock creep, 204
Rock unit, 256
Rocks, 158–165
Rocky mountains, 263, 265
Rossi-Forel scale, 195
Rotation, 37
RR-Lyrae variables, 86
Russell, Henry, 82, 87

Sahara desert 211, 213
Salinity, 397–399
Saltation, 211, 231
San Andreas fault, 190, 193, 194
Sandage, A., 102, 112
Sand dunes, 213
Sandstone, 161, 162
Santa Ana, 307
Santa Barbara channel, 442
Sargasso sea, 394
Saturn, 11, 13, 16, 18, 48, 62
Saturn's rings, 16, 62
Savanna, 339
Scarps, 136, 190
Schaefer, Vincent J., 370
Schiaparelli, G., 53
Schmidt, M., 102, 107
Schwabe, H., 29
Science, definition, 3
Sea floor spreading, 139, 141, 193
Sealab, 393
Seamounts, 136
Seasons, 331–333

Secchi, Angelo, 81
Secchi disk, 408
Secondary (S) wave, 194, 195
Sedimentary rock, 161–163
Seine river, 229
Seismographs, 195
Seismology, 192
Selenite, 145
Shale, 161, 162
Shapley, Harlow, 93, 99
Shatter cones, 70
Shields, 258
Shield volcanoes, 182
Shooting star, 69
Sial, 134
Sidereal day, 37
   month, 39
   year, 38
Siderite, 69
Siderolite, 69
Sierra Nevada mountains, 266
Sigma-t, 406
Silicates, 155
Silurian period, 260, 261, 262
Sima, 134
Sink, 242
Singh, Awtar, 208
Sirius, 10, 76
Sirius B, 79
61 Cygni, 73
Sliding, 204–206
Slipher, V. M., 102
Slip-off slope, 233
Slow flowage, 203–204
Smith, William, 256
Smog, 274, 360
Socrates, 19
Soil, 170
Solar constant, 33, 277
   day, 37
   flares, 30
   granulation, 27
   wind, 31, 277
Solstice, summer, 331
   winter, 332
Solution, 168, 231
Source region, 308
Specific gravity, 152, 403
Spectral class, 82
Spiral galaxy, 96
Spits, 246
Spring tide, 423
Squall line, 314
Stalactite, 243
Stalagmite, 243

Index

Stationary front, 311, 316
Steady state, 113–114
Stellar classification, 81–82
   density, 78–79
   diameter, 77–78
   distance, 72–75
   evolution, 87–89
   Magnitude, 75–77
   mass, 78–79
   motion, 79–81
   temperature, 77
Steno, Nicolaus, 128, 177
Stock, 185
Storms, 320–327
Stratopause, 275
Stratosphere, 275
Stratus, 293
Streak, 150
Strike-slip fault, 190
Stromboli, 182
Submersibles, 391
Suess, E., 134, 137
Sulfates, 154
Sulfides, 154
Sulfur dioxide, 360, 365, 367
Summer solstice, 331
Sun, 11, 17, 25–33, 76, 87, 94, 95
Sunshine switch, 289
Sunspots, 16, 28
Supergiants, 83, 85, 88
Supernova, 86, 88
Suspension, 231
Swallow buoy, 417
Swells, 429
Swift, Jonathon, 58
Syncline, 186
Synodic month, 39
Synoptic weather report, 350
System, drainage, 229

Taconian orogeny, 260
Talus, 204
Tarn, 222
Tau Ceti, 117, 118
Telescope, 15
Temperature inversion, 281
Terminal moraine, 225
Terminator, 41
Tertiary period, 265
Tetragonal system, 147
Thales, 192
Therapsid, 262
Thermocline, 402
Thermohaline circulation, 419
Thermometer, 284

Thermosphere, 275
Thoreau, H., 438
Thrust fault, 190
Thunderstorm, 325
Tidal bore, 428
Tides, 423, 427, 429
Tigrus river, 238
Till, 223
Titan, 62
Tombaugh, Clyde, 64, 65
Tornado, 323
Torrey Canyon, 442
Torricelli, E., 286
Trade Wind Cell, 299
Travertine, 163
Trellis pattern, 230
Triangulation, 73
Triassic period, 262, 263
Triclinic system, 148
Trilobite, 261, 262
Triton, 64
Trojan group, 67
Tropical cyclone, 320
Tropical year, 38
Tropopause, 275
Troposphere, 274, 275
Tsunami, 197–198
Tufa, 163
Turbidity currents, 206
Turtle mountain, 207
Type locality, 256
Typhoon, 320
Tyrannosaurus rex, 264

Umbra, 28, 41
Unconformity, 253
Uniformitarianism, 130
Upper cold front, 319
Upper zone, 402
Upwelling, 422
Ural mountains, 176
Uranus, 17, 18, 48, 62–63

Valley glaciers, 216
Valley wind, 306
Van Allen radiation belts, 277
Variable stars, 85, 88
Varve, 226
Vega, 38, 81
Venera IV, 50
   VIII, 50
Vent, 177
Venus, 11, 13, 15, 17, 18, 48, 50–52, 66, 76, 115
Veraart, August W., 369
Vernal equinox, 38, 332

Vesta, 67
Vesuvius, 182
Volcanic ash, 170
  blocks, 179
  bombs, 179
  cinders, 179
  dikes, 184
  domes, 182
  gases, 180
  neck, 184
  plug, 184
  sill, 184
Volcanism, 177–186
Voltaire, 58
Vonnegut, Bernard, 370
von Weizacker, C. F., 24
Vredefort Ring, 70
Vulcan, 29

Warm front, 307, 311, 315–316
Washington, George, 239
Water erosion, 228–247
Water table, 241
Wave action, 246–247
Wave cut cliffs, 246
Wave cut terrace, 246
Wave cyclone, 316
Wave height, 431
Wave period, 431
Wave refraction, 433
Wave speed, 431

Wavelength, 431
Waves, 423–435
Weather fronts, 310–319
Weather forecasting, 348–352
Weathering, 165–170
Weather modification, 368–375
Weather vane, 288
Wegener, Alfred, 137
Well logging, 252
Wentworth, C. K., 162
Westerlies, 302
Wet and dry bulb, 284
White dwarf, 83, 88
Wilson, J. Tuzo, 141
Wind erosion, 210–215
Wind velocity, 288
Winter solstice, 332
Wisconsin stage, 217
Wolf Trap, 20
World Meteorological Organization, 388, 410
World Weather Watch, 410
Wright, Thomas, 92, 93

X-ray, 19, 277

Ylem, 112
Yosemite National Park, 221
Youthful stage (river), 233

Zeeman effect, 29
Zone of aeration, 241